Eco-city Planning

Tai-Chee Wong · Belinda Yuen
Editors

Eco-city Planning

Policies, Practice and Design

Springer

SINGAPORE INSTITUTE OF PLANNERS

In Association with the
Singapore Institute of Planners

Editors
Dr. Tai-Chee Wong
Nanyang Technological University
National Institute of Education
Nanyang Walk 1
637616 Singapore
Singapore
taichee.wong@nie.edu.sg

Dr. Belinda Yuen
Singapore Institute of Planners
Singapore
belyuen8@gmail.com

ISBN 978-94-007-9708-6 ISBN 978-94-007-0383-4 (eBook)
DOI 10.1007/978-94-007-0383-4
Springer Dordrecht Heidelberg London New York

Cover illustration: Figure 4.12 from this book

Printed on acid-free paper

Springer is part of Springer Science+Business Media (www.springer.com)

Foreword

Eco-city planning is putting the emphasis on the environmental aspects of planning while sustainable planning treats equally the economic, social and environmental aspects. Eco-city planning and management are based on the principle of a cyclical urban metabolism, minimizing the use of land, energy and materials, and impairment of the natural environment, ultimately leading to zero carbon settlements. This principle is illustrated by Hammarby Sjöstad (Stockholm)[1], as indicated by the editors in their book's introductory chapter (see Brebbia et al. 2010).[2] The book starts with a historic account of eco-city planning. Seven thousand years of urban civilization and planning history have clearly more to tell us than a century of functionalist planning, which leaves a questionable legacy of economic, social as well as eco-city planning.

The division of the book into three parts allows an encompassing coverage of the main components of eco-city planning according to the scale of observation: macro-level policies issues, practice and implementation experiences, and micro-level sustainable design. It is indeed the scale of observation that determines the observed phenomena from diversified perspectives.

The geographic coverage is truly worldwide, with cases from all continents, both in industrialised countries and developing countries. Both positive and less positive examples are described in each level of observation. Regional observation is applied to places such as Malaysia (Iskandar). Urban observation is ranging from the emblematic Curitiba city taken as a whole (land use and transport) down to Nairobi (Umoja Neighbourhood) and to Istanbul (Büyükdere Avenue). Micro level observation includes the indoor ambient air quality, analyzing the effects of air conditioning. At this point, the work of Belinda Yuen about perception of high-rise living by Singapore inhabitants comes to mind. Another special chapter in this volume is devoted to "Eco-cities in China: Pearls in the Sea of Degrading Urban Environments" by Tai-Chee Wong.

Angles of observation are equally diverse, including the specific issue of tourism. Tourism too often kills what it feeds on. Short term interests favour numbers, long term interests favour stewardship and preservation. Eco-tourism keeps rural populations in their traditional settlements while giving them opportunities for external contacts and added value for their products.

As the editors point out the eco-city planning has to be quantified in order to be comparatively assessed. Green labels are generously given to regions, cities, neighbourhoods and individual buildings. Calculation methodologies and their implementation is a new and promising field for eco-planning assessment.

As an example of attempt towards quantification at city level one could mention the European Green City Award. Stockholm was selected as the 2010 European Green Capital, through an evaluation based on a 13 areas list of eco-city parametres including quality of life indicators, among others, as follows:

- Emissions
  - CO_2 equivalent per capita, including emissions resulting from use of electricity;
  - CO_2 per capita resulting from use of natural gas;
  - CO_2 per capita resulting from transport; and
  - CO_2 per kWh use.
- Annual mean concentration of NO_2 and PM10.
- Transport modal split – share of population living within 300 m of a public transport stop.
- Percentage of green areas (public and private) in relation to the overall area and specific percentage of areas set aside to protect urban nature and biodiversity.
- Share of population exposed to noise values of L (day) above 55 dB (A)/of L (night) above 45 dB (A).
- Amount of waste per capita; proportion of total/biodegradable waste sent to a landfill, percentage of recycled municipal waste.
- Proportion of urban water supply subject to water metering; water consumption per capita; water loss in pipelines.
- Energy consumption of public buildings, per square metre.

Each of these indicators has to be scrutinised as to the methodology of calculation. For example, the GHG emissions calculation methodologies at city level were surveyed by the College of Europe in Bruges. Seven standard methodologies were assessed, resulting in widely different per capita figures.

This book volume has also mentioned international "green" evaluation systems for individual buildings, mostly commercial. Among these systems are the BRE Environmental Assessment Method (BREEAM) used in United Kingdom; the Leaders in Energy and Environmental Design (LEED) applied in the United States; the Comprehensive Assessment System for Building Environmental Efficiency (CASBEE) of Japan; and the Green Star of the Green Building Council of Australia (GBCA) (WGBC 2010). The authors have pointed out the diversity of assessment criteria, of which emphasis can vary from energy consumption, water consumption, wastes treatment, building to service materials or indoor environmental quality.

More recently, recycling friendliness has been added to be another assessment criterion, using the "cradle to cradle" approach.

The book *Eco-City Planning: Policies, Practice and Design* gives a number of glimpses about the multiplicity of eco-planning assets. It constitutes a welcome addition to the literature about eco-city planning and opens important perspectives for further research.

Kortenberg, Belgium

Pierre Laconte
President, International
Society of City and
Regional Planners, 2006–2009

Notes

1. Hammarby Sjöstad is Stockholm's largest urban development project whose work began in the early 2000s. It is developed from a disused industrial brownfield and a waterfront harbour site and it is to be transformed into an Ecocycle city by 2015.
2. Brebbia, C. A., Hernandez, S. & Tiezzi, E. (Eds). (2010). *The sustainability city VI: urban regeneration and sustainability*. Ashurst (UK): WIT Press.

Preface

From the Kyoto Protocol, Copenhagen Accord to the current Cancun Conference in Mexico, international concern has been expressed on how best to combat global warming effects to achieve a more sustainable environmental development. Despite differences in commitments and responsibilities from participating countries, the common goal is to protect our mother Earth and our common future. As environmental sustainability becomes a core value of urban development, practising professionals in land use planning versed with ecocity planning ideals will have a great role to play and in contributing towards this common goal.

In this book, more than 12 leading experts, urban planners and academics have collectively expounded, shared their concerns and strategies on the new eco-city urbanism movement in our world today. It will be a "must read" book for a wide market spectrum, including city decision makers, academics and researchers, the public, private sector professionals such as planners, architects, engineers, landscape designers, geologists and economists, etc.

I read with interest the visions of eco-city and the emerging trends of tailor-made eco-towns and cities that are fast transforming scores of new cities in China, including Tianjin Eco-City development by the governments of China and Singapore; United Kingdom's plan to build 10 eco-towns across the country, and the world's first ambitious multi-billion dollar carbon neutral city in Masdar, Abu Dhabi in the Middle East, etc.

As President of the Singapore Institute of Planners with an energetic and ambitious Council, I hope that we shall embark on more publications to showcase the excellent works of Singapore planners and those of the city-state of Singapore reflecting her great effort to build a sustainable and eco-friendly living environment. It is my great pleasure to present to you this book, which is comprehensively loaded with key aspects on eco-city planning. The book shares the world's aspiration in the search for a sustainable solution to the newly emerging urbanism towards building a better urban habitat.

Singapore

William HL Lau
President, Singapore Institute of Planners, 2010–2012

Contents

Contributors

Carlos H. Betancourth Independent International Consultant, chbetanc@msn.com

Hoong-Chor Chin Department of Civil Engineering, National University of Singapore, Singapore, ceechc@nus.edu.sg

Scott Dunn AECOM Technology Corporation, Singapore, scott.dunn@aecom.com

Wee-Kean Fong CTI Engineering International Co., Ltd., Tokyo, Japan, fwkeanjp@yahoo.co.jp

Peter Head Arup (International Consultancy Firm), London, UK, peter.head@arup.com

Chin-Siong Ho Universiti Teknologi Malaysia, Johor Bahru, Johor, Malaysia, ho@utm.my

Walter Jamieson College of Innovation, AECOM Technology Corporation, Singapore, wjtourism@hotmail.com

Asfaw Kumssa United Nations Centre for Regional Development (UNCRD) Africa Office, Nairobi, Kenya, asfaw.kumssa.uncrd@undp.org

Debra Lam Arup (International Consultancy Services), London, UK, debra.lam@arup.com

Gissella B. Lebron Natural Sciences & Science Education, National Institute of Education (NIE), Singapore, gissella.lebron@nie.edu.sg

Steffen Lehmann Research Centre for Sustainable Design & Behaviour, University of South Australia, Adelaide, SA, Australia, steffen.lehmann@unisa.edu.au

Reuben Mingguang Li Institute of High Performance Computing, Agency for Science, Technology and Research, Singapore, lirm@ihpc.a-star.edu.sg

Eleanor Smith Morris Commonwealth Human Ecology Council, London, UK, emorrischec@yahoo.co.uk

Selin Mutdoğan Hacettepe University, Ankara, Turkey, selinse@hacettepe.edu.tr; smutdogan@gmail.com

Isaac K. Mwangi United Nations Centre for Regional Development (UNCRD) Africa Office, Nairobi, Kenya, isaac.mwangi.uncrd@undp.org

T.L. Tan Natural Sciences & Science Education, National Institute of Education (NIE), Singapore, augustine.tan@nie.edu.sg

Meine Pieter van Dijk Water Services Management, UNESCO-IHE Institute for Water Education, Delft, The Netherlands; Urban Management, ISS, Erasmus University in Rotterdam, Rotterdam, The Netherlands, m.vandijk@unesco-ihe.org

Tai-Chee Wong National Institute of Education, Nanyang Technological University, Singapore, taichee.wong@nie.edu.sg

Belinda Yuen Singapore Institute of Planners, Singapore, belyuen8@gmail.com

About the Editors

Tai-Chee Wong received his BA and MA from University of Paris (Urban & Regional Planning), and PhD from the Department of Human Geography, Research School of Asian and Pacific Studies, Australian National University. He is currently Associate Professor at National Institute of Education, Nanyang Technological University, Singapore. He teaches urban geography and planning courses, and was Visiting Professor to Institute of Geography, University of Paris IV-Sorbonne in 2007. His main research interests are in urban and regional issues on which he has published books and many articles in international journals. His five latest books are: *Four Decades of Transformation: Land Use in Singapore 1960–2000* (Eastern University Press, Singapore 2004) and *A Roof Over Every Head: Singapore's Housing Policies between State Monopoly and Privatization* (Sampark and IRASEC 2005); Edited volume with B. J. Shaw & K-C Goh, *Challenging Sustainability: Urban Development and Change in Southeast Asia* (Marshall Cavendish Academic, 2006); Edited volume with B. Yuen & C. Goldblum, *Spatial Planning for a Sustainable Singapore* (Springer, 2008); Edited volume with Jonathan Rigg, *Asian Cities, Migrant Labour and Contested Spaces* (Routledge, 2010).

Belinda Yuen is council member, Singapore Institute of Planners. She has been President, Singapore Institute of Planners (2005–2008), Vice-President, Commonwealth Association of Planners (2006–2008; 2010–2012), member of United Nations Commission on Legal Empowerment of the Poor Working Group and advisory board member of several UN-HABITAT flagship urban publications and research network. Belinda is a qualified urban planner. She has a MA (Town and Regional Planning), University of Sheffield, UK and PhD with focus on environmental planning, University of Melbourne, Australia. Belinda has served on various local planning committees of Singapore including as Planning Appeals Inspector, subject group of Singapore Master Plan 2003, Concept Plan 2011, Action Programme Working Committee of Singapore Green Plan 2012. Her research includes spatial planning and urban policy analysis, most recently on planning livable, sustainable cities and vertical living. Belinda is on the Editorial Board of *Asia Pacific Planning Review; Regional Development Studies; Cities; Journal of Planning History*.

About the Authors

Carlos H. Betancourth is a PhD candidate at Columbia University and has been working since 2003 as an independent international consultant on Sustainable Urban Re-design. His current research and work aims at filling in an important gap on urban sustainability, namely, the crucial importance of networked urban infrastructures and their re-design as weaves of eco-infrastructures for the development of feed-back loop urbanisms and networks of zero carbon settlements, as strategic responses for the ecological sustainability and safety of cities in the context of climate change, resource scarcity and risk. Carlos has been working internationally with various communities, governments and companies on Sustainable Urban Development for the European, American and Latin-American Regions. He is currently involved in many collaborative projects on eco-infrastructures, adaptation planning in Belize Mexico, Spain and New Mexico (USA). His latest publications include: *Urban responses to climate change: Creating secure urbanities through eco-infrastructures; self-enclosed spaces and networks of zero carbon settlements*: the case of Cartagena, Colombia. World Bank, Fifth Urban Research Symposium, Marseille (2009).

Hoong-Chor Chin is Associate Professor and Director of Safety Studies Initiative, Department of Civil & Environmental Engineering, National University of Singapore. He holds a PhD in Transportation Engineering from University of Southampton. His areas of specialization include transportation planning, transport systems modelling and transportation safety and is consultant to the Asian Development Bank and Cities Development Initiative for Asia in several regional transportation planning and safety projects. He has also undertaken numerous traffic planning and safety studies in Singapore. Among his publications are chapter contributions, "Urban Transportation Planning in Singapore" in the book on Infrastructure Planning for Singapore and "Modeling multilevel data in traffic safety: A Bayesian hierarchical approach" in the book "Transportation Accident Analysis and Prevention". He won the UK "Institution of Civil Engineers" Webb Prize for his innovative work on Benchmarking Road Safety Projects, and has been on several government committees to review land transport policies in Singapore.

Scott Dunn is the Regional Managing Director for Planning, Design and Development (PDD) at AECOM Technology Corporation in Southeast Asia. Scott leads multidisciplinary teams of design and planning professionals on projects ranging from large-scale resort developments to mixed-used new communities and high-density master plan developments across Asia and the Middle East. Scott has won numerous awards in master planning and architecture, was published in several design magazines and is a highly-regarded speaker on thought leadership. He is also active in lecturing and teaching on issues of sustainable resort development and community building in Southeast Asia, India, Korea and Hong Kong. Over the past 17 years, Scott has been involved in various golf community developments such as the Shenzhou Peninsula Golf Community project and has also been in charge of numerous resort planning projects, including the award-winning Subic Resort Master Plan in the Philippines.

Wee-Kean Fong holds a Bachelor Degree in Urban and Regional Planning from the Universiti Teknologi Malaysia, Master of Engineering and Doctor of Engineering from the Toyohashi University of Technology, Japan. He is a Senior Associate at the China Office, World Resources Institute (WRI) where he leads WRI China's works in city-level greenhouse gas accounting program and low-carbon city planning with his extensive experience in these areas. Before joining WRI, Fong was affiliated with a Tokyo-based international consulting firm and was involved in a number of Japanese official development assistance (ODA) projects. He has gained international project experience in several Asian countries including Malaysia, where he built his strong technical background in environmental management and urban and regional planning.

Peter Head is Consultant at Arup, an international consultancy firm involved in designing eco-cities globally, and a champion for developing practices in promoting sustainable development principles. He has won many awards for his work, including the Royal Academy Silver Medal, Award of Merit of IABSE and the Royal Academy of Engineering Sir Frank Whittle Medal for innovation in the environment. He joined Arup in 2004 to create and lead their planning and integrated urbanism team. He was appointed in 2002 by the Mayor of London as an independent Commissioner on the London Sustainable Development Commission and led the planning and development subgroup of the Commission. He was also project director for the planning and development of the Dongtan Ecocity on Chongming Island in Shanghai and other city developments in China for the client Shanghai Industrial Investment Company. He supported the development of a Zero Carbon housing project in Thames Gateway and now Chairs the new Institute for Sustainability nearby. He was awarded a CBE in the Queen's New Year Honours for services to civil engineering and the environment. In 2008 he was nominated by Time Magazine as one of 30 global eco-heroes.

Chin-Siong Ho is currently the Deputy Director of the Office of International Affairs and Professor of Faculty of the Built Environment at Universiti Teknologi Malaysia, as well as Senate Member of the University. He received his B.A in

Urban and Regional Planning from Universiti Teknologi Malaysia, MSc from Heriot Watt University, Edinburgh, UK and Doctor of Engineering from Toyohashi University of Technology, Japan in 1994. He is registered member of the Board of Town Planning Malaysia (MTPB) and corporate member of Malaysian Institute of Planning (MIP). He was a post-doctoral fellow under Hitachi Komai Scholarship to Japan in 1995 and Royal Society of Malaysia/Chevening Scholarship to United Kingdom in 2005. His research interests are in urban sustainable development, energy-efficient city, low carbon city planning, and Built Environment education. His published books include: *Introduction to Japanese City Planning UTM* (2003), *Encyclopedia of Laws and Planning Administration of Town and Country Planning Malaysia* (2003 in Malay language) and *Best Practice of Sustainable Development* by Asian Development Bank (2006).

Walter Jamieson holds a PhD from the University of Birmingham, England, M.Sc. from Edinburgh College of Art/Heriot-Watt University, Scotland, and M.E.S. from York University, Toronto, Ontario. He has been involved in academia in Canada, Thailand and the United States as well as consultancy activities in over 20 countries over the last 35 years. He presently is the Sustainable Tourism Planning and Development Specialist for AECOM in Asia. Formally Dean of the School of Travel Industry Management at the University of Hawai'i at Manoa and prior to that member of faculty and administration within the Faculty of Environmental Design at the University of Calgary. His consultancy activities include working with the United Nations World Tourism Organization, ESCAP and UNESCO. He has published and presented widely for over 135 papers, and lectures. His latest publication is Managing Metropolitan Tourism: An Asian Perspective published by the United Nations World Tourism Organization.

Asfaw Kumssa is the coordinator of the United Nations Centre for Regional Development (UNCRD) Africa Office, Nairobi, Kenya. He earned his M.S. in national economic planning from Odessa National Economic Planning Institute, Ukraine, and a M.A. and Ph.D. from Graduate School of International Studies, University of Denver, U.S.A., where he was subsequently an adjunct professor of economics and political economy. Kumssa has published in *International Journal of Social Economics, International Review of Administrative Sciences, the Journal of African Studies, Journal of Social Development in Africa, Social Development Issues, Regional Development Studies;* and *Regional Development Dialogue*. He co-edited a book with Terry G. McGee, *Globalization and the New Regional Development*, Vol. 1. (2001) and co-edited another book with John F. Jones, *The Cost of Reform: the Social Aspect of Transitional Economies* (2000).

Debra Lam graduated in Foreign Service at Georgetown University and has a graduate degree in public policy at University of California, Berkeley. She is a senior policy consultant at Arup (International Consultancy Services). Having several years of international experience in governance, sustainable and strategic development, best practices, government analysis and policy, and project management, Debra now works on bringing governance, policy and sustainability into city planning, stakeholder engagement, local capacity building, and project implementation. Her ongoing research and analysis includes low carbon strategies for new built and retrofit, climate change adaptation and mitigation, and overall resilience of local governments. She works closely with local stakeholders in assessing key issues; strengthening their governance, process, and policy; and coordinating key roles/responsibilities towards implementation, operation and monitoring and evaluation.

Gissella B. Lebron received her Bachelor of Science degree in Chemistry from De La Salle University in Manila, Philippines where she also completed the required academic coursework leading to a Master of Science degree in Physics. In 2007 she took the nationwide Licensure Examination for Teachers in the Philippines and ranked 8th overall. She had worked as a secondary school Physics teacher, a college instructor, a textbook writer and editor in the past. Presently, she is working as a full-time research assistant while pursuing her Master of Science degree by Research at the National Institute of Education, Nanyang Technological University, Singapore under the supervision of Dr Tan. Together, she and Dr Tan have published journal articles.

Steffen Lehmann received his doctorate from the Technical University of Berlin and is Professor of Sustainable Design and Director of the Research Centre for Sustainable Design and Behaviour (sd+b), at the University of South Australia, Adelaide. Since 2008, Steffen holds the UNESCO Chair in Sustainable Urban Development for the Asia-Pacific Region. He is currently the General-Editor of the US-based Journal of Green Building. Over the last 15 years, he has presented his research at over 350 conferences in 25 countries. His research includes sustainable design for high performance city districts and buildings, design strategies for green urbanism and healthy cities, as well as urban regeneration through the reuse of buildings and materials. Besides winning architectural awards, he is acknowledged as a leader in the emerging field of green urbanism and regularly consults with companies and governments on issues of sustainable design, integration of technology and the built environment. His latest books include: "Back to the City", Hatje Cantz Publisher (Stuttgart, 2009); "The Principles of Green Urbanism", Earthscan (London, 2010); and the forthcoming book: "Designing for Zero Waste", Earthscan (London, 2011). See also: www.slab.com.au.

Reuben Mingguang Li received his Bachelor's degree from the National University of Singapore with First Class Honours in Geography. He won the President's Honour Roll of the University Scholars Programme for outstanding academic results and co-curricular contributions in the course of study. He is currently a Research Officer with the Institute of High Performance Computing under the umbrella of the Agency for Science, Technology and Research, Singapore. He concurrently pursues postgraduate studies at the National University of Singapore focusing primarily on the spatio-temporal variability of the urban thermal environment of Singapore. His main research interests are in spatio-temporal studies, climate change modelling, urban climatology, and applied Geographic Information Systems (GIS) and remote sensing.

Eleanor Smith Morris is currently Chairman, Executive Committee, Commonwealth Human Ecology Council, London. She was Visiting Professor of Urban and Environmental Planning, Clemson University, South Carolina (USA) in 2003. Previously Eleanor had been the Academic Director of the Centre for Environmental Change and Sustainability, Faculty of Science, Edinburgh University, Scotland, U.K. and had been Lecturer in Urban Design and Regional Planning, University of Edinburgh. She received her Doctorate from Edinburgh University, her Master's Degree from the University of Pennsylvania and an A.B. Architectural Sciences (Hons) from Harvard University and is a member of both the American Institute of Certified Planners and the Royal Town Planning Institute. She was Chairman of the Royal Town Planning Institute of Scotland (1986–1987) and served on the Council, Executive and Buildings Committees of the National Trust for Scotland. She has published over 100 articles and reports on urban design and town planning, written for BBC television and organised over 14 town planning conferences. Her publications include: British Town Planning and Urban Design (Longman, Harlow, 1997) and James Morris, Architect and Landscape Architect, (Royal Scottish Academy, Edinburgh, 2007).

Selin Mutdoğan received her Bachelor Degree from Faculty of Art, Design and Architecture, Department of Interior Architecture and Environmental Design in 2001 from Bilkent University, Turkey. She received her M.A. degree with the thesis titled "Analysis of interior spaces of contemporary housing according to psychosocial determinant" in Hacettepe University, Department of Interior Architecture and Environmental Design in 2005. She is currently studying for her PhD in the same department. Her research topic is related to the sustainable design and sustainable strategies for high-rise housing units especially for interiors. Since 2005, Mutdoğan is a research assistant in Hacettepe University.

Isaac K. Mwangi is curriculum, research and capacity building expert at UNCRD Africa Office. He earned his B.Sc (Hons.) and M.A. (Planning) from the University of Nairobi, Kenya and a PhD from School of Planning, University of Waterloo, Ontario, Canada. His teaching and research experience at the University of Nairobi include the areas of planning law, urban development administration and region development planning. Mwangi is a Fellow of the Kenya Institute of Planners of which he is the founding Vice-Chairman. He has served as researcher and consultant in urban and regional planning and development for the UN-HABITAT, International Finance Corporation (IFC) and Netherlands Development Organization (SNV). He has published in the *UN-HABITAT Publications, Eastern and Southern Africa Geographical (ESAG) Journal, ACTS Research Programme, Plan Canada, Spring Research Series,* and other international journals.

T.L. Tan is Associate Professor in Physics in Natural Sciences & Science Education, National Institute of Education (NIE), Singapore. He received his PhD by research in Physics from the National University of Singapore in 1993. He spent his post-doctoral training in Steacie Institute of Molecular Science (formerly Herzberg Institute of Astrophysics), Ottawa, Canada and in University of Washington, Seattle, USA, in high-resolution Fourier transform infrared (FTIR) spectroscopy of gases of atmospheric interests. Later, he worked for four years as a senior research engineer in Hewlett-Packard, Singapore, specialising in material characterization techniques such as FTIR and Raman spectroscopy, scanning electron microscopy (SEM), transmission electron microscopy (TEM), and x-ray fluorescence (XRF). To date, he is the author or co-author of 120 papers in international journals of USA and Europe, and referees papers for several journals. His present research extends to the studies of toxic gases in indoor air quality (IAQ) of buildings using infrared techniques. Since 2005, he is appointed Associate Dean for Academic Research, Graduate Programmes and Research Office, NIE, Singapore.

Meine Pieter van Dijk (PhD Economics Free University Amsterdam) is an economist and professor of Water Services Management at UNESCO-IHE Institute for Water Education in Delft, professor of entrepreneurship at MSM and professor of Urban management at the Institute of Social Studies and at the Economic Faculty of the Erasmus University in Rotterdam (EUR), all in the Netherlands. He is member of the research schools CERES and SENSE. He worked on and in developing countries since 1973 and as a consultant for NGOs, the Asian Development Bank, the Inter-American Development Bank, the World Bank, different bilateral donors and UN agencies. He recently edited a volume of the International Journal of Water on the role of the private sector in water and sanitation. His recent books are on the new presence of China in Africa (Amsterdam: University Press, 2009); Managing cities in Ethiopia (eds, with J. Fransen, Delft: Eburon, 2008); Managing cities in developing countries, the theory and practice of urban management (2006, Cheltenham: Edgar Elgar) and with C. Sijbesma (eds., 2006): Water in India (New Delhi: Manohar). Since 2000, he published, among many others, these books: with E. Guiliani, R. Rabelotti (eds, 2005): Clusters facing competition: The importance of external linkages. Aldershot: Ashgate; 2006, Managing cities in developing countries: urban management in emerging economies, in Chinese. Beijing: Renmin University Press; with Tegegne G/Egziabher (eds, 2005): Issues and challenges in local and regional development, Decentralization, urban services and inequality. Addis Ababa: University RLDS; with M. Noordhoek and E. Wegelin (eds): Governing cities, New institutional forms in developing countries and transitional economies. London: ITDG.

Chapter 1
Understanding the Origins and Evolution of Eco-city Development: An Introduction

Tai-Chee Wong and Belinda Yuen

1.1 Introduction

The world is increasingly urban. Since 2008, more than half of the world's population is living in urban areas. The number of urban residents is expected to continue to grow, especially in developing countries. In Asia, some 1.1 billion are anticipated to move to cities in the next 20 years (Kallidaikurichi and Yuen 2010). This includes 11 megacities, each with a population exceeding 10 million, for example, Beijing, Shanghai, Kolkata (Calcutta), Delhi, Jakarta and Tokyo. With the exception of Tokyo, the rest are in developing countries. The expanding urban population will require a whole range of infrastructure, services, housing and jobs, not to mention land. The urban land expansion could threaten agricultural land supply, cause growth in traffic volumes and increased pressure on the environment, and be massively unsustainable for the country and the rest of the planet. It is vital that sustainable urban development be pursued as cities continue to grow.

Dramatic urban demographic expansion and keen competition with globalization have called for urgent actions in the management of the human–environment interactions especially in the wake of rising consumerism. Consumerism has added to the worsening conditions of environmental degradation in the developed world and is spreading to the developing world, especially the fast growing economies of China and India in recent decades. To make matters worse, the global shift of manufacturing industries from advanced nations (since the oil crisis in the mid-1970s) to developing countries is also transferring sites of industrial and household wastes, and carbon emissions to the developing world (Randolph 2004, Jayne 2006, Roberts et al. 2009, Dicken 2005). For the latter the urge to use domestic consumption as a means to bolster economic growth, their more rapidly rising urban population, relatively low levels of environment-led technologies, management and civic awareness in environmental protection all contribute to the urgency for action.

T.-C. Wong (✉)
National Institute of Education, Nanyang Technological University, Singapore
e-mail: taichee.wong@nie.edu.sg

T.-C. Wong, B. Yuen (eds.), *Eco-city Planning*, DOI 10.1007/978-94-007-0383-4_1,

Indiscriminate material consumption patterns if unchecked can contribute to large amounts of wastes and unsustainable development of cities (Girardet 1999). Mounds of solid wastes on dump sites of many cities in developing countries visibly illustrate this challenge. Wastes of plastic materials, for instance, are durable and resistant to natural processes of degradation as their total natural decomposition may take hundreds or thousands of years. Furthermore, burning plastics could produce toxic fumes and manufacturing of plastics often creates chemical pollutants. The cycle of modern production, consumption and disposal which motivates urban metabolism must be re-examined from a new perspective.

Ecologists have long argued for equilibrium with basic ecological support systems, and since the 1987 Brundtland Commission, the notion of sustainable development has taken on renewed and urgent currency (Daly 1991, United Nations 1987, Silvers 1976). The notion of sustainable development enjoins current generations to take a systems approach to urban growth, and to manage resources – economic, social and environmental – in a responsible manner for their own and future generation's enjoyment in line with the Earth's carrying capacity. Over the years, various writers from a range of disciplines have expounded the concept, and suggested ways to measure, monitor and implement sustainability (see, for example, Aguirre 2002, Kates et al. 2005, Hasna 2007, Boulanger 2008). In the main, the objectives have been to direct urban development towards minimizing the use of land, energy and materials, and impairment of the natural environment while maximizing human well being and quality of life. The implication is that settlement patterns need to be liveable, attractive while sustainable, and this can be achieved through ecological planning.

Urban land use planning can no longer afford to be merely anthropogenic (human-centred). Instead, it has to also consider environmental issues including the interdependency of human and non-human species and the "rights" and "intrinsic values" of non-human species in our pursuit for a sustainable ecosystem. It has to be ecological. Ecological planning involves conceptual thinking in environmental urban sustainability, land use allocations, spatially designed and distribution patterns that contribute and lead to achieving such objectives of ecological balance. Yet, the in-principle outcome should not be detrimental to aggregate economic development without which environmental sustainability efforts might remain a lip-service. In other words, the logic and *modus operandi* of ecological planning should be also contributing to economic progress. How this is effectively done will be a challenging task ahead.

This book explores one of the widely emerging settlement patterns of eco-city. The premise, origin and evolution of the notion of eco-cities are examined in this chapter.

1.2 The Visions of Eco-city

Appealing to live harmoniously with nature is nothing new in human history. Ancient philosophers and thinkers in both Western and Oriental civilizations observed the omnipresent mightiness of natural forces in influencing human habitation and cultural life. More than 2,500 years ago, Lao Zi had propounded the Taoist concept – Dao (the path), laying the core regulatory rule that stresses the essence of balanced and interdependent developments of Heaven, Earth and humans (Zhan 2003). Taoist thoughts giving due respects to nature are generic, universal, albeit aspatial in implication, and remain influential in modern societies where Taoism are practiced, for example, China and Taiwan.

An eco-city by its very appellation is place-specific, characteristically spatial in significance. It suggests an ecological approach to urban design, management and towards a new way of lifestyle. The advocacy is for the city to function in harmony with the natural environment. This implies that cities should be conceptualized as ecosystems where there is an inherent circularity of physical processes of resources, activities and residuals that must be managed effectively if the city's environmental quality is to be maintained. As Wolman (1965) suggested, there are major physical inflows to the city and outflows from it that should be accounted for, and more importantly, integrated to the rest of the biospheric web. To this recognition, eco-cities are designed with consideration of socio-economic and ecological requirements dedicated to the minimization of inputs of energy, water and food, and waste output of heat, air pollution, etc so as to create an attractive place to live and work.

The term "eco-city" is widely traced to Richard Register's (1987) book, *Ecocity Berkeley: Building cities for a healthy future*. Register's vision of the eco-city is a proposal for building the city like a living system with a land use pattern that supports the healthy anatomy of the whole city, enhances biodiversity, and makes the city's functions resonate with the patterns of evolution and sustainability. Some of the strategies used to manage this balance include building up instead of sprawling out, giving strong incentives not to use a car, using renewable energy and green tools to make the city self-sustaining. Eco-cities would characteristically comprise compact, pedestrian-oriented, mixed-use neighbourhoods that give priority to re-use of land and public transport. Since then, several similar themes such as "eco-neighbourhoods", "urban eco-village" and "eco-communities" have emerged, all emphasizing ways of making the city more environment-friendly and sustainable (Roseland 1997, Barton 2000). It should be stated that notions of ecological planning and design are not new in the planning literature.

Tracing the more recent Western civilization following the Industrial Revolution, a review of the work of nineteenth century planning pioneers such as Frederick Law Olmsted, Patrick Geddes and Ebenezer Howard would indicate views of landscape as a living entity and their concern for preservation of nature beauty and ecological function with planning tasks (Hall 1996, Welter and Lawson 2000). Geddes, for example, proposed the idea of a bioregion where he highlighted the importance of a comprehensive consideration of the interrelationship between cities and their surrounding ecosystem. He gave emphasis to survey-analysis-plan, in particular, that a regional scale survey of the ecological environment and the place-work-folk relationships should be conducted before developing any planning concept and development project. Similarly, Howard in his influential garden city concept argued for the importance of bringing nature back to cities, and suggested the need of decentralization and urban containment for managing urban growth.

Into the twentieth century, these early ideas were expanded on by Lewis Mumford and Clarence Stein, leading to the development of several greenbelt towns in USA (Parsons 1990, Luccarelli 1995). Mumford (1961, 2004) identified the unsustainability of urban development trends in the twentieth century, arguing for "the development of a more organic world picture, which shall do justice to all the dimensions of living organisms and human personalities" (p. 567). In his work published in 1938 "The Culture of Cities", Mumford (1997) associated cities as "a product of Earth [and as] a fact of nature". For him, urban culture was faced with crises, harmful to the local community culture. Urban sprawl accompanying massive suburbanization was particularly seen as having created a series of new social problems.

Moving on, others such as Ian McHarg (1969) have developed the concept of ecological planning, proposing the theory and methodology of ecological land use planning that explicitly connected ecology theory to planning and design practice and laid a new integration of human and natural environments. Urban ecological concerns of McHarg's *Design with Nature* published in 1969 spread fast in practical terms to continental Europe, especially the Netherlands. In Utrecht and Delft wetland layout, nature-imitating features (logs, stones, wild rose) were landscaped around office and housing blocks. Some old buildings in The Hague were dismantled and replaced with cuddle garden for children (Nicholson-Lord 1987: 110–111). Quite uniquely, the Dutch experience reflected a social-cum-human driven response with an artificial but natural setting to fit harmoniously into their habitat of dyke, polder and reclaimed land on which concrete structures have been introduced! It also had strong influence in North America on *New Urbanism*.

Several other planners and designers have also worked on applying the theory of landscape ecology to land use planning (see, for example, Dramstad et al. 1996), and developed new urban design theories related to *New Urbanism* (see, for example, Calthorpe 1993) in which they try to integrate an array of related concepts including ecology, community design and planning for a liveable and walkable environment. *New Urbanism* emerged in the 1980s as a strategy with new typologies in land use to deal with the ecological weakness arising from the massive scale of post-war sprawling suburbanization, which has led to a landscape of low-density, single family dwellings, almost totally automobile dependent lifestyle. With no intention to replace the low-density suburbia prevalent in the United States, a group of young American architects initiated building designs that capitalized on natural resources in constructing environmentally sustainable buildings.

A key development strategy is to promote sensitive urban development that preserves open space and ecological integrity of land and water, that is, a balance of city and country. These qualities may be achieved through a wide variety of means including urban consolidation, various methods to reduce traffic and urban heat island effect, encourage greater use of renewable energy, green roofs and public transport, a holistic approach to nature, history, heritage, health and safety, and a life cycle approach to energy, resources and waste. Much of the elements highlighted in *New Urbanism* such as transit, walkability, environmental sustainability and social integration came close to the present-day eco-city notions. Led by pioneers, Andres

Duany and Elizabeth Plater-Zyberk, this *New Urbanism* model which combined the "green design" ethic and individualistic home ownership "doctrine" of the American dream tradition gained acceptance in Kentlands, Maryland and Windsor in Florida, United States (see Kelbaugh 2002).

Another important source of thinking that has contributed to the conceptualization of eco-cities is indisputably the environmental ethics.

1.2.1 Environmental Ethics

History of environmental ethics could be traced to 1962 when Rachel Carson (1962) published her book *Silent Spring* that revealed the harmful effects of pesticides to humans and other creatures. With an initial concern over the death of birds, she showed how farming practices using DDT as a pesticide could affect the food web, and hence the living and public health. Despite being attacked for exaggerating the impact, her thinking and ideas were seen to have set the cornerstone of modern environmentalism. Her love of nature, especially birds and natural plants challenged the anthropocentric development practices that put humans as the central figure that count on Earth. Richard Routley (1973) followed suit by addressing the issue of human chauvinism in which humans were treated as a privileged class; all other species had been discriminated against. Again, this would not be helpful to ecological balance. During the 1970s, there were ethical, political and legal debates to support animal rights in the ethical thinking. The rise of "Green Parties" in Europe in the 1980s further condemned the anthropogenic approach that had contributed substantially to environmental devastation, and rising levels of pollution.

The key interpretation of the anthropogenic approach is that it serves human-centred instrumental values of identified ends but neglects the intrinsic values of all living things in existence that forms the basis of interdependent ecosystems. For example, trees with little or no commercial value are not looked upon as useful and therefore should be disposed of though their contribution to the ecological balance is considerable. Arguably, as humans have no ecological superiority compared to other non-human species on Earth and since the latter's extinction can affect human species' own existence, an anthropogenic approach is self-destructive.

Quite along the same line, the works of Naess (1973, 1989) in the 1970s and late 1980s exposed the aims of the deep ecology movement that supports the "biospheric egalitarianism". This egalitarianism stipulates principles that all living things are alike in having value in their own right, independent of their usefulness to others. Naess' idea has been interpreted as "an extended social-democratic version of utilitarianism", which counts human interests in the same calculation alongside the interests of all natural things in the natural environment. Nevertheless, the deep ecology theory was criticized as being inadequate, acting as "a disguised form of human colonialism", unable to give nature its due status, and being elitist serving "a small selected well-off group" (see Stanford Encyclopedia of Philosophy 2008).

Taken as a whole, activists promoting ethical environmentalism have acted as a counter force against the Western traditional ethical theories such as utilitarianism which are associated with the values (pleasure) and disvalue (pain) (ibid, Nash 1989). Whilst utilitarian followers are more inclined to support anthropogenic sources of pleasure and have little concern to non-sentient beings (for example, plants, mountains, rivers), ethical environmentalists attribute more intrinsic values to the natural environment and its inhabitants. The latter's environmental ethics correspond with the objective of eco-city promoters, and they share in many aspects the urgent need to manage production and consumption in a sustainable way.

1.3 Towards Sustainable Production and Consumption

Modern urban-industrial consumption patterns and habits differ in essence from those of the pre-industrial and feudal times characterized by low-productivity and consumption levels meeting largely basic needs. Not only is the modern industrial age much more productive in producing daily needs, but the consumer goods designated for the market place involve use of unnatural sources often harmful to the ecological system. More significantly, the prevailing market economy relies on large scales of consumption to justify its profitability and corporate survival or expansion.

Consumption cultures based on material possessions have increasingly been related to fashionability rather than durability. Consumerism and consumer ethic, according to Corrigan (1997, cited in Jayne 2006: 27), first developed among the aristocrats during the sixteenth century Elizabethan period but only blossomed after the Industrial Revolution in late eighteenth century with the advancement in industrial capitalism and its production technologies, that enabled consumption of rare consumer goods to reach a much larger cohort of consumers and could render them to show social prestige and status. A sharp turn took place in the post-World War II period. With further technological progress, aided by the Fordist mode of production and world-scale marketing strategies, consumer goods became highly accessible in developed countries, especially private automobiles. Today, in the midst of environmental preservation, consumerism has become a collective consumption lifestyle in the developed world and has also spread to the more affluent social groups in the developing world. In the face of increasing environmental degradation, unsustainable consumerism is being questioned and sustainable consumption is being elevated to the international forum as a balancing force.

Unsustainable consumerism in daily practice is inherently distinct from environmental ethic in theory discussed earlier. However, individuals with an environmental ethic and awareness could be contradictory in actions if consumption is seen as an individual's lawful right and he/she is not prepared to give up his/her preferences. Environmental ethical consciousness or citizen preferences, as Sagoff (1988) suggested, are judgments about what one should do whereas consumer preferences mean to do what one desires to possess or consume. Satisfying individuals' massive scales of desires could be ecologically disastrous in some cases. But if enormous economic sacrifice is needed to achieve insignificant pollution or contamination control, the role of sustainable production and consumption acting as a compromising agent is very useful.

What is sustainable production and consumption? Sustainable consumption must be matched by sustainable production regulated by demand management which does reliable valuations of natural resources and arouses public awareness in recycling, reduction and reuse of materials. Technologies employed in the sustainable production processes are those that protect the environment, are less polluting and handle all residue wastes in environment-friendly ways. The methods of production would use much fewer resources and generate close to zero waste (Newman and Jennings 2008: 188–189, White 2002). In light of the large gaps between affluent nations and poor countries, meeting the basic needs of the latter is crucial to ensure environmental, economic and social sustainability which are interdependent and mutually reinforcing. For the urban poor in many African and Asian countries, for instance, sustainable consumption implies not so much material consumption of consumer goods but more the safeguarding of their living environment often built precariously on poorly serviced quarters of the cities.

The future direction of sustainable consumption would need to promote conserver lifestyle yet maintaining a high quality of life. Looking from the perspective of more developed societies, Newman and Jennings (2008: 191–198) have conceived a series of sustainable consumption strategies, as listed below:

(1) Voluntary simplicity strategy
Disapproving consumerism and viewing overconsumption as an illness in society, this strategy aims to assist people to find alternative ways to satisfy their needs and promote simple ways of living;

(2) Demand management strategy
Education is sought to educate consumers the ways in which to meet one's needs without consuming much non-renewable resources. The premise is however that reducing resource use should not mean lowering quality of life. Application of this strategy needs to be adopted at both household and corporate levels, in order to achieve a meaningful reduction as a consequence.

(3) Sustainable procurement strategy
Government and institutions, together with households should adopt purchasing programmes using the notion of sustainability. This sustainable shopping behaviour should build up more sustainable markets by consuming less. More attention should be directed towards more environmentally sound products.

(4) "Slow movement" strategy
"Slow food", "slow cities" and "slow traffic" are three elements of the "slow movement" strategy that are anticipated to help cut down consumption. "Slow food" is to counteract fast food and fast life in an attempt to rediscover the real taste of authentic and local/regional food sources and quality pace of life. "Slow cities" place emphasis on small towns and cities, with preference modelled after

the European late medieval and renaissance era. "Slow traffic" calls for traffic calming in favour of small road capacity emphasizing walking, cycling and transit.

The above strategies are apparently more relevant to more developed societies. Most of these societies are in post-industrial stage of development where material shortage is not a major issue. The notion of "small cities" appears idealistic and nostalgic in sharp contrast to the current global trend of mega-urbanization, taking place at grandiose scale globally. Given the diverse socio-economic backgrounds between the developed and developing worlds, it is understandable and logical that the strategies of sustainable eco-city development must follow the specificities and circumstances of the adopting countries.

1.4 Emerging Trends: Building One's Own Tailored-Made Eco-towns or Cities

Cities are different. Serving the identical purpose of environmental sustainable development, different countries have adopted different approaches in implementing their own eco-city development programmes or schemes. Criteria used and standards set would be localized in accordance with financial and technological capabilities that one could afford. The eco-city index system worked out recently by a group of Chinese researchers, for instance, has taken into consideration the local urban physical features as a basis of implementation reference (see Li et al. 2010). In approach, the "one size fits all" equation must be ruled out when dealing with environmental sustainable issues.

Eco-cities are on the rise in different parts of the world. In the Middle East, Abu Dhabi in 2006 has initiated a US$22 billion project to build the world's first carbon neutral city, Masdar. The city is planned on a land area of 6 km^2 for a population of 45,000–50,000, setting new standards in green living including clean power, desalinization plant run on solar power, magnetic trains for transportation (cars are not welcome), and 100% waste recycling. In the United Kingdom, the Prime Minister, Gordon Brown, announced in 2007 the building of 10 eco-towns across the country (BBC News 24 Sep 2007). A new planning policy statement was published on 16 July 2009 setting out the standards that eco-towns will have to meet.[1] The intention is to offer an opportunity to promote sustainable living and zero carbon development while also maximising the provision of green space and potential for affordable housing.

In China, eco-city building is proposed for not just the big cities like Beijing, Shanghai and Tianjin but also the small- and medium-sized cities of Yuxi, Wehai, Rizhao and Changshu, among others. Under the State Environmental Planning Agency *Guidelines for the Building of Eco-communities* (1996–2050), the intention is to promote the planning and construction of eco-communities across China. The objective is to apply sustainable planning and design principles in the building of new communities. Since then, many countries have offered to help China develop eco-cities. The most advanced of these developments is the Tianjin eco-city developed by the governments of China and Singapore.

In 2007, the Chinese and Singapore governments announced the signing of a collaborative framework to plan and develop a 30 km^2 eco-city at Tianjin. By 2010, the basic infrastructure for the start-up area (4 km^2) has been completed. Development projects with a total gross floor area of more than 800,000 m^2 are under construction. Key performance indicators comprising both short-term and long-term targets for key aspects of the eco-city development such as water and waste management, air and water quality, green buildings and transportation, resource usage and conservation, public housing have been established. The aim is to achieve harmonious living with man, economy and environment. The Sino-Singapore Tianjin eco-city is planned with several distinguishing features including the use of clean, renewable energy; 100% green buildings, an efficient and easily accessible public transport system, extensive greenery, heritage conservation, water recycling and more efficient use of water resources, integrated waste management, development and strengthening of social harmony among residents and specialization in service industries.

Other Chinese cities have followed suit. In January 2010, Kunming (China) was honoured by the United Nations to be the "most leisure and liable green eco-city in China and United Nations liveable eco-city". Endowed with pleasant climate all year round and locational advantage, Kunming has become known as the Chinese brand of model eco-cities (ECN News 2010).

Recently, in 2009, the World Bank has launched the Eco2Cities program, containing many of the world's best practices as well as a comprehensive financial support, analytical and operational framework to help cities adopt the ecological approach as part of their city planning (Suzuki et al. 2009). Some of these best practices include Stockholm – how integrated and collaborative planning and management on the principle of a cyclical urban metabolism can transform an old inner city industrial area (Hammerby Sjostad) into an attractive and ecologically sustainable neighbourhood; Curitiba – how innovative approaches in urban planning, city management and transport planning (such as Bus Rapid Transit) are an investment in the city's economy and welfare; Yokohama – how an integrated approach in waste management combined with stakeholder engagement could significantly reduce solid waste; Vancouver – how a set of basic land use planning principles and inclusive planning can help to create a highly liveable city and region.

Environmentally, eco-city development is used as a new environmental paradigm to counter global warming, ecological degradation and unsustainable resource exploitation. Within this paradigm, ideas of green urbanism, sustainable building design or architecture, promoting more compact cities to fight sprawling are subsumed. Economically, building eco-cities as a green infrastructure has inevitably to be used as a form of new business opportunity serving the objectives of economic sustainability. In the eyes of Richard Register (2006: 214), developing green technologies and turning them to serve a vital economy would help us win a tough and expensive ecological war. Socially, eco-cities have to be made implementable and applicable globally to be effective in countering environmental degradation, even in varied forms and standards. Implementing countries have to consider implementing it against their own budget constraints, key social concerns and development priorities.

1.5 Organization of the Book: The Chapters

The rest of this book is divided into three parts, covering (a) macro-level policies issues, (b) practice and implementation experiences, and (c) micro-level sustainable design and management measures. The intent is to provide both big picture as well as issue-specific discussion on eco-city planning, development and management. Each chapter is written by specialist authors.

"Part I: Macro Strategic Planning: Policies and Principles" comprises four chapters that primarily address some of the key policies and principles relating to eco-city planning and development, illustrated with case examples. Beginning the discussion is Peter Head and Debra Lam who in Chapter 2 have used a generic, strategic and policy-driven approach to examine "How Cities Can Enter the Ecological Age". In particular, they examine the ways in which eco-cities would continue to serve urban residents with clean and healthy necessities such as water and air. They believe feasible policy measures could be put in place through international and cross-border co-operations in low, middle, and high income countries. Eco-friendly-oriented business models will have potential to restrict ecological footprint and take humanity into the future.

Meine Pieter van Dijk's Chapter 3 "Three Ecological Cities, Examples of Different Approaches in Asia and Europe" explicates the interest of developing and developed economies in building eco-cities. Since the 1990s, different urban planning approaches have been used to create eco-friendly neighbourhoods within cities. Three cities are examined in this chapter – Shanghai's Dongtan, Singapore and Rotterdam. These cities offer examples of promising eco-city practices that address the negative effects caused by widespread pollution and mounting waste problems.

In Chapter 4, Carlos Betancourth in his "Eco-infrastructures, Feedback Loop Urbanisms, and Networks of Energy Independent Zero Carbon Settlements", using the context of Latin American cities posits a different urban growth approach based on eco-infrastructures. He argues that urbanization can be a sustainable process through an eco-infrastructure approach that seeks to reduce urban vulnerabilities and apply a series of strategic responses including feedback-loop urbanisms and networks of zero carbon settlements powered by renewable energies.

Scott Dunn and Walter Jamieson in Chapter 5 look at "The Relationship of Tourism and the Eco-cities Concept". Arguably, with rising numbers of cities over one million and tourists, urban tourism will not only imprint a deeper ecological footprint in high density urban agglomerations, but also will be a dynamic sector of hospitality activities. In Asia and elsewhere, eco-tourism has been developed to meet the needs of local residents and tourists, and to protect heritage and environmental values. The planning and development process involves therefore policy measures that develop innovative sustainable tourism in line with the fundamental concepts of eco-cities.

"Part II: Implementation and Practice" contains five chapters. Its thematic focus is on the implementation process and practice of eco-city development from around the world – United Kingdom, China, Singapore, Malaysia, Kenya. Eleanor Smith Morris begins with the complex implementation process of the politically sensitive British eco-towns (Chapter 6). She reviews the ups and downs of eco-town proposals during 2009–2010. Having a rich tradition of new town development in the immediate post-war era, British new towns had brought little success in creating local employment that made public authorities suspicious of the prospects of the proposed eco-towns. Debates on the pros and cons of the proposals were on the agenda of both the Conservative and Labour Parties. The new Coalition Government of Conservatives and Liberal Democrats decided to keep four of the proposed eco-towns, and the general consensus is that eco-towns should be situated adjacent to existing centres of population, transport, infrastructure and employment. In terms of sustainability, the proposed British eco-towns are being tested if they could achieve zero carbon building development, as a source of housing supplier in offering affordable housing, and as a green infrastructure capable of managing waste effectively.

Tai-Chee Wong, in Chapter 7, focuses on the implementation of "Eco-cities in China" whilst he inquires whether eco-cities are merely "Pearls in the Sea of Degrading Urban Environments". Over the last 30 years, economic reforms have created tremendous amounts of material wealth accompanied by unprecedented level of consumption, particularly in the cities. Pollution hazards are so serious that China has now become the largest carbon emitter in the world. This chapter investigates the difficulties in developing an environmentally sustainable urban system via eco-city development while seeing its great potential as an instrument to improve the environment. Eco-city norms and standards such as energy saving, use of renewable energy, public transport, reforestation, recycling of water and other materials are expected to lead a new development path towards a more sustainable urban future in China.

Moving on to Chapter 8, Steffen Lehmann explores the ways in which greenery and green urbanism is being incorporated in city development. In his "Green Urbanism: Holistic Pathways to the Rejuvenation of Mature Housing Estates in Singapore", he argues for more compact, polycentric mixed-use urban clusters, supported by a well integrated public transport network. In mature and aged housing estates, however, rejuvenation and retrofitting by breathing in new air of sustainability is most appropriate. Management of waste, energy, water, public transport, materials and food supply must be done in an integrated manner by bringing in eco-city planning concepts. Further adaptation is required for cities such as Singapore situated in the humid tropical zone. He concludes that good urban governance and leadership is crucial to the success of eco-city development.

Asfaw Kumssa and Issac Mwangi address the sustainable housing problem in urban Africa, a basic need of eco-city development (see World Bank 2010). In their "Challenges of Sustainable Urban Development: The Case of Umoja 1 Residential Community in Nairobi City, Kenya (Chapter 9), they draw on rich local lessons to identify the causes of ineffective planning and implementation. Problems specific to the Umoja 1 Residential Plan include too low capacity of infrastructure provided to meet the residents" demand, poor standards of maintenance, and unreliable supply of clean water supply. Moreover, local interest groups have not actively participated in the communal affairs. Substantial improvement is thus needed.

Chapter 10 prepared by Chin-Siong Ho and Wee-Kean Fong investigates the potential of achieving environmental sustainability in a new growth area in Malaysia. In their "Towards a Sustainable Regional Development in Malaysia – The Case of Iskandar Malaysia", they explore if this economic-driven region in the southern tip of West Malaysia could combine the objective of economic sustainable development with that of environmental sustainable development. This chapter also refers to the success cases of low carbon cities elsewhere and examines the scenarios of transforming the Iskandar economic region into an environmentally sustainable urban region.

"Part III: Design and Micro Local Planning" consists of studies relating to ecological footprint, indoor air quality management and building design approach prepared in three respective chapters. Hoon-Chor Chin and Mingguang Li examine in Chapter 11 the methods of presenting ecological footprint information, a key source of measuring the carbon impact on the environment. Lately, the ecological footprint concept has been a useful tool to measure environmental impact and assess sustainability levels. The authors re-examine the notion of ecological footprint, arguing for a different approach to ecological footprint analysis, with results that help to identify several shortcomings, upon which site improvements could be made.

In Chapter 12, Selin Mutdogan and Tai-Chee Wong examine the efforts made by the Istanbul municipal government to construct a green building environment. In "Towards Sustainable Architecture: The Transformation of the Built Environment in Istanbul, Turkey", they first review international efforts, supported by technological innovations and rising environmental consciousness that had made contribution to building designs. By referring to sustainable architecture and green design in Istanbul, the study uses a chosen set of evaluation criteria to assess the green building standards that the central city buildings along Büyükdere Avenue might have achieved. Results revealed that though standards achieved were low, they reflected a progressive initiative to move towards a high level of urban ecological protection.

The final chapter (Chapter 13) is by Tan and G. B. Lebron who look at indoor air quality control of city buildings acting as shared public spaces in their joint research "Urban Air Quality Management: Detecting and Improving Indoor Ambient Air Quality". As a source of public health hazards, the "sick building syndrome" captures increasing public concern. For example, carbon monoxide is emitted at high concentration levels in buildings through burning of tobacco and incense, and its decay rates in air can be measured using the Fourier transform infrared spectroscopy. The research uses many air-conditioned buildings in Singapore as test samples and basis of analysis.

The collection of papers in this volume provides but a glimpse of the many complex, sometimes inter-related issues of planning and implementing eco-city, a settlement type that is rapidly being created in both developed and developing countries. There is no singular recipe but a range of strategic responses and tools that cities and planners will need to examine and adapt to their own local circumstances in dealing with unsustainable consumption and growth. Eco-city development is not a fad. It is our future.

Note

1. Because of the protests from environmental groups and local residents who questioned the impact of eco-towns on the planning system, transport links, jobs opportunities and the environment, the building programme was scaled down and confirmed to four eco-towns in July 2009 (BBC News 16 July 2009).

References

Aguirre, M. S. (2002). Sustainable development: why the focus on population? *International Journal of Social Economics*, 29(12): 923–945.

Barton, H. (2000). *Sustainable communities*. London: Earthscan.

BBC News (2007). "Eco-towns" target doubled by PM. http://news.bbc.co.uk/2/hi/uk_news/politics/7010888.stm. Accessed 28 August 2010.

BBC News (2009). Four sites to become eco-towns. 16 July 2010 http://news.bbc.co.uk/2/hi/uk_news/8152985.stm. Accessed 20 October 2010.

Boulanger, P. M. (2008). Sustainable development indicators: a scientific challenge, a democratic issue. SAPIENS 1(1), http://sapiens.revues.org/index166.html. Accessed 20 October 2010.

Calthorpe, P. (1993). *The next American metropolis: ecology, community and the American dream*. New York, NY: Princeton Architectural Press.

Carson, R. (1962). *Silent spring*. Boston, MA: Houghton Mifflin.

Corrigan, P. (1997). *The sociology of consumption*. London: Sage

Daly, H. E. (1991). *Steady-state economics*. Washington, DC: Island Press.

Dicken, P. (2005). *The global shift: mapping the changing contours of the world economy* (5th ed). London: Sage.

Dramstad, W. E., Olson, J. D. & Forman, R. T. (1996). *Landscape ecology principles in landscape architecture and land use planning*. Washington, DC: Island Press.

ECN News (2010). Kunming honored liable eco-city by the United Nations. http://www.ae-eco-city.net/ae_ecocity/News/Industry%20news/20100120/58966.shtml & http://www.cityup.org/ae_ecocity/News/index.shtml. Accessed 01 October 2010.

Girardet, H. (1999). The metabolism of cities. In D. Banister, K. Button & P. Nijkamp (Eds.), *Environment, land use and urban policy* (pp. 352–361). Cheltenham: Edward Elgar.

Hall, P. (1996). *Cities of tomorrow: an intellectual history of urban planning and design in the twentieth century*. Oxford: Blackwell.

Hasna, A. M. (2007). Dimensions of sustainability. *Journal of Engineering for Sustainable Development*, 2(1): 47–57.

Jayne, M. (2006). *Cities and consumption*. London: Routledge.

Kallidaikurichi, S. & Yuen, B. (Eds.). (2010). *Developing living cities*. Singapore: World Scientific.

Kates, R. W., Parris, T. M. & Leiserowitz, A. A. (2005). What is sustainable development?. *Environment: Science and Policy for Sustainable Development*, 47(3): 8–21.
Kelbaugh, D. (2002). The new urbanism. In S. S. Fainstein & S. Campbell (Eds.), *Readings in urban theory* (2nd ed, pp. 354–361). Oxford: Blackwell.
Li, S.-S., Zhang, Y., Li, Y.-T. & Yang, N.-J. (2010). Research on the eco-city index system based on the city classification. Bioinformatics and biomedical engineering (ICBEE) 2010 4th international conference. Chengdu, pp. 1–4.
Luccarelli, M. (1995). *Lewis Mumford and the ecological region*. New York, NY: The Gulford Press.
McHarg, I. (1969). *Design with nature*. Dockside Green, VIC: Wiley.
Mumford, L. (1961). *The city in history: its origins, its transformations and its prospects*. New York, NY: Harcourt Brace and World.
Mumford, L. (1997). *The culture of cities*. London: Routledge/Thoemmes Press.
Mumford, L. (2004). Cities and the crisis of civilization. In S. M. Wheeler &T. Beatley (Eds.), *The sustainable urban development reader* (pp. 15–19). New York, NY: Routledge.
Naess, A. (1973). The shallow and the deep, long-range ecology movement. *Inquiry*, 16: 151–155.
Naess, A. (1989). *Ecology, community, lifestyle*. Cambridge: Cambridge University Press.
Nash, R. (1989). *The right of nature: a history of environmental ethics*. Madison, WI: University of Wisconsin Press.
Newman, P. & Jennings, I. (2008). *Cities and sustainable ecosystems*. Washington, DC: Island Press.
Nicholson-Lord, D. (1987). *The greening of the cities*. London: Routledge & Kegan Paul.
Parsons, K. C. (1990). Clarence Stein and the greenbelt towns settling for less. *Journal of American Planning Association*, 56(2): 161–183.
Randolph, J. (2004). *Environmental land use planning and management*. Washington, DC: Island Press.
Register, R. (1987). *Ecocity Berkeley: building cities for a healthy future*. Berkeley, CA: North Atlantic Books.
Register, R. (2006). *Rebuilding cities in balance with nature*. Cabriola Island, BC: New Society Publishers.
Roberts, P., Ravetz, J. & George, C. (2009). *Environment and the city*. London: Routledge.
Roseland, M. (1997). Dimensions of the ecocity. *Cities*, 14(4): 197–202.
Routley, R. (1973). Is there a need for a new environmental ethic? Proceedings of the XVth world congress of philosophy, held on September 17–22, 1973 at Varna, Bulgaria.
Sagoff, M. (1988). The allocation and distribution of resources. In M. Sagoff (Ed.), *The economy of the earth* (pp. 50–73). Cambridge: Cambridge University Press.
Silvers, R. (1976). *The sustainable society*. Philadelphia, PA: Westminster Press.
Stanford Encyclopedia of Philosophy (2008). Environmental ethics. http://plato.stanford.edu/entries/ethics-environmental. Accessed 09 September 2009.
Suzuki, H., Dastur, A., Moffat, S. & Yabuki, N. (2009). *Eco2Cities: ecological cities as economic cities (conference edition)*. Washington, DC: The World Bank.
United Nations (1987). *Our common future: report of the world commission on environment and development*. New York, NY: United Nations.
Welter V. M. & Lawson J. (Eds.). (2000). *The city after Patrick Geddes*. Oxford: Peter Lang.
White, R. (2002). *Building the ecological city*. Cambridge: Woodhead.
Wolman, A. (1965). The metabolism of cities. *Scientific American*, 213: 179–190.
World Bank (2010). Eco2 cities: ecological cities as economic cities – synopsis. http://siteresources.worldbank.org/INTURBANDEVELOPMENT/Resources/336387-1270074782769/Eco2Cities_synopsis.pdf. Accessed 25 September 2010.
Zhan, S. (2003). *Fifteen lectures on the Taoist culture*. Beijing: Peking University Press (in Chinese).

Part I
Macro Strategic Planning: Policies and Principles

Chapter 2
How Cities Can Enter the Ecological Age

Peter Head and Debra Lam

Abstract The aim of eco-cities is to build a viable future for humanity with a healthy planet where the Earth, water and air will continue to support our complex solar-powered ecosystems. Presently, our over-dependence on depletable resources is destabilising the planet's life-support systems. Three key issues that have exacerbated our problems are: (a) the continued growth of population; (b) the rapid growth of resource consumption associated with urbanization, especially in emerging economies; and (c) climate change. Against this background, this paper analyses current global knowledge and examine if and how we can reach a sustainable future. The authors believe that this is feasible if cities, driven by urbanization, population growth, and climate change, can lead the way. Working together globally and with the supporting policy framework in low, middle, and high income countries, and new eco-oriented business models, cities can reduce their carbon emissions, retain a limited ecological footprint, and improve their human development to enter the ecological age.

2.1 Introduction

In recent decades it has dawned on many of us that there can be no viable future for humanity without a healthy planet. Earth, water and air support the existence of an immensely complex living system, powered by the sun. We are part of this web of life. But within a few generations, we are using up most of the Earth's stored fossil fuel resources and their transfer from the Earth to the atmosphere is significantly altering its composition. Our globalising, resource over-dependent path is destabilising the planet's life-support systems. The total global resource consumption has gone up substantially, with nearly all of it from non-renewable sources. The direct impacts of this on human development, plus increase in population; rising

D. Lam (✉)
Arup (International Consultancy Services), London, UK
e-mail: debra.lam@arup.com

T.-C. Wong, B. Yuen (eds.), *Eco-city Planning*, DOI 10.1007/978-94-007-0383-4_2,

food and resource costs mean that traditional economic growth is rapidly becoming unsustainable and a global transition is underway to the ecological age of human civilization.

Three key issues that exacerbate our problems are: (i) the continued growth of population – it is predicted to reach 9 billion by 2050; (ii) the rapid growth of resource consumption associated with urbanization, especially in emerging economies; and (iii) climate change. The year 2008 marked the first time in history that half of the population lived in urban areas. The world urban population is expected to nearly double by 2050, increasing from 3.3 billion in 2007 to 6.4 billion in 2050 (United Nations 2008). As for climate change, even if we were to stabilize carbon emissions today, increases in temperature and the associated impacts will continue for many decades. And given the outcome of the Copenhagen Accord, pending expiration of the Kyoto Protocol and mixed national commitments, carbon emissions are not likely to stabilize soon.

The drivers for urbanization are strong, with the potential for better living standards, improved health, higher education, and greater gender equality. But this current model is unsustainable. Life in high income urban areas gives rise to a large proportion of CO_2 emissions and subsequent climate change impacts. It is also dependent on outside resources shipped in, and wastes shipped out. Seeing only the economic success of high income countries, low and middle income countries have followed the same fossil-fuel dependent route, and accelerated inefficient resource consumption. The rapid economic development of China, with over 800 million people living in cities by 2020 (People's Daily 2004) – 60% of its population – has alarmed many. There would be insufficient resources if every Chinese wanted to live the same high and inefficient standard as an American.

Urban centres and cities of the future need to be refashioned to enable people to live much more lightly on the planet with a huge reduction in greenhouse gas emissions and resilience to climate change impacts. Especially for low and middle income areas, there are opportunities to leapfrog the problems of the current high income world, making much more efficient use of their resources, following the new ecological age model.

2.2 Ecological Age Performance Measurements

This chapter carefully analyses current global knowledge in an attempt to see if and how we can reach a sustainable future. The conclusion is that we could move to a sustainable way of living within environmental limits over the next few decades, allowing for continued human development and population growth, whilst adapting to climate change impacts. Clear objectives are set out for 2050 Ecological Age, using three performance measurements:

- *CO_2 Reduction*: 50% average from 1990 levels by 2050
- *Ecological Footprint Decrease*: Within the Earth's biocapacity of 1.44 gha/person, based on a projected global population in 2050

- *UN Human Development Index Improvement:* Raise overall wellbeing in GDP/capita, life expectancy, and education.

"Between 2000 and 2005, emissions grew four times faster than in the preceding 10 years, according to researchers at the Global Carbon Project, a consortium of international researchers. Global growth rates were 0.8% from 1990 to 1999. From 2000 to 2005, they reached 3.2%" (New Scientist 2006). We need to decrease our carbon emissions or risk greater and more frequent impacts of heat waves, drought, typhoons, etc. However, decreased carbon emissions are not enough to transition towards an Ecological Age. We need to ensure that we continue to grow and develop, but within our resource constraints and improve our living standards.

Ecological footprint was developed by William Rees and Mathis Wacknernagel, and is a resource measurement tool similar to a life-cycle analysis. It attempts to account and compare human's demand for ecological resources, and the planet's ability to supply that demand and regenerate. Its methodology involves calculating "the area of productive land and sea needed to provide a given quantity of energy, food and materials for a defined population in a given land mass, and the area of land required to absorb the emissions" (Global Footprint Network 2005) – in other words, nature's ability to provide for our lifestyle consumption, or biocapacity. In 1998 WWF started publishing a biennial Planet Report, which in 2006 showed that we are now living in severe ecological overshoot. Worldwide, the report says that we are consuming 25% more resources than the planet can replace and are drawing down the stock of natural capital that supports our lives (World Wildlife Fund 2006).

The UN Human Development Index measures overall well-being in three basic dimensions of human development: a long life, formal education, and average per capita income of GDP (UNDP Human Development Report 2007–2008). It has been used by the United Nations since 1990 as an indicator of human well-being beyond sheer economic growth. Together these three objectives serve as our guide in entering an Ecological Age and future ecological age cities. Each indicator alone has weaknesses, but together, they provide a holistic assessment of where cities should strive for. The three keep us in balance with nature while continuing to promote our growth and development. Happiness will not be attained with material accumulation, but rather in a change in our living conditions and thinking.

2.2.1 Different City Conditions

Recognizing the different performance levels in each city– along with local conditions and policies – we aim to set recommendations that are relevant to each context while promoting an overall transition towards an Ecological Age. Existing urban centres are simplified into three basic models (Table 2.1).

The first type- emerging economy- focuses on the expansion or creation of urban areas, while the final two look into retrofitting existing areas. The emerging economy's goal is to avoid an increase in ecological footprint as it continues to grow

Table 2.1 City models

Urban centre models	Main characteristics	Ecological footprint (gha/capita)	Human development index	Example locations
Emerging economy	Dense living, growing population	1–2	0.4–0.8	Africa, Latin America, Eastern Europe, China, India
European	High density, low car use	4–8	> 0.8	Western Europe, Japan, Korea, Singapore
USA	Sprawl, high car use	8–15	> 0.8	North America, Australia

Source: Collated by authors from various sources

and improve its human development index. The European and USA models aim to decrease their ecological footprint while maintaining a high human development index.

Low and middle income cities need to develop in a way that improves quality of life and creates jobs and opportunities within the new global economy where resource efficiency underpins development. The planning, design and investment model will be a new one following the long term lessons from cities. For these low and middle income economies this approach can be thought of as a way of leapfrogging from the Agricultural Age to the Ecological Age.

At the same time high-income countries need to rebase their paradigms around city living, rural food production, water management, energy supply and manufacturing to take advantage of the ecological age economy. They need to avoid the ravages of inflation and political risks of shortages of basic needs that result from a continued focus in industrial production. This will require investment to transform existing cities along the lines of the London Climate Change Action Plan and various One Planet Living studies by WWF. We call this retrofitting and envisage this will be carried out at a regional scale of communities of at least 50,000–100,000 people.

2.2.2 Climate Change Resilience

At the same time, cities are retrofitting or developing anew, they will be facing greater and more frequent climate change impacts. There are an increasing number of natural disasters caused by climate change. The growing populations – particularly in coastal areas have increasing exposure to cyclones, droughts and floods – are affecting food production and prices and higher summer temperatures are creating dangerous conditions for the elderly and infirm. This is caused by the higher intensity of storms. Fires are also becoming more frequent.

Many of the cities most at risk from the impacts of climate change are low and middle income nations that have contributed very little to greenhouse gases. These cities are not the best equipped with the skills and resources to combat climate change impacts or to prevent its occurrence. The number of deaths in certain countries is decreasing thanks to early warning systems that trigger mass evacuations to shelters, but the social and economic impacts are terrible. Overall the impacts are already becoming very serious.

Adaptation must be a priority and should go hand in hand with mitigation. Costs can be reduced by combining infrastructure investments to serve both purposes. For example, we can plan urban areas to take advantage of natural cooling through green roofs and parks, combining lower greenhouse gas emissions and hence lower heat stress for residents. Emergency preparation plans can be part of the city's sustainable development programme and supported by a communications systems for up-to-date, accurate information for the residents.

2.2.3 Sustainable Urban Design and Transport

Competition for land in most urban areas is driving up the land part of house prices. This means that rising land value can be used to underpin investments in improved efficiency. Inequalities are widening however, especially between homeowners and renters. For most, the ambitions of those moving to urban centres globally are not being realised. As the former Executive Director of UN Habitat, Anna Tibaijuka, notes, "People move to the cities not because they *will* be better off but because they *expect* to be better off" (BBC News 2006). These members of the population find it hard to find the economic opportunities they envisioned. Their dire financial situation and lack of affordable housing, exacerbated by rising fuel and food costs, is leading to homelessness and slum housing. The slum population is forecast to reach 1.4 billion by 2020, with Africa most affected.[1]

The approach to city living needs to change radically to a much more efficient use of land if we are to live within the carrying capacity of the planet. Ecological footprint is changed fundamentally by the level of urban density, food and goods selection, energy supply efficiency, fuel choice, and transport. Food and goods are consumer choices while urban density, supply efficiency, and fuel choice are largely planning decisions. Good urban design and planning is therefore a key to a successful change of direction and clarity of legal structure for land use planning is critical.

One of the largest differentiators in the ecological footprint of cities is the relationship between urban density and transport energy use. An average urban dweller in the United States consumes about 24 times more energy annually in private transport than a Chinese urban resident (Kenworthy 2003). There is a sweet spot of urban density of 75 persons/hectare in which transport energy use is reduced through the economic provision of public transport and there is still ample room for urban parks and gardens (Newman and Kenworthy 2006). Higher urban density combined with good public transport and a switch to use of fuel efficient and renewable energy powered vehicles can decrease transport-related energy use and improve liveability.

Opening up the city roads to walking, biking improves air quality, reduces traffic congestion, and enhances community and healthy living. Real time information can support greater public transport use and scheduling. Intercity connections can rely on high speed rail, waterways, and green logistics services from freight hubs for goods delivery. Better transport options also improve other infrastructure. A simple example is that the use of quiet electric vehicles and pedestrianised streets can mean the facades of buildings can be lighter in weight with the need for less noise attenuation, therefore consuming fewer resources; or that choosing more sustainable building material results in lower CO_2 emissions. Improved air quality from non-polluting vehicles can facilitate natural ventilation of buildings, saving energy costs and improving residents' health.

Increasing biodiversity with green roofs, urban parks and tree planting along streets will reduce the heat island effect and give benefits of improved health through lowering heat stress and improving mental health (Mind 2007). The link between biodiversity and health can be illustrated by Singapore's visionary approach to biodiversity management in parks. Dragonfly habitats are being introduced to try to help control mosquitoes and the problem of dengue fever in the city. Melbourne also uses species planting to create an eco-system in which mosquitoes do not proliferate.

There is a virtuous cycle between the biodiversity of a city, and therefore living in harmony with nature, and the energy consumption and quality of life. There is strong evidence that access to green space increases demand for developments and opens the door for funding through land value uplift. It will also benefit the natural systems that maintain life. Trees and vegetation also help with water-management, slow down water run-off and improve air quality. There is also a need to restore rural and aquatic bio-diversity outside urban areas. Future urban centres can be transformed to reflect places where we live in harmony with nature in all its forms.

2.2.4 Urban Agriculture

Food security and self-sufficiency have become a problem in many areas due to decreasing supply from agriculture, livestock, and aquaculture, and increased consumption and higher demands of energy-intensive goods. Food demand globally is expected to rise by 70–90% by 2050 due to population growth and higher standards of living (Varma 2008). Productive agricultural land area is generally decreasing in part because of urbanization, pollution and climate change impacts. Deterioration of soil quality and overgrazing are reducing food productivity on the shrinking land area, requiring ever increasing use of chemical fertilisers and yet more non-renewable energy consumption and associated carbon emissions. Water resources are also becoming depleted. Three-quarters of the world's fish stocks are fully exploited, overexploited or depleted. It is forecast that most fish stocks will collapse by 2050 (UNEP GEO-4 2007). Tropical forests, important to ecosystems preservation and efficient stores of carbon are being destroyed to make way for food and bio-fuel production.

The 850 plus million hungry people will continue to grow, while others will be forced to change their spending and give up other necessary goods or services, such as healthcare and education (Varma 2008). Josette Sheeran, Executive Director of the United Nations' World Food Programme (WFP), notes with alarm:

> For the middle classes, it means cutting out medical care. For those on $2 a day, it means cutting out meat and taking the children out of school. For those on $1 a day, it means cutting out meat and vegetables and eating only cereals. And for those on 50 cents a day, it means total disaster (The Economist 2008a).

Food is becoming a larger part of one's budget. "The average Afghan household now spends about 45% of its income on food, up from 11% in 2006" (Ban 2008). As a result, people buy less and cheaper foods. But cheaper foods, such as processed or packaged goods are usually less nutritious and require more energy.[2] The rising middle class faces a different situation with food. As living standards rise we are consuming more resource intensive foods. For example, moving from cereals to meat results in 2.5–3.5 times more land required for food production (UNEP GEO-4 2007). This is most acutely seen in China as its increased living standards have resulted in a 2.5 times increase in meat consumption in less than 30 years.

We actually produce enough food now to feed every child, woman and man and could feed up to 12 billion people. But in reality, while 850 million people (mostly women and children) remain chronically hungry there are 1.1 billion people who are obese or overweight (Economist 2008b). Our food supply is unequally distributed.

Diet, food production efficiency and distribution are key elements of resource efficiency and these are issues that can be tackled. For example, it is likely that we will need to turn to new low energy processes of building and balancing soil fertility and this can be assisted by closing the resource loops between urban living and rural food production. Research is being carried out into food production in buildings in which artificial light is used together with hydroponics culture and nutrient recycling from city waste streams to grow green vegetables and fruit. This takes advantage of new LED lighting technologies and plant science and recognises that plants only need a proportion of the white light spectrum to grow healthily. It is likely that by 2050 a proportion of food can be grown commercially by supermarkets within their existing facilities in towns and cities and sold directly to customers with low ecological footprint as long as a supply of renewable energy is available. Control of nutrient supply to plants grown in this way will also enable the mineral balance in the food chain to be improved.

Urban co-operative gardens and urban agriculture can be an important contributor to food supply in cities. Storing and transporting food from the rural areas not only widens the rural-urban divide but also creates a dependency relationship on cities. Green walls and roofs can produce agriculture, improve air and water quality and engage the community. There is also substantial opportunity is the growing of food in urban areas using hydroponics and nutrients recovered from the waste stream and the recycling of carbon from energy consumption in urban areas back to the productive land. This could also free up land for new forests to create other additional carbon absorption capacity and to improve biodiversity.

2.2.5 Total Water Resource Management

Freshwater resources are fundamental to agriculture, food production and human development. The UN Environmental Programme reports that "if present trends continue, 1.8 billion people will be living in countries or regions with absolute water scarcity by 2025, and two thirds of the world population could be subject to water stress (UNEP GEO-4 2007)." This is caused primarily by over-abstraction, inefficient/inequitable use, man made pollution and damage to the eco-system by deforestation. There is also an overconsumption by the agricultural sector and draw-down of most aquifers, largely from inefficient water pricing (Timmins 2004).

There are major opportunities to use recycled water. This can be from urban development to give efficient irrigation of surrounding farmland and to collect and store water run-off in cities and use it as grey water for secondary uses. These lead to a reduction in the demand for potable water and the associated energy needed for treatment. It can also help mitigate climate change impacts of increased storm rainfall intensity on flooding.

Likewise wastewater can be separated, and treated for reuse, and for conversion to energy. All the technology that would allow us to do this is on the market and is not excessively expensive, especially if the urban economies of scale are taken into consideration.

2.2.6 Energy Efficiency and Renewable Energy

If current trends continue the world's primary energy demand will more than double by 2030; almost half of that will be accounted for by energy demand in India and China alone (International Energy Agency 2007). Currently two thirds of potential energy is wasted through inefficiency in generation, distribution, supply and usage (The Economist 2008c). Demand for all fuels is predicted to rise.

Coal consumption is rising faster than oil and gas with global demand forecast to jump 73% between 2005 and 2030 (International Energy Agency 2007). Coal powered stations are being built all over the world despite the threat of emissions caps because coal is now the cheapest most plentiful fossil fuel we have left and could last beyond oil and gas. Current resource estimates assume consumption at present rates- not increasing consumption, but official coal reserve estimates may not be as high as believed so there may not be the 150 years of reserves some have estimated (Strahan 2008).

Oil prices have increased fourfold in 7 years (The Economist 2007). The cost of traditional production has changed little but deeper wells and a transition to more inaccessible sources is having an impact. Most of the increase in prices is down to the classic economic model of supply unable to meet demand. The inelasticity of oil demand means that the price must get high before demand is "killed". The second report of the UK Industry Taskforce on Peak Oil and Energy Security finds that oil shortages, insecurity of supply and price volatility will destabilise economic, political and social activity potentially by 2015 (Industry Taskforce on Peak Oil and Energy Security 2008). An increase in oil prices not only affects our energy costs, but trickles through to the costs of other goods and services. Particularly on necessities, those who are less able to afford it will feel the largest impact.

Improving the energy efficiency is one of the cheapest and easiest ways to conserve energy sources. Behaviour changes, and smart energy monitoring in buildings and homes can reduce the need for excess energy. Work has shown that improvements such as insulation, efficient water heating and use of energy efficient appliances and lighting can reap rapid cost benefits to most householders. Cities can also look towards combined heat and power and local heat and power grids to supply their energy. They can take advantage of the waste to energy links and use secondary biomass for energy and products, including biofuel for transport.

For low and middle income countries, the rapid uptake of the use of microfinance to install photovoltaic panels, local energy from waste facilities and solar powered irrigation pumps shows that, at current oil prices, the use of local renewable sources of energy is much more attractive for human development in remote inaccessible areas than expensive centralised power supply. This could also extend into transport once economic electric vehicles are available, and can already be seen in the use of electric bicycles.

Energy from renewable energy sources such as solar, wind, tidal stream and wave power are greatly underutilized. We already see that development will move forward with a greater consumption of renewable resources (with non-renewables gradually being priced or regulated out as they become more scarce) and will be underpinned by greater efficiency, lower environmental pollution and an emphasis on improving the effectiveness of human development through the transition. For example we now see the increasing sales of energy efficient and renewable resource products and services. Renewable energy is the primary job creator in Germany with 100,000 new jobs expected by 2020, largely as a result of government policy (The Climate Group 2007). In Japan, new building energy codes for residential and commercial buildings will save US$5.3 billion in energy costs and 34 million tonnes of CO_2 annually (The Climate Group 2007).

There is much more solar energy available in the desert regions of the world than we are currently generating from fossil fuels. According to the 2006 United Nations Environment Report, an area of 640,000 km^2 could provide the world with all of its electricity needs (the Sahara is more than 9 million km^2 in size) (United Nations Environment Report 2006). But climate change could also turn valuable deserts and their solar resource uninhabitable with unbearable rising temperatures and water scarcity. We have to be willing to build the infrastructure to transfer the desert power into our urban areas. This is one of the greatest opportunities for new technology to help solve our problems.

Emerging technologies will also be important to future energy supply. This will be an important component to research institutes and academia, translating that knowledge first into demonstration projects, and gradually into wider use and decrease costs. Carbon capture and storage, plus new coal gasification technologies offer the opportunity to reduce emissions from coal power stations. The costs are high, however, because of the need to liquefy and store the carbon dioxide gas. Other new technologies are in sight for creating short carbon cycles in urban areas by absorbing carbon dioxide at local power stations into different algae forms and using the algae and by-products as a local fuel with the carbon being returned to the land.

2.3 Smart Responsive Simplicity

All of these systems are connected and form virtuous cycles that integrate the environmental, economic and social performance of different components of built environment so that change in the design of one can lead to benefits in another.

The stacking of problems has led to a complexity of infrastructure with high maintenance costs. A clear vision is now emerging that the way forward is one of smart responsive simplicity rather than rigid complexity. For example, in a new compact mixed use development, people can easily go to work, school, shops and leisure facilities by walking, cycling or by public transport; the residents save money and travel creates less pollution from car exhausts. This leads to better health, lower social care costs and creates a more desirable place to live in and a higher return for the developer. Local utility systems for energy, water and waste management should be integrated to allow cooperation, shared land use and shared resources. Retrofitting of new sustainable systems need to follow this model too. Typical examples are energy from waste anaerobic digestion plants for both municipal waste and sludge from sewage treatment viable, particularly when they interconnect new and existing rail routes.

Recent surveys in many countries have shown that people are prepared to live differently and willing to make lifestyle changes (BBC News 2007). Diversity of cultures, ages and family groups in local accommodation can greatly assist human development through mutual support systems which are "bartered" within communities. All of these point to high quality urban design for compact mixed use which includes access to education, leisure and parks as well as work to help human development.

2.3.1 Policy Framework

When countries are not able to provide basic necessities like food for their people, social disruption results. We have seen this in many African countries and increasingly in other low- and middle-income countries. "Food riots have been reported from Kolkata to Namibia, Zimbabwe, Morocco, Uzbekistan, Austria,

Hungary and Mexico" (Varma 2008). Similarly, disputes and frustrations have erupted over land, water, and energy. For cities to effectively provide reliable, consistent and cheap services to their citizens, we need to have strong policies and an economic model that promotes resource efficiency, sustainable development and climate change resilience.

First, policies which drive towards the sustainable or optimal scale need to address the limiting of scale and the fact that previously free natural resources and services have to be declared scarce economic goods. Once they are scarce they become valuable assets and the question of who owns them arises and therefore the issue of distribution must be addressed, for example:

- Energy feed-in legislation.
- Polluter pays taxes introduced progressively, with proceeds used to drive public sector investments which help the private sector.
- Tradable permits with quotas set so that the marginal social and business costs are equal to the societal benefits.

Second, as sustainability is the criterion for scale, justice is the criterion for distribution to ensure that there is fairness across society and globally, for example:

- National and regional land use plans.
- Land value taxation to redistribute value to the community.
- Bartering of human development benefits against environmental clean up benefits.
- "Contract and Convergence" for carbon and "Shrink and Share" for ecological footprint.

Thirdly, policy needs to ensure that allocation of resources is as efficient and cost effective as possible, for example:

- National resource efficiency targets and circular economy laws to incentivize symbiotic manufacture.
- National policy to manage the rebound effect of improved resource efficiency.

2.3.2 Economic Model

Most detailed carbon emissions reduction studies like the Stern Review (see UK Chancellor of the Exchequer 2007) and a major report from McKinsey Company (2007) say that the transition costs are within our means and will not hurt economic growth. The McKinsey report says that the United States can reduce greenhouse gas emissions by one-third to a half by 2030 at manageable costs to the economy. In reality if the broader canvas of resource efficiency was adopted, as suggested here, then the economic drivers would be even clearer. Failure to act now means that the cost of climate change, especially its impacts will be much higher.

The economics of scale are bringing down the costs of low carbon technologies. Already in the United States, studies of energy efficient buildings designed and built to LEED standards have shown that initial increases in costs have disappeared as the numbers have increased and substantial energy performance improvements compared to non-LEED buildings (Turner and Frankel 2008). A combination of top down policy and individual action is needed to enable the direction of development to change. A major obstacle is the fact that culturally, we have convinced ourselves that human development cannot occur without resource consumption.

The financing solutions will require long-term infrastructure partnerships between public and private sectors and community groups and NGOs and we can expect to see these emerging at a regional level and to include mitigation and adaptation. Partnerships are necessary because often land ownership will be in both public and private sector hands. Pension funds have a significant interest in this area of investment. Risks of losses of value will be mitigated and so partnerships with insurance companies are also likely to be productive as will partnerships with mortgage lenders for the upgrading of homes and surrounding infrastructure to enable occupiers to see cost reductions quickly. Microfinance and micro-insurance schemes that deal with both adaptation and mitigation are emerging quickly and these can operate at a local community or regional scale in low and middle income countries to manage and share risks over the long term.

2.4 Conclusion

This is a first glimpse of a way forward and a credible vision of the future for ecological age cities, but it is only a modest start for a long journey. There is clear evidence that first movers in this transition are gaining benefit both at a regional economic level and at a business level. Cities, driven by urbanization, population growth, and climate change can lead the way towards an ecological age. Despite different local conditions, and various economic and environmental levels, the principles of urban design and resource management are universal. But they can also be localized to fit the context. Planning for new and retrofitting high, middle, and low income cities can transform the way we manage our water and waste; feed our community; supply our power; travel to places; and live each other and with nature. Together, with the supporting policy framework and new business model, cities can reduce their carbon emissions, retain a limited ecological footprint, and improve their human development to enter the ecological age.

Notes

1. UN Habitat characterizes slum housing as lack of durable housing material, insufficient living area, and lack of access to clean water, inadequate sanitation, and insecure tenure.
2. The cost of fresh vegetables and fruit (lower calorie foods) are rising fastest.

References

BBC News (2006, 16 June). Report reveals global slum crisis. http://news.bbc.co.uk/1/hi/world/5078654.stm#slums. Accessed 12 April 2010.

BBC News (2007, 9 November). Most people ready for green sacrifices. http://news.bbc.co.uk/1/hi/world/7075759.stm, and http://news.bbc.co.uk/1/shared/bsp/hi/pdfs/09_11_2007bbcpollclimate.pdf. Accessed 12 April 2010.

Ban, K.-M (2008, 12 March). The new face of hunger, Washington Post. http://www.washingtonpost.com/wp-dyn/content/article/2008/03/11/AR2008031102462.html?hpid=opinionsbox1. Accessed 12 April 2010.

GEO-4 (2007). United Nations Environmental Programme. http://www.unep.org/geo/geo4/media/p 142. Accessed 12 April 2010.

Global Footprint Network (2005). Definition in multiple sources and reports. http://www.footprintnetwork.org/en/index.php/GFN/page/frequently_asked_questions/. Accessed 2005.

International Energy Agency (2007, 7 November). World energy outlook 2007 – China and India insights, http://www.iea.org/w/bookshop/add.aspx?id=319. Accessed 12 April 2010.

Kenworthy, J. R. (2003). Transport energy use and greenhouse gases in urban passenger transport systems: a study of 84 global cities. Presented to the international third conference of the Regional Government Network for Sustainable Development, Notre Dame University, Fremantle, Western Australia, during September 17–19, 2003. http://cst.uwinnipeg.ca/documents/Transport_Greenhouse.pdf. Accessed 20 April 2010.

Mckinsey & Company (2007, December). Reducing greenhouse gas emissions: how much at what costs? US Greenhouse Gas Abatement Mapping Initiative Executive Report. http://www.mckinsey.com/clientservice/ccsi/pdf/US_ghg_final_report.pdf. Accessed 20 April 2010.

Mind Organisation (2007, May). Ecotherapy: the green agenda for mental Health. http://www.mind.org.uk/mindweek2007/report/. Accessed 22 April 2010.

New Scientist (2006). http://www.newscientist.com/article/dn10507-carbon-emissions-rising-faster-than-ever.html

Newman, P. & Kenworthy, J. (2006). Urban design to reduce automobile dependence. *Opolis: An International Journal of Suburban and Metropolitan Studies*, 2(1): 3. http://repositories.cdlib.org/cssd/opolis/vol2/iss1/art3. Accessed 03 March 2010.

Peak Oil Group (2008). Industry taskforce on peak oil & energy security. http://peakoiltaskforce.net/. Accessed 20 April 2010.

Peoples' Daily Online (2004). China's urban population to reach 800 to 900 million by 2020. http://english.people.com.cn/200409/16/eng20040916_157275.html, and http://www.newscientist.com/article/dn10507-carbon-emissions-rising-faster-than-ever.html. Accessed 12 April 2010.

Strahan, D. (2008, 19 January). Coal: bleak outlooks for the black stuff. New Scientist. http://environment.newscientist.com/data/images/archive/2639/26391802.jpg. Accessed 20 April 2010.

The Climate Group Report (2007, August). In the black: the growth of the low carbon economy – summary report. http://theclimategroup.org/assets/resources/TCG_ITB_SR_FINAL_COPY.pdf. Accessed 20 April 2010.

The Economist (2007, 15 November). Shock treatment. http://www.economist.com/finance/displaystory.cfm?story_id=10130655. Accessed 12 April 2010.

The Economist (2008a, 17 April). The new face of hunger. http://www.economist.com/world/international/displaystory.cfm?story_id=11049284. Accessed 12 April 2010.

The Economist (2008b, 10 May). The elusive negawatt. http://www.economist.com/displaystory.cfm?story_id=11326549. Accessed 20 April 2010.

The Economist (2008c, 3 November). An expensive dinner. http://www.economist.com/world/international/displaystory.cfm?story_id=10085859. Accessed 12 April 2010.

Timmins, C. (2004). Environmental resource economics. Volume 26, Number 1, sourced from, http://www.springerlink.com/content/rp22580246p40t34. Accessed 12 April 2010.

Turner, C. & Frankel, M. (2008, 4 March). Energy performance of LEED for new construction buildings, US Green Building Council. http://www.usgbc.org/ShowFile.aspx?DocumentID=3930. Accessed 20 April 2010.

UK Chancellor of the Exchequer (2007). Stern review: the economics of climate change. http://www.hm-treasury.gov.uk/independent_reviews/stern_review_economics_climate_change/stern_review_report.cfm. Accessed 12 April 2010.

United Nations Development Programme (2007–2008). Human development report, http://hdr.undp.org/en/media/hdr_20072008_en_complete.pdf. Accessed 20 April 2010.

United Nations Environment Report (2006). The environmental effects assessment panel report for 2006. http://www.unep.ch/ozone/Assessment_Panels/EEAP/eeap-report2006.pdf. Accessed 20 April 2010.

United Nations (2008). World urbanization prospects: the 2007 revision, UN publication, 26 February 2008. http://www.un.org/esa/population/publications/wup2007/2007WUP_Highlights_web.pdf. Accessed 20 April 2010.

Varma, S. (2008, 13 March). Hunger is set to grow as global food stocks fall. The Times of India. http://timesofindia.indiatimes.com/Hunger_is_set_to_grow_as_global_food_stocks_fall/articleshow/2859771.cms. Accessed 12 April 2010.

World Wildlife Fund (2006). Living planet report. http://assets.panda.org/downloads/living_planet_report.pdf. Accessed 12 April 2010.

Chapter 3
Three Ecological Cities, Examples of Different Approaches in Asia and Europe

Meine Pieter van Dijk

Abstract Developing countries and emerging economies have been active in creating ecological cities. An analysis of some Asian cases will be presented to show that the reasons to create a new neighbourhood or to introduce a different approach to urban planning are mainly environmental considerations. Since the 1990s a number of cities have created new neighbourhoods taking environmental factors into consideration. More recently Shanghai announced plans to build the city of the future on an island at the mouth of China's Yangtze River, in the same way Singapore has planned new ecological neighbourhoods. These examples and one from the Netherlands (Rotterdam) will be reviewed to answer three questions like what would the ecological city of the future look like and what can we learn from these experiences for the ecological city of the future? Pollution, solid waste and wastewater problems, all aggravated by climate change require a different approach to urban management to build the ecological city of the future!

3.1 Introduction

There is not one definition of ecological or for short eco-cities that is generally accepted. Different authors have very different views of what makes a city an ecological city (van Dijk 2009b). Table 3.1 provides a number of examples of terms that are used to refer to ecological cities. One can conclude that people are driven by ideals (to create heaven on Earth) or needs (to deal with climate change). Some sources express the importance of having a livable (and economically vibrant city; van Dijk 2006), others stress a more green (more trees and parks) city. Finally some sources stress the ecological and others the sustainability aspect: the initiative should carry

M.P. van Dijk (✉)
Water Services Management, UNESCO-IHE Institute for Water Education, Delft, The Netherlands; Urban Management, ISS, Erasmus University in Rotterdam, Rotterdam, The Netherlands
e-mail: m.vandijk@unesco-ihe.org

T.-C. Wong, B. Yuen (eds.), *Eco-city Planning*, DOI 10.1007/978-94-007-0383-4_3,

Table 3.1 Examples of terms referring to more ecological cities

Titles for eco-city initiatives	Source
a. Eco-heaven, a model city near Shanghai	a. International Herald Tribune (24-6-2008)
b. Garden city brochure	b. Suqian city Jiangsu province China
c. Sustainable urbanization or cities	c. Van Dijk (2007b)
d. Sustainable urban development network	d. SUD-network UN Habitat Nairobi
e. Cities of the future	e. Switch project Delft, the Netherlands[a]
f. Livable and vibrant cities	f. National library Singapore
g. Sustainable living, bringing together best practices	g. Ministry of Housing, Spatial Planning & Environment in the Netherlands
h. Green urbanites	h. *Strait Times* (21-6-2008)
i. Climate resilient cities	i. World Bank primer, Washington Oct. 2008
j. Rotterdam, climate proof	j. Rotterdam municipality the Netherlands
k. Keeping cities alive	k. *Strait Times* (21-6-2008)
l. Green city philosophy, cooperation between Thailand and the Netherlands	l. Dutch Ministry of Agriculture, Nature and Food Quality
m. Eco systems and biodiversity, the role of cities	m. UNEP & United Nations Habitat brochure

[a]The Switch project (Sustainable Water Improves Tomorrow's Cities' Health) with support from the European Union (EU) is seeking a paradigm shift in urban water management. Its purpose is to make water treatment more sustainable and protect the quality of drinking water sources. In addition, it wants to reduce risks such as water related diseases, droughts and flooding. www.switchurbanwater.eu

on, because economically, environmentally and institutionally it is durable. An easy definition of an ecological city would be one emphasizing what should be or should not be there. The positive points of environmentally friendly cities are: they are livable and energy saving, promote integrated water and sanitation, better urban waste collection and processing, more gardens and trees, bio diversity and better public transportation and deal with climate change. On the negative side one could mention: do away with air, water and soil pollution, congestion, flooding and the lack of green areas. This paper deals with three questions:

a. What would the ecological city of the future look like?
b. To what extent do these examples satisfy the criteria for sustainable urban development formulated in the literature?
c. What can we learn from these experiences for the ecological city of the future?

Besides the "what is an ecological city" question, the why question will also be answered and some examples of the how will be given.

3.2 What Would the Ecological City of the Future Look Like?

To explore how the ecological cities of the future would look like it is necessary to assess what the relevant dimensions for an ecological city are. What makes a city an ecological city and what does sustainable urban development really mean? An overview of experiences with stimulating more sustainable urban development will be given. We will first review why we are concerned about more ecological cities and what sustainable urban development means. Subsequently we introduce the approach of the Switch project, which embodies a more ecological attitude towards water and environmental issues.[1] This will also mean a discussion about a theoretical framework for sustainable cities and explaining what following a more integrated approach to environmental problems means. Ten dimensions for sustainable city development in the Third World were developed by Kenworthy (2006: 67). They will be presented as a possible analytical framework to decide whether certain initiatives qualify for the ecological city label. His ten dimensions for sustainable city development in the Third World give a good impression of the issues at stake. A sustainable city is characterized by:

a. A compact, mixed urban form that protects the natural environment, biodiversity and food-producing areas
b. The natural environment permeates the city's spaces and embraces the city, while the city and its hinterland provide a major proportion of its food needs
c. Freeway and road infrastructure is deemphasized in favour of transit, walking and cycling infrastructure, with a special emphasis on rail. Car and motorcycle use is minimized
d. There is extensive use of environmental technologies for water, energy and waste management – the city's life support systems become closed loop systems
e. The central city and sub-centers within the city are human centers that emphasize access and circulation by modes of transport other than the automobile, and absorb a high proportion of employment and residential growth
f. The city has a high quality public culture, community, equity and good governance. The public realm includes the entire transit system and all the environments associated with it
g. The physical structure and urban design of the city, especially its public environments are highly legible, permeable, robust, varied, rich, visually appropriate and personalized for human needs
h. The economic performance of the city and employment creation is maximized through innovation, creativity and uniqueness of the local environment, culture and history, as well as the high environmental and social quality of the city's public environments
i. Planning for the future of the city is a visionary debate and decision process, not a predict and provide computer-driven process
j. All decision making is sustainability-based, integrating social, economic, environmental and cultural considerations as well as compact, transit-oriented urban form principles. Such decision making processes are democratic, inclusive, empowering and engendering of hope.

We would like to use a list of characteristics to find out if certain initiatives in Asian cities qualify for the label *ecological city* or ecological neighbourhood. Kenworthy's principles are quite broad and come from someone with a transport background (points c, e, f and j). There is a vision behind these dimensions and an integrated strategy is necessary to implement the envisaged solutions for the implicit problems. The importance of appropriate technologies for water and sanitation is only mentioned under point four.

Urban management should help take steps towards more ecological cities. My definition of a more ecological approach to urban development, based on the existing literature, would be that such a city requires a strategy combining:

a. Integrated water resources management: closing the water cycle
b. Energy management, reducing greenhouse gases
c. Waste minimization and integrated solid waste management
d. But also a different approach to sanitation
e. Integrated transport policies
f. A policy dealing with pollution issues
g. Anticipation of climate change
h. A different housing policy
i. Objectives concerning justice, for example promoting an equal distribution of the benefits
j. Integration in the framework of sustainable urban management, while also managing urban risks.

3.3 Three Levels of Eco Practices

Ecological initiatives can be taken at three levels. In the first place at the level of the city, a new town, or a neighbourhood would be an example. We will point to all kinds of ecological neighbourhoods appearing. Secondly at the level of buildings one notes ecological villas, blocks of houses, or apartment buildings with common heating/cooling systems or a shared grey water re-use facility. Finally individual initiatives can be noted at the household level, spontaneously or triggered by incentives or price increases. There are a number of Chinese eco-city initiatives that are interesting and have been studied. The reaction of these cities to climate change should be evaluated.

Examples of some Chinese eco-cities and eco-provinces (Wang 2006) will be presented to show how China, a country which is considered to grow at the expense of its environment, deals with urban environmental issues. Are these cities introducing a very different, more integrated approach to a number of related environmental issues? Urban environmental policies in Asia are also illustrated by the positive example of Singapore, where special attention is paid to the issue of building in a sustainable way. Subsequently we will present how Rotterdam in the Netherlands tries to deal with climate issues in its plan Rotterdam Climate Proof (Stadshavens Rotterdam 2008b), before some conclusions will be formulated.

The possibilities for Third World cities, particularly those located in Deltas, to prepare themselves for climate change, will be identified. We focus on what it means for water and sanitation, but drainage and flooding would also need to be considered.

Integration could take place in the framework of urban management (van Dijk 2006). Issues discussed are the integration of the different sectoral interests, the role of planning and management, the importance of economic, financial, social and environmental criteria (and how to combine them), who are the decision makers and how do we deal with the strict and the loose meaning of sustainable urbanization? After all sustainable development is a normative concept. In 1987 the World Commission on Environment and Development provided a definition of sustainability that is still often used. Brundland (1987) defines sustainable development as development that meets the needs of the present generation without compromising the needs of future generations.

The literature keeps struggling over what to put into the sustainability concept, while the environment continues to degrade. Mohan Munasinghe, Vice-Chairman of the United Nations (UN) Intergovernmental Panel on Climate Change (IPCC) tried to bring together the economic, human and environmental aspects of development. His analytical framework is called sustainomics (Munasinghe 2007). Through sustainomics he offers alternative mechanisms to help us bring environmental degradation and social cost into the analysis and applies his methodology to greenhouse emissions and the transport sector in Sri Lanka. At the same time he criticizes the traditional cost benefit analysis. Earlier we suggested letting the weight of the issues play a role in the definition of urban sustainability (van Dijk and Zhang 2005).

There are definitional problems as shown in the literature (Finco and Nijkamp 2001). One can find very idealistic, very sectoral, or issue based definitions of ecological cities and sometimes norms and values play a role such as the distributional issue: should the Chinese be denied the level of energy consumption of average citizens in the United States? Sen (2009) provides a theory of comparative justice, judgments that tell us when and why we are moving closer to or farther away from realizing justice in the present globalized world.

3.4 Why More Ecological Cities?

Not only higher energy prices and increased emissions of carbon dioxide (CO_2) force a reconsideration of the priorities for the future of cities in developing countries. Besides traditional urban environmental issues such as urban pollution, traffic congestion and inappropriate waste collection, the results of rapid urbanization and of climate change force cities to think more about their future.

Water stress can be noted in many countries (Seckler et al. 1998). A deteriorating environment accelerates the trend towards a gradual shortage of fresh water. While freshwater supplies are clearly limited, for most people water scarcity is caused by competition between water uses and by political, technological and financial barriers that limit their access to water (Falkenmark and Lundqvist 1998). The Switch program intends to generate new efficiencies from an integration of actions across the urban water cycle in order to improve the quality of life in cities.[2] It also promotes urban agriculture projects, as part of the integrated approach to water use and reuse (Box 3.1).

Box 3.1 Switch Project on Ecological Cities of the Future

The UNESCO-IHE Institute for water education in Delft in the Netherlands carries out a European Union supported Switch project on ecological cities, where sustainability is defined as the process and the ecological city as the result. Global changes such as climate change and volatility, urbanization and industrialization, population growth, urban sprawl, and rural-urban migration put pressure on cities. A sustainable urban water system is a basic feature of an ecological city, but is it enough?

The Switch project according to the proposal, intends to improve water governance and to translate scientific innovations into improvements of day-to-day management of urban water and sanitation. The approach is focused on closing the urban water cycle, defined as the link between the resource, its use for drinking water and eventual reuse to allow the water to flow back into the resource. From the literature we know that reuse is currently at a price of 30–40 euro cents per m^3, while desalinated water may cost around one euro per m^3. Unfortunately the latter is always produced at sea level, implying transportation costs in most countries.

3.5 Monitoring Sustainable Urban Development

Achieving sustainable urban development also includes considering water and sanitation integral parts of urban infrastructure planning. The Switch vision relates to storm water to drinking water and waste water treatment. It emphasizes:

a. Thinking in terms of systems of interrelated components (system engineering)
b. A more ecological approach to sustainable urbanization
c. A more integrated approach to different water related issues.

Part of the first approach would be developing indicators to monitor constantly the score of city with respect to the quality of the aquatic urban environment and to take corrective actions if certain variables reach threshold levels. Modeling the system and emphasizing decision support systems is inherent to this vision.

A more ecological approach to sustainable urbanization implies moving from traditional environmental technologies to more ecosan options in the ecological city of the future. It will be necessary to focus more research on the topic of ecological cities, to study certain phases in the process of becoming more environmentally conscious as well as finding ways to interest some of the major urban actors in these issues. However the coordinating role of local governments and urban managers should not be underestimated. In fact it is their task to coordinate a multiplicity of actors. That is the essence of urban management: participatory, inclusive and with all actors concerned taking into consideration equality, the environment and economic development.

It needs to be clear what will be integrated, how and by whom? Integrated Urban Water Management (IUWM) can be achieved in each of the cities if we work towards a plan. A major assumption of this approach is that if we follow a holistic approach we will have better results. We assume that such an integrated policy will be the result of scientific research, rather than consultancy reports. Such integrated environment friendly urban development plans may be too ambitious for big cities like Beijing and we may have to content ourselves with providing strategic direction for moving towards a more ecological city.

Strigl (2003) stresses that a real improvement in eco-efficiency requires a fundamental change in culture, structure (institutions) and technology. Switch intends to develop, apply and demonstrate a range of scientific, technological and socio-economic solutions that will be tested to determine their contribution to the achievement of sustainable and effective urban water management schemes. It implies a multi-disciplinary approach for Switch that is the integration of the technological means, socioeconomic aspects, environmental concerns and health considerations.

Each flash in the figure represents a point where costs are made and revenues can be obtained. It is also possible to deal with the water cycle process in an integrated way, as is done in Singapore. In that case the costs and charges could also be integrated in one exercise (for the costs) and one bill for the customers (Fig. 3.1)

How do we hope to achieve all this in the Switch project? Learning alliances have been created consisting of interested stakeholders to discuss the issues and to identify directions for research (van Dijk 2008). The researchers hope to provide a broader perspective to the members of the learning alliance and to increase the range of options between which they can now make an informed choice. Why is Switch different? Because the project promotes sustainable and integrated urban water management, to make the city a better place to live. It suggests closing the urban water cycle for the city of the future.

The point of departure is closing the urban water cycle. In Singapore no water gets lost between the resource, the use for drinking water and the treatment and

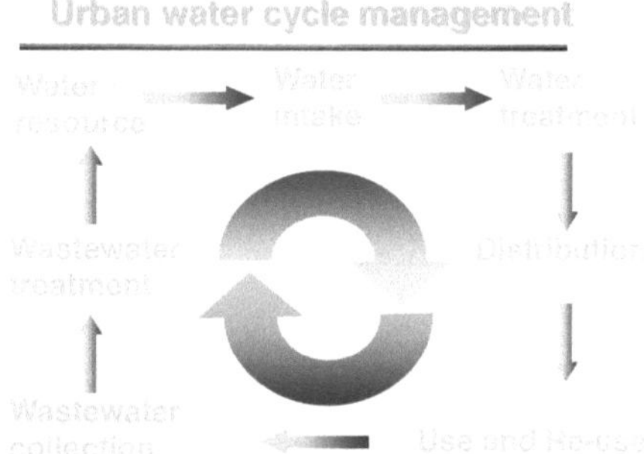

Fig. 3.1 The closed urban water cycle. *Source*: van Dijk (2007a)

reuse. This is the work of NEWater, a company with the mission to go from "Sewage to Safe". The city is also using reverse osmosis technology for the process of transforming sea water into drinking water, if additional water is necessary in the closed urban water cycle.

3.6 Examples of Urban Environmental Policies in Asia

We will provide some Asian examples of urban environmental policies and ask the question: do these examples satisfy the criteria for sustainable urban development formulated in the literature (for example Kenworthy 2006)? Special attention is paid to a number of Chinese cities and in Singapore to the issue of building in a sustainable way and the role of housing finance. At the end we briefly present how Rotterdam in the Netherlands deals with these issues (Stadshavens Rotterdam 2008b), before drawing some conclusions.

3.6.1 Ecological Initiatives in China

A number of ecological initiatives have received support from the Chinese government. They range from alternative building methods (emphasizing the need to isolate houses better) to promoting other ways of dealing with drinking water and sanitation. The question is to what extent these disjointed initiatives also contribute to building the much-needed ecological city of the future.

In the integrated urban water cycle, managing water resources, drinking water supply and wastewater treatment are three important stages, each with specific problems in China. The risks in the water cycle are substantial. The water situation in northern China can be described by the term, water scarcity. In the north, there is not enough water for the different types of use and for the big cities, which have high per capita consumption figures, probably due to substantial water loss. For that reason, China has embarked on a number of river linking projects (WWF 2005). The main problems with water and pollution in China can be summarized as follows:

a. Water prices are not realistic (*Financial Times* 20-3-2003), but efforts to increase water prices by 30% have not been approved by the Municipal Commission of Development and Reform (*China Daily* 2-7-2004). Recently a small increase was announced.
b. The river transfer project is extremely costly (*Financial Times* 20-3-2003)
c. Pollution has led to algae in the Yellow Sea (*NRC* 17-6-2004)
d. The Three Gorges Dam may cause serious ecological risks (Financial Times 2009).

There are numerous problems with the water cycle in China. Just to mention some examples: the impact of climate change on water resources and development in China (*China Daily* 2-7-2004), as well as the risks linked to the current practice of water management for Chinese rivers (CICED 2006). Flooding is common just like pollution. However, the river is also important for irrigation, drinking water, transport and fishing activities. In the northern port city of Tianjin, the river became polluted and consequently the population could not drink the water for weeks. This is a big city and the impact of upstream pollution was enormous. The risks this time are not so much the risks of flooding, but of not supplying clean drinking water to the big cities on the coast (Pahl-Wostl and Kabat 2003). There were reservoirs to serve Tianjin and Beijing, but the water was not available at the crucial moment. Currently the city is using a desalination plant, but it will also benefit from the south-north river linking program, which connects the northern Yellow and southern Yangtze rivers.

There are a number of other eco-city initiatives in China, ranging from simple water and sanitation technologies for the western part of the country (through a project financed by the Netherlands) to sophisticated ecological projects in the framework of the 2008 Olympic Games in Beijing. The Chinese authorities exhibit a preference for large modern high tech solutions; even if they know they cannot always manage the technology properly. They are less willing to pay for management support, training or software; while given the high energy use per unit of Gross Domestic Product (GDP) and the huge water consumption in per capita terms, there is scope for improvement of the efficiency of the system through better management.

There is in China this trend to focus on obtaining the most advanced technology, counting that this will be sufficient to deal with the issue. The emphasis is on the hardware and not enough attention is paid to managing water systems in a more optimal way. Not enough attention is paid to managing existing water supply and waste water treatment systems properly. Hence many water resources are polluted, drinking water is scarce and the quality of the water produced by the waste water treatment plants is not always appropriate. Environmental norms have been put at a high level in China, unfortunately the strict norms are not always implemented seriously. The State Environmental Protection Agency (SEPA) is not very powerful, compared to Provincial governments, or the Ministry of Construction, which is responsible for the construction of water and sanitation facilities. SEPA will be upgraded and obtain the status of a Ministry, which will make it easier to deal with the environmental issues in different Chinese provinces, because they would be administratively at the same level.

The goals to be achieved in the water sector according to China's 11th Five-Year Plan are ambitious. The planners want to reduce for example water consumption per industrial unit by 30% and to increase the coverage for water and sanitation facilities in line with the Millennium Development Goals (MDGs). The governance structure to achieve this is relatively simple because the governance structure in China is still highly centralized and hierarchical. The role of the Ministry of Construction and the corresponding line offices at the city and district levels is very important. However, this excludes broader participation of all stakeholders. Water and sanitation facilities are for example not owned and managed by local authorities, which hinders innovative local solutions, innovative ways of financing.[3] It also means that very little cost recovery is achieved. It is normally possible to recover the cost for water treatment through the drinking water bill, however. The current price per m^3 is only 3.5 Yuan, of which 0.5 Yuan is for wastewater treatment, which is much too low.[4] Unfortunately in the case of ecological initiatives taken at the neighbourhood level to recuperate grey (lightly polluted) water, the treatment charge will not be repaid to the inhabitants, while they do pay the 0.5 Yuan for large scale treatment (Liang and van Dijk 2010).

3.6.2 Examples of Chinese Eco-cities

Chinese cities are facing the pressure of a water crisis. More than 400 cities are lacking water resources and more than half the rivers are polluted. In 2004, 5.548×10^{12} m^3 water was used for agriculture, industry and domestic activities. Meanwhile 6.930×10^{8} m^3 waste water was discharged from Chinese cities, but no more than half the amount of wastewater is subject to secondary treatment (China Bulletin of Water Resources 2004). We will look at initiatives in Beijing, Shenzhen, Shanghai (a new neighbourhood and the Thai lake) and Wuhan in that order.

Beijing is the capital of the People's Republic of China, lies in the northern part of the country and is geographically on the edge of a desert. Because of its geography, Beijing has low average rainfall. Beijing's total precipitation is 640 mm per year, 80% of which is concentrated during the period of June to September. The population of Beijing is 15.38 million, of which 3.2 million people reside in the peri-urban districts and rural counties of the metropolitan area. Because of the dramatic economic development during the last 20 years, Beijing has been urbanizing rapidly, with an average annual official population increase of 2.48%. Ground water is the primary source of water for agriculture and industry, and recently has shown a gradual decrease. Water scarcity, depletion of underground water stocks and environmental degradation are the main problems faced by Beijing. Given the negative effects on the environment, Beijing has decided to direct businesses, which utilize large amounts of water, out of the city (*China Daily* 10-4-2004).

In Beijing there are thousands of ecological initiatives and other Chinese cities are also doing their best. The question is, whether this is enough to counter a looming environmental crisis. Praising sustainable development is a beginning, but not enough. One example is the development of urban agriculture in Beijing. Beijing being a metropolitan region has large rural areas as well and urban agriculture has a very specific background with practices that can be repeated elsewhere. The projects are examples of eco sanitation (re-using urine and compost for urban agriculture) and could be elements of a more ecological city.

Given a number of the problems are related to water governance, an institutional analysis is required to identify the different bottlenecks. In the example of Tianjin in the north where the river became significantly polluted because of an upstream industrial accident, there was no riverbank infiltration system to mitigate the negative effects of the pollution. In addition constructed wetlands, which help to clean the water, were not used in Beijing as this approach required too much land.

Riverbank infiltration projects may be an alternative for constructed wetlands, which require much space, while river banks are available for this purpose. The model of Singapore, closing the urban water cycle completely, may also be an appropriate option and could help to economize the expenditures for this kind of projects. The example of Beijing shows how difficult it is to be an ecological city.

Shenzhen is another example of a major Chinese city trying to become an ecological city (see Box 3.2).

Box 3.2 Is Shenzhen Already an Ecological City?

In 2002, the State Environmental Protection Administration (SEPA) and the Ministry of Construction jointly formulated a series of standards and rules on the construction and recognition of ecological cities, which are related to economic development, environmental protection and social progress. All detailed standards are published on the website of SEPA www.sepa.gov.cn. SEPA is the decision-maker to approve or disapprove cities' applications. On June 2, 2006, SEPA for the first time awarded the title of the ecological city to the following cities: ZhangJiaGang City, ChangShu City, Kunshan City and JiangYin City of Jiangsu Province.

The city has set this target for the year 2010. Shenzhen's urban greening ratio has reached 51.1%, with 16.01 m^2 of green area per person, ranking top among other cities of the country. With 218 parks and 5,000 ha of ecological scenic forests, Shenzhen takes the lead in both land area and quantity of greening compared to other cities. The City has been awarded titles including "China's Best 10 Cities for Greening", "National Garden City", "Nations in Bloom", "National Greening Pioneer". At present, Shenzhen is on her way of thriving development with the aim of building itself into an "ecological city with high tastes".

Source: Taken from the website of the Shenzhen Bureau of Trade and Industry.

Box 3.3 summarizes the initiative in Shanghai to create an environmental neighbourhood.

Box 3.3 Shanghai's an Environmental Neighbourhood

Shanghai plans to build on an island at the mouth of the Yangtze River the city of the future (*Economist* 23-9-2006; *Trouw* 9-11-2007; *Financial Times* 15-9-2006). The idea is that the city will be self-sufficient in energy and water and will generate almost no carbon emissions. Petrol and diesel vehicles will be banned in favour of solar-powered boats and fuel-cell-driven buses, according to the *Economist*. The city should number around 500,000 inhabitants in 2040 and will house an agro park of 27 km^2 to grow food in a sustainable way (*Trouw*, 9-11-2007). Finally the *Financial Times* describes energy conservation at the level of the house and shows the use of water conservation (rain water harvesting). The houses will use only one third of the energy consumed by a normal house, while the energy will be renewable, for example through windmills. The project received attention and press coverage, but the question is how to diminish pollution in neighbouring Shanghai city, with 20 million inhabitants and many polluting industries.

In 2003 an environmental study of Tai Lake near Shanghai carried out by a Dutch consulting firm together with UNESCO-IHE showed the seriousness of pollution of the water resources and the need to introduce wastewater treatment plants. What has been done so far and to what extent the risk of pollution of the water resources have been limited by treating used water properly is not clear. It is our experience that the Chinese started building water treatment plants before the feasibility study was finished. Now they are not always working at full capacity nor turning out the expected quality of water. Recently another effort to clean Tai Lake was announced. Ten billion euro will be spent to clean it (*De Pers* 29-10-2007). According to these plans, it would take 5 years to clean the lake while the problem would be totally solved in 8–10 years.

Another example is Wuhan, one of the largest cities in China, with total area of 8,494 km^2 and a population of 8.3 million.[5] Unlike Beijing, Wuhan has much richer water resources, ranking first among the largest Chinese cities. Called *water city* in China, Wuhan is located about halfway along the several thousand kilometres reach of the Yangtze River and has nearly 200 lakes of various sizes. The water area makes up 25.8% of Wuhan's entire territory. Although Wuhan has abundant water resources, the Yangtze River and many lakes suffer from serious pollution. In 2000, Wuhan's wastewater discharge totalled about 2 million cubic metres per day with domestic sewage and about 25% of that was industrial wastewater. Water quality in Wuhan has significantly decreased over the last 15 years, making the concern for sustainable urban water management in this city greater than in other cities.

Other Chinese provinces want to get the eco-province label and take initiatives to achieve this. In China this usually means that competition is created and a prize may be given to the most ecologically friendly province or city. The Jiangsu province is an example that is implementing a policy for sustainability. It will implement the Jiangsu Eco-Province Plan with the Nanjing Eco-city Project as a major component.

3.6.3 Initiatives at the Level of Buildings

Although many initiatives are taken at the level of the city, the real promotion of ecological innovations comes from the national level through subsidies. For example a

30% subsidy of the construction cost is possible in the case of an ecological housing project. An interesting case of an ecological neighbourhood can be found in Wuhan and concerns a project of about ten buildings with seven or eight floors per building. The project would receive a 30% subsidy for using energy saving techniques, but one of the conditions was that the project would also recycle their grey water.[6] Energy savings is based on double-glazing and the use of ground source heat pumps. The geothermal heat pump uses a system of pipes absorbing latent heat from the ground and transferring it to the home's heating and hot water systems. The details are provided in Box 3.4.

Box 3.4 The Taiyue-Jinhe (Tai) Residential Project in Wuhan

The Taiyue-Jinhe (Tai) project is about establishing an ecological residential area with low energy consumption and a water recycling system. It is located in Jinyin Hu district, which is a suburban area of Wuhan city. Because there are two big lakes: Jin Lake and Yin Lake, the district is called Jinyin Hu (lake). Jinyin Hu district was an agricultural production field 20 years ago, mainly for rice production. Presently Jinyin Hu district is being developed as a residential space and ecological park.

The Tai project began in 2006, and the residential building was completed and sold out in 2007. The water recycling system was estimated to be completed in 2008 (see Fig. 3.2). The Tai project is involved in a national level energy saving program (initiated by the Ministry of Construction) on the condition that energy saving and water recycling systems are included. This program was organized by the Chinese Ministry of Construction which also issues permits to build water recycling systems. Moreover the Tai project could get a subsidy from the Ministry of Construction. At present there is no policy on water reuse system construction in Wuhan.

There are two main parts to water recycling: water reuse and rainwater harvest. The water reclamation technology used by the Tai project is Membrane Bio-Reactor (MBR) with wetlands. Two pipes are constructed in the residential buildings to collect wastewater: one for grey water and another for black water. Only grey water is recycled, the black water goes directly to the municipal sewage system. The MBR method is the first step and wetlands is the second step for wastewater cleaning. Rainwater is collected through drainage pipes in the buildings and beside the paths. After the rainwater is collected, it moves directly into the wetlands. Finally the reused water is pumped from the wetlands and used to water the green areas and wash cars.

The wetlands consist of three lakes: North Lake, Middle Lake and South Lake, which are shown in Fig. 3.3. The water moves from the south to the north due to water level differences. In the middle, there is a windmill, which transfers the water from outside the lake into the wetland in order to keep enough water in the wetland. There are several pumps in the northern lake to transfer reused water. Unfortunately we found during our fieldwork in October 2007 that the houses were almost finished (to be occupied in December 2007), but the grey water treatment facility was not yet built. The question is whether this will still happen, since the project developer considered thermal isolation more important and expected to get the subsidy anyway. When we checked in the summer of 2008 it had still not been finished. For the apartment buyers thermal isolation is an asset, but they were not very interested in separating grey and black (heavily polluted) water, since this would incur additional cost and they would not get the money back.

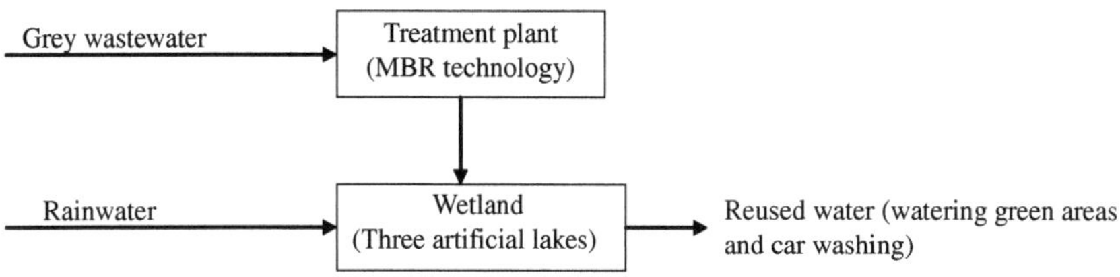

Fig. 3.2 Water recycling in the Taiyue-Jinhe project. *Source*: Interview with the manager of the Tai project

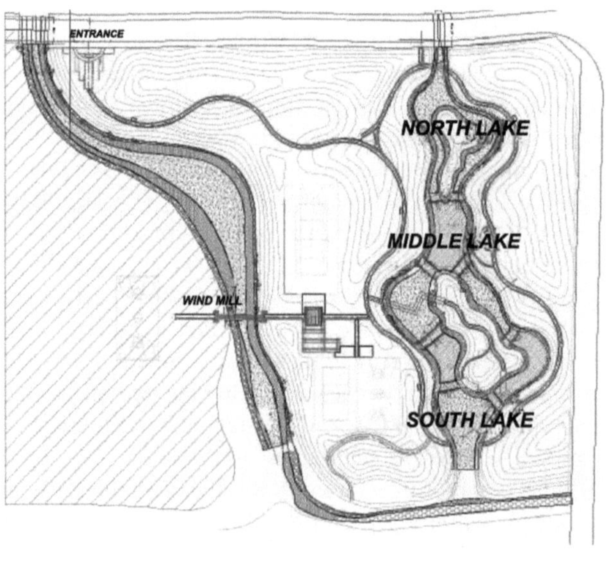

Fig. 3.3 The wetland. *Source*: The Tai project introductory document

Our research aims at completing a financial and economic analysis of the decentralized system of urban water management (also Zhang 2006). The expected outcome of the research may contribute to developing and selecting sustainable plans for urban water management by:

a. Determining costs and benefits for the alternative systems from the point of view of social economics
b. Financially appraising the alternative systems
c. Exploring the sustainable financing plans
d. Comparing the economic competitiveness of the alternative systems with that of the existing centralized system.

3.6.4 Initiatives at the Household Level

Finally individual initiatives can be noted, spontaneously or triggered by incentives. Environmental awareness may not yet be very developed in China and more time and policies that raise the consciousness of the people may be needed to achieve more activities at this level. However, people may save energy and tend to use less water than in developed countries, but this is partly due to the level of development, availability and price. Individual households usually install water heaters on the roofs of houses. In certain cities this is becoming a trend; the question is whether the systems are efficient enough to be recommended to large numbers of people and to have a substantial impact.

3.6.5 The Example of Singapore

Singapore is a city-state, an island of 20 × 30 km counting currently 4.5 million inhabitants. Its government has the ambition to almost double this number in the next 50 years. Singapore became independent in 1965 and started as an Asian tiger producing low-tech labour intensive products. In the 1980s it deliberately increased wages substantially, since it wanted to become an economy based on technologically more advanced products. Currently a third transformation is envisaged where Singapore wants to become a high-tech service economy in Southeast Asia. Yuen (2006: 414) notes that the "planning, design, and management of the urban environments are much admired by other Asian nations".

In the remarks of Roberts and Kanaley (2006), the country is also presented as an example of good practice, in particular its approach to sustainable urban development. Singapore is one of the four Chinese cities/regions figuring on the list of most competitive locations in Asia (on the list also figures Taiwan, Hong Kong and Zhejiang Province in China). There are some of the reasons for Singapore's economic success since its independence:

a. Political stability
b. Long-term vision and a development strategy
c. Leadership
d. Strategic location with a booming port, which is first in the world in terms of throughput of containers (measured in TEU)

Singapore is a kind of laboratory for housing and environmental policies in Asia. It also shows a coordinated effort to become a green city. The Ministry of Information of Singapore (2008a) published a brochure on "Green Singapore" and one on "Sustainability" (Ministry of Information of Singapore 2008b). The first publication details Singapore's urban planning and community involvement to make it a green city. In Singapore the shortage of water led to integrating wastewater treatment in an innovative way in the drinking water cycle, under the lead of the Singapore Public Utilities Board.[7] Having learned from this experience Singapore now wants to become a hydro-hub.

3.7 Rotterdam in Europe: Different Approaches to Urban Water Management

Rotterdam (in the Netherlands) is also an example of a city trying to become more ecological. It takes part in the Clinton initiative and is currently considering storing carbon dioxide in its port area. Rotterdam wants to become a climate proof city by 2020 (Rotterdam 2008a). Every city needs enough water for its population and industries, and hence it needs water resources. However, a city also needs institutions that secure good use of water. The current set-up in the Netherlands is complicated and the fragmentation of institutions makes integrated water management at the city level difficult. Given the need for a city like Rotterdam to deal with the risks involved in urban water management, we suggested three alternative approaches (van Dijk 2007a).

The first option is an integrated approach to water management, combining drinking water and surface-water management perspectives, which are currently institutionally separated in the Netherlands. However for such an approach, the current institutional context is too complicated and not appropriate for the problems Rotterdam is facing. Integrating the production of drinking water with surface water management was the option chosen by another Dutch city, Amsterdam. The authorities announced a merger between the water board and the municipal water company, which would lead to water chain management, where the customer would eventually pay only one bill for all water related services.

The second alternative is closing the water cycle to deal with water in a more efficient way. Closing the water cycle means not losing any of the scarce resource and controlling the quantity and quality constantly. Such an approach would favour integrating the management of the whole water cycle. Singapore has managed for example to close the water cycle and in principle, no water gets lost between resource and users. All of it is cleaned and made available for reuse. In the Dutch context this would mean a closer cooperation between the water utilities and the water boards. It would also imply a different role for the municipalities. However this may be easier than continuing to clean dirty water from the rivers to discharge it again after treatment to the North Sea.

The third option is to strive for a more ecological city, where integrated water management would be part of a broader approach to the urban environment. The term ecological city could be used as an approach to urban management that combines water with environmental management and focuses on long-term urban sustainability. The perspective is broader than just water related environmental issues. Examples in the European context are Hanover and Hamburg and invite debate on the ecological city of the future.

Considering these options, a more effective management of the water system and making it more sustainable is needed. Water management can be undertaken by central government or by communities. In Europe the task is usually allocated to the city level, which makes it interesting for Rotterdam as they develop plans to deal with water in a different way (van den Berg and Otgaar 2007).

3.8 What Can We Learn From the Ecological City Experiences for the Future?

What can we learn from these different experiences to build the ecological city of the future? There is currently no definition of what an ecological city would really be, so we need to agree on what we consider the important criteria for sustainability and I would go for stakeholder planning to assure that all partners will work together for the common future of the city. Stating that it requires an integrated approach is not enough, because this could mean integrating the analyses of the issue (look at them in relation to each other). But also an integrated approach to deal with the issues can be chosen and finally the activities undertaken to solve the problems can be integrated.

Ecological cities are more than ecologically managed closed urban water systems. Sustainable urban water management is just the beginning. Changes in the behaviour of consumers will be required, just like a combination of better water management, collection and treatment of solid waste and striving toward integration (van Dijk and Oduro-Kwarteng 2007). Water demand management may be a good start at the household level, just like separation at source and composting at home is a good start for ecologically friendly solid waste management.

In China the initiatives are undertaken at three distinct levels, but there is no real integrated approach at the provincial (Fujian province for example, *China Daily*, 27-8-2002), or at city level. The institutional framework of provinces taking the initiative, provincial capitals trying to do something and a state level Ministry of Construction to approve projects are in place, while the state level Environmental Protection Agency that does the regulation, does not function properly at the moment.

Consultancy firms claim that sustainable urban development starts with integrated design (DHV 2007). However what's important is convincing people that it is essential to do something to improve one's environment. As the Dutch government claimed in a media campaign: The environment starts at home. More is necessary than consultancy reports. Good research shows what works and why help is needed with realistic suggestions for ecological cities of the future. Private developers are looking for new ideas, but they are also mainly interested in cost savings and offering attractive alternative options for the customers for their projects. In Europe we may need besides a "cultural capital" an annual example of a good eco-city initiative.

3.9 Conclusions

Urban development means forging new partnerships between parties that have not often worked together: government officials, non-governmental organizations (NGOs) and private sector businessmen. This requires "organizing capacity" (van den Berg et al. 1996) and the ability to develop an integrated approach to the key issues facing the city. This is the job of an urban manager (van Dijk 2006). Ideas about ecological cities change over time and this affects the design of policies and projects to improve the urban environment in which we live. We assume the ideas will change again, once the consensus thinking of the 1990s will start to fall apart because we will start to realize that countries, cities and wards differ from one part of the world to another, as anthropologists, non-western sociologists and geographers keep telling us. Pollution, solid waste and wastewater problems, all aggravated by climate change require a different urban management approach to build the ecological city of the future!

However, the eco-city of the future is not just about dealing with environmental issues. Such a city will also need a sound economic basis, appropriate solutions for its transport systems and requires urban amenities. The presence of sufficient amenities is an important factor to make a city attractive and receives more attention because it is contributing to the quality of life in cities. In the European Union this element is emphasized in its program of choosing periodically "a cultural capital of Europe". This is usually an opportunity for such a city to show what it has to offer and to make additional investments to increase its attractiveness. In the future we may need to choose an eco-city as well.

Notes

1. The paper is based on research carried out in the framework of the Switch project, where work on modelling floods has been quite prominent, but where also the question has been asked how to link the results of such modelling efforts to the existing decision making structures (Van Dijk 2009a).

2. Nine cities around the world serve as demonstration cities and a learning alliance framework will be established in each demo city. Through the learning alliance platform, the barriers to information sharing are broken down and the process of technological and institutional innovation is sped up.
3. There are even some Build, Operate and Transfer [BOT] projects of local investors in this sector.
4. The current rate is 10 yuan to the euro.
5. The case study has been undertaken in Wuhan in November 2007 with a doctoral student, Mrs. X. Liang.
6. Grey water is wastewater generated in households, excluding water containing human excreta or urine, but including water from kitchens, bathrooms and laundry rooms.
7. The water utility in Singapore functioned already well for a long time, although it was only a municipal department, not even corporatized to separate its finance from the regular municipal finance. However, the authorities did not interfere!

References

Berg, L. vanden, Braun, E. & van der Meer, J. (1996). *Organizing capacity of metropolitan cities.* Rotterdam: EURICUR.

Brundland, G. (1987). *Our common future.* New York, NY: United Nations.

China Bulletin of Water Resources (2004). *An overview of waste water treatment.* Beijing: Ministry of Construction.

CICED (2006). *Lessons learned for integrated river basin management.* Beijing: China Environmental Science Press.

DHV (2007). *Integraal ontwerpen is de sleutel tot duurzaamheid* (pp. 2–3). Amersfoort: DHV Times.

Falkenmark, M. & Lundqvist, J. (1998). Towards water security: Political determination and human adaptation. *Crucial, Natural Resources Forum*, 21: 37–51.

Finco, A. & Nijkamp, P. (2001). Pathways to urban sustainability. *Journal of Environmental Policy and Planning*, 3: 289–302.

Kenworthy, J. R. (2006). Dimensions for sustainable city development in the third world. *Environment Urbanization*, 67–84.

Liang, X. & van Dijk, M. P. (2010). Financial and economic feasibility of decentralized waste water reuse systems in Beijing. *Water Science and Technology*, 61(8): 1965–1974.

Ministry of Information of Singapore (2008a). *Green Singapore.* Singapore.

Ministry of Information of Singapore (2008b). *Sustainability.* Singapore.

Munasinghe, M. (2007). *Making development more sustainable, sustainomics framework and practical applications.* Colombo: Mind Press.

Pahl-Wostl, C. & Kabat, P. (2003). *New approaches to adaptive water management under uncertainty.* Brussels: EU.

Roberts, B. & Kanaley, T. (Eds.). (2006). *Urbanization and sustainability in Asia, good practice approaches in urban region development.* Manila: ADB Cities Alliance.

Rotterdam (2008a). *Rotterdam, climate proof.* The Netherlands: Rotterdam Municipality.

Rotterdam, S. (2008b). *1600 hectares, creating on the edge, five strategies for sustainable development.* Rotterdam: Projectbureau Stadshavens Rotterdam.

Salih, M. A. M. (2009). *Climate change and sustainable development, new challenges for poverty reduction.* Cheltenham: Edward Elgar.

Schouten, M. & Hes, E. (Eds.). (2008). *Innovative practices of African water and sanitation providers.* Johannesburg: Sun Media.

Seckler, D., Amarasinghe, U., De Silva, R. & Barker, R. (1998). *World water demand and supply, 1990–2025: Scenarios and Issues.* Colombo: IWMI.

Sen, A. (2009). *The idea of justice.* Cambridge: Harvard University Press.

Strigl, A. W. (2003). Science, research, knowledge and capacity building. *Environment, Development and Sustainability*, 5(1–2): 255–273.
Urbano, F. (Ed.). (2009). *Building safer communities*. Amsterdam: IOS Press.
van Dijk, M. P (2006). *Managing cities in developing countries, the theory and practice of urban management.* Cheltenham: Edward Elgar.
van Dijk, M. P. & Oduro-Kwarteng, S. (2007). Urban management and solid waste issues in Africa. Contribution to ISWA World Congress September, Amsterdam.
van Dijk, M. P. (2007a). Water management in Rotterdam: towards an ecological city? In L. van den Berg & A. Otgaar (Eds.).
van Dijk, M. P. (2007b). Urban management and institutional change: an integrated approach to achieving ecological cities. Contribution to an international seminar on sustainable urbanization in Tripoli, Libya, Hotel Bab Africa, 30 June and 1 July.
van Dijk, M. P. (2008). Turning Accra (Ghana) into an ecological city. In M. Schouten & E. Hes (Eds.). pp. 99–114.
van Dijk, M. P. (2009a). Urban water governance as part of a strategy for risk mitigation, what is different in Third world cities? In F. P. Urbano (Ed.). pp. 182–200.
van Dijk, M. P (2009b). Ecological cities, illustrated by Chinese examples. In M. Salih (Ed.). pp. 214–233.
van den Berg, L. & Otgaar, A. (Eds.). (2007). *Rotterdam, city of water.* Rotterdam: EURICUR.
van Dijk, M. P. & Zhang, M. (2005). Sustainability indices as a tool for urban managers, Evidence from four medium-sized Chinese cities. *Environmental Impact Assessment Review*, 25: 667–688.
Wang, R. (2006). *Integrating of eco-industry, eco-scape and eco-culture, a case study of Hainan eco-province planning*. Beijing: Academy of Sciences, Research Centre for Eco-environmental Sciences.
WWF (2005). Linking rivers. www.riverlinkinsdialogue.org. Accessed 10 January 2010.
Yuen, B. (2006). Innovation, key to sustainable urban development in Singapore. In B. Roberts & T. Kanaley (Eds.), *Urbanization and sustainability in Asia, good practice approaches in urban region development*. Manila: ADB, pp. 414–417.
Zhang, C. (2006). An assessment of centralized and decentralized wastewater reclamation system in Beijing. Wageningen: Unpublished MSc thesis.

Chapter 4
Eco-infrastructures, Feedback Loop Urbanisms and Network of Independent Zero Carbon Settlements

Carlos H. Betancourth

Abstract More than half the world's population now lives in cities, and the rate of urbanization is accelerating. Cities are major sources of greenhouse gas (GHG) emissions. They are vulnerable to climate change. The limited success of the December 2009 Copenhagen climate negotiations heightens the urgency of cities' efforts to adapt and mitigate to climate change. Urban growth in the developing countries of Latin America, India and China is fundamentally changing the lives of hundreds of millions of people. So far, these urbanization processes have dramatically increased developing countries' environmental damage and vulnerability to climate change. This paper aims to show that urbanization can be a sustainable process capable to create secure urbanities through an eco-infrastructure approach for reducing urban vulnerabilities that explores a series of strategic responses in a weave of eco-infrastructures, feedback-loop urbanisms and networks of zero carbon settlements powered by renewable energies.

4.1 Introduction

Climate change impacts such as increases in global temperatures, loss from flooding and hurricanes accompanied by rising sea levels are becoming an all too frequent occurrence in many countries, particularly in cities where people and assets are concentrated. This is generating uneasiness over the environmental security to maintain and enhance economic growth at the national scale. In a context of resource constraints and climate change, questions of environmental, social and economic reproduction become strategically entangled at the city level. It is expected that increasing concerns over the environmental security of cities will give rise to attempts to protect their critical infrastructures. Cities need to actively engage in developing strategic responses to the opportunities and constraints of

C.H. Betancourth (✉)
Independent International Consultant
e-mail: chbetanc@msn.com

T.-C. Wong, B. Yuen (eds.), *Eco-city Planning*, DOI 10.1007/978-94-007-0383-4_4,

climate change and resource constraints. This means that urban centres must be prepared with a knowledge base of climate projections and specialized tools to deal with these impacts to look after their critical infrastructures through the protection of flows of ecological resources, infrastructure and services at the urban scale.

Moreover, given the potential devastation associated with future climate change-related disasters, it is vital to change the way we build and manage our cities, through new strategies to reconfigure them and their infrastructures in ways that help secure their reproduction. The spatial planning of cities requires the consideration of climate change impacts as vital components of urban development. In order to start to build up the case for the strategic relevance of the city in generating responses to climate change, it is important to design tools for local governments and their communities to better understand the concepts and consequences of climate change and resource constraint; how their impacts generate urban environmental in-security; and what needs to be done to build ecological secure urbanities. In this paper, we begin to put together a framework for a tool-based process that takes into consideration the limited resources that characterize cities in developing countries (as well as the uncertainties and risks that characterize the complexity of climate change), and start to build a knowledge base that will inform and support the design of strategies to protect cities through comprehensive adaptation programs and plans.

Based on the case of Latin America in general and in particular on the case of informal settlements in the city of Cartagena (Colombia), I outline the challenges posed and the responses required by the environmental security of cities. I propose an eco-infrastructure approach for reducing urban vulnerabilities and start to explore a series of strategic responses which I characterize as a weave of eco-infrastructures which points in the direction of a new logic of infrastructure provision. It is critical that the definition of urban infrastructure must be expanded from just basic services to include climate change impact and hazard management investments for a secure built environment. The argument is developed in five sections.

The second section following the introduction presents the case of climate change impacts in Latin America and how these ongoing changes are already affecting ecosystems and social systems. The third section provides an overview of the concept of environmental security and formulates the need for new planning and design tools. The fourth section elaborates and presents the tools and their components. This section makes the case for the strategic involvement of cities in addressing climate change. It provides a preliminary evaluation of impacts on ecosystems and eco-infrastructures at the regional scale. It maps information regarding potential hotspots of vulnerability at the regional scale. It introduces some of the steps necessary to create a knowledge base for the city and its citizens, and explores responses for secure urbanities through eco-infrastructures, enclosed spaces, and networks of zero carbon settlements. The final section summarizes the key findings of the paper and makes some general suggestions for further research and investigation.

4.2 Climate Change Impacts in Latin America

The observed changes in the global climate suggest that warming of the climate system is undeniable (IPCC 2007, Stern 2008). 2010 is becoming the year of the heat wave, with record temperatures set in 17 countries. The recorded temperature for Colombia, in January 2010, was, 42.3°C (Guardian 2010a). The rise in global temperatures is impacting Latin America's cities including low-lying cities located in the Colombian Caribbean coast. Temperatures in Latin America increased by about 1°C during the twentieth century, while sea level rise reached 2–3 mm/yr since the 1980s. The IPCC's Fourth Assessment Report predicts that under business-as-usual scenarios, temperature increases in Latin American countries with respect to a baseline period of 1961–1990 could range from 0.4 to 1.8°C by 2020 and from 1 to 4°C by 2050 (Magrin et al. 2007). The effects from a rise of two degrees-modifying weather patterns, which in turn affect temperatures, precipitation patterns, sea levels, storm frequencies and floods will be felt by every town and city, especially those in coastal zones. Changes in precipitation patterns have been observed, with some areas receiving more rainfall, and others less. Extreme weather events have become more common in several parts of the region, including more and/or stronger storms (Raddatz 2008, Hoyos et al. 2006, Webster et al. 2005).

Climate change is likely to cause severe impacts on ecosystems and species such as the bleaching of coral reefs; the damage of wetlands and coastal systems and the risk of forest degradation in the Amazonan basin as well as on socio-economic systems and cities of the Latin American region (Milly et al. 2005, Ruiz-Carrascal 2008, Coundrain et al. 2005). It is expected that the agricultural sector will suffer direct and large impacts from gradual changes in temperatures and precipitation (Mendelsohn 2008, Medvedev and van der Mensbrugghe 2008). Cities and localities will also suffer serious economic and social impacts: the expected increase in the frequency and/or intensity of hurricanes and tropical storms will impact coastal cities, their livelihoods, infrastructures and biodiversity (Curry et al. 2009, Toba 2009); the expected disappearance of tropical glaciers in Los Andes (Bradley et al. 2006) and changes in rainfall patterns will have economic consequences on water supply and the availability of water for use and consumption in Andean cities, in agriculture, and in hydroelectric production. The increase in the rate of sea level rise will economically damage coastal areas and cities through the loss of land, of tourism infrastructure, of buildings (UNFCCC 2006, Dasgupta et al. 2007). Climate change could also have multiple impacts on health (Confalonieri et al. 2007), such as increase in malnutrition and mortality, cardio respiratory diseases from reduction in air quality, and an increase in water-borne diseases-such as malaria in rural areas and dengue in urban areas.

The evidence indicates that climate change and resource constraints will impose significant costs on Latin American cities and eco-systems (De La Torre et al. 2009). However, current efforts to address climate change focus mainly on attempts to *mitigate* climate change and on reducing GHG emissions of greenhouse gases as well as on endeavors to *adapt* through reducing the vulnerability of communities at risk by improving and building hard and grey infrastructures. This paper attempts to move beyond adaptation and mitigation and aims at long-term climate resilience (the physical and the institutional capacity to absorb the long-term trends and near-term vagaries of climate while maintaining risk at socially acceptable levels) of cities and their critical infrastructures, and sets out an argument for including an eco-infrastructure-based approach in strategies to address climate change. As these ecosystems have a critical role to play in building resilience and reducing vulnerabilities in cities, communities and economies at risk, the enhanced protection and management of ecosystems, biological resources and habitats can mitigate impacts and contribute to solutions as nations and cities strive to adapt to climate change. This proposal for an informal settlement located in the Delta City of Cartagena Colombia proposes an eco-infrastructure approach to climate change as a supplement to national, regional and local strategies. Such eco-infrastructures based strategies can offer sustainable solutions contributing to, and complementing, other national and regional adaptation strategies, and facilitate a transition of informal settlements from "a slum condition" to a living laboratory of eco-infrastructure landscapes for low carbon growth and development. This requires a transition from mono-functional grey infrastructures to a network of multi-functional eco-infrastructures and living spaces that all work together as a connected system to conform an integrated habitat.

In a context of resource constraints and climate change risks (floods, droughts, heat stress, diseases, loss of infrastructure and lives, displacement of people), a series of new environmental, socio-economic and political problems (energy security, scarcity of water resources) is forcing issues of environmental security up the agenda of national governments (UNEP 2007, Pirages and Cousins 2005, Hodson and Marvin 2009, Giddens 2009). Major and emerging environmental changes (such as depletion of fresh water supplies, fisheries, biodiversity, agriculture lands, food and health safety, stratospheric ozone and global warming) can lead to environmental conflicts (Betancourth 2008a), and to short and long term decreases in environmental security. Resource constraints and climate change can be characterized as problems of environmental security. This in turn invites to rethink the concept of security. Addressing environmental insecurity requires collective and preventive action (through re-design) and a transition to alternative models of development and economic growth where the sustainable use of natural resources and joint efforts to protect the environment can contribute to environmental security and conflict prevention.

4.3 The Need for New Tools

This issue of environmental security can be formulated at different but interrelated spatial scales: global, national, sub-national, regional, urban, metropolitan and local. The implementation of the mitigation and adaptation policies necessary to successfully address the climate change challenge will only be achieved, and sustained, through involvement and commitment at all these levels of decision-making. In particular, the full engagement of the sub-national scales is important to move the climate change and alternative development agendas forward. Their decisions can influence GHG emissions and most site-specific adaptation initiatives as well as to promote long-term planning and to incorporate climate change considerations into decision-making. Adaptation to climate change is very site specific, and local planning decisions will be critical to tailor almost every single adaptation action to the conditions in which it will take place. The relevant questions at this local scale are: how do cities and regions prevent their reproduction in conditions of environmental insecurity? Which are the strategic responses and which insights, capacities and new tools are needed for successful decision-making? To elaborate on these questions requires formulating a new agenda for urban development.

4.3.1 The Agenda

This agenda is built around the following problems and themes:

1. the problem of the environmental security of cities as protecting flows of environmental resources at the regional and urban scales;
2. the strategic importance of cities in developing responses to climate change and resource constraint for the production of secure urbanities;
3. the reorientation of the management, growth, and development of the city to climate change and resource constraint; by building on the synergy and interdependence of ecological and economic sustainability;
4. the reconfiguration of the city and its infrastructures in ways that help to secure their environmental, social and economic reproduction, around the following responses:
   (a) improving the strategic protection of cities through:

- the redesign of layers of eco-infrastructures (the environment as infrastructure);
- enclosed and autonomous urban spaces (feedback loop urbanisms); and
- networks of zero carbon settlements.

(b) reducing the sensitivity of citizens to climate hazards by using the sustainable management of ecosystems and of eco-infrastructures to:

- expand livelihood assets; and
- enable economic development through enterprise development related to ecosystems management.

(c) improving adaptive capacity through eco-infrastructure governance that builds:

- adaptative/mitigative planning;
- flexible and coordinated institutions; and
- learning and dissemination of knowledge needed to empower people in planning and decision-making related to adaptation.

In order to gather social actors around this agenda, we need to offer them new insights, guidance and tools as they seek to take steps to adapt and mitigate to climate change. In what follows I will briefly present some of these tools.

4.3.2 The Tools and Their Components

4.3.2.1 Constraints of Key Local Institutional Players

There are a number of barriers that need to be recognized and overcome to enable key local actors to play a critical role in addressing climate change. First, there is an increasing body of scientific literature on global climate change impacts but a lack of knowledge at the local level. Second, in a new field like climate change, local public authorities may have limited technical and financial capacities. Third, knowledge sharing is limited by the varying roles and responsibilities of regions and cities. If local authorities are to succeed in their efforts to address climate change, effective partnerships must be formed with a variety of social actors – their constituencies, the national government, international donors, the academic community, technical centres of excellence, and the private sector, who share common interests in addressing climate change. Fourth, the preparation of integrated urban and regional climate change plans can remove some of these barriers above. Such plans will require a rethinking of the development processes and the formulation of strategic approaches and innovative policy development and planning instruments to promote long term planning and to incorporate climate change considerations into decision-making (UNDP 2009).

4.3.2.2 Changing Needs and Uncertainty

Climate change is unequivocal. Less certain is the timing and magnitude of climate change. Climate change represents a dramatic increase in uncertainty and new decision-making methods will be required to cope with it. Many infrastructure investments and planning decisions, such as water and transportation infrastructure, building design and urban/land-use planning, require substantial lead-time for conception and implementation. By the end of this century, investments may have to cope with climate conditions that will be radically different from current ones, otherwise they risk to be obsolete or sustaining damages due to the impacts of climate change. Different climate change models could predict a full range of possible future climates for one and the same region and city. These entails that infrastructures could face different and opposite climate change scenarios.

Water infrastructure could, over its lifetime, face a significant drying, a progressive wetting, or even an initial wetting period followed by a significant drying. Water engineers can easily design water infrastructure adapted to a progressive drying or wetting but it will be more difficult to design infrastructure adapted to a full range of possible future climates. Thus, infrastructures would need to be designed and re-designed so that they can be adapted not just to one scenario or the other, but to a full range of possible future climates. While it is known that our climate will change over the long-term, decision-makers are confronted with a situation where the direction of change is not fully clear at this stage. The chain of causality between emissions today and the future impacts of climate change has many links, and there is a great deal of scientific uncertainty involved in moving from each one to the next. Yet decision-makers will still need to make investment decisions today, with incomplete and imperfect information to estimate both the costs and benefits of such decisions. It is very hard to quantify the probabilities associated with specific climate impacts. Thus, policy makers are confronted not only with risk-randomness with known probabilities, but also with uncertainty (Knight 1921).

The risk of simply reacting to changes in the short- or medium-term could result in poor investment decisions, the cost of which could exceed the direct costs of global warming. These considerations of risk and uncertainty may make it prudent for policymakers to adopt an approach based on precaution, in which a large weight is assigned to the objective of avoiding such events. Addressing environmental insecurity requires acting preventively through re-design and a transition to alternative models of development and economic growth. Therefore, it is important to design strategies which can cope with climate change uncertainty regardless of how the local climate will change. In what follows, I will be exploring strategies for risk-informed mobility, multiple land use planning and risk informed water management through the concepts of eco-infrastructures; of feedback loop urbanisms and of networks of zero carbon settlements. But we will present first some preliminary prospective techniques and scenario based approaches that can help us overcome some of the constraints posed by the lack of information and help local and regional decision-makers deal with climate uncertainty and complexity.

4.4 Evaluating the Vulnerability of Ecosystems and Eco-infrastructures

Adaptation to climate change calls for a new paradigm that manages risks related to climate change by considering a range of possible future climate conditions and associated impacts, some well outside the realm of past experience, trends, and variation. This means not waiting until uncertainties have been reduced to consider adaptation actions. Mobilizing now to increase the city's security and adaptive capacity can be viewed as an insurance policy against an uncertain future. Reducing vulnerability to climate change requires to consider a combination of: (a) the nature and magnitude of the changes experienced and reduced exposure to hazards; (b) underlying social, cultural, economic, geographic, and ecological factors that determine sensitivity to climate change, and reduced sensitivity to the effects of climate change; and (c) the city's ability to prepare for, and respond to impacts on ecological, economic, and human systems, and increased adaptive capacity (See agenda cited above in Section 4.3). This process can be described as a series of stages: a first step is to identify current and future climate changes relevant to the city under consideration; the second step is to assess the vulnerabilities and risk to the system; the third step is to develop an adaption strategy using risk-based prioritization schemes; the fourth step is to identify opportunities for co-benefits and synergies across sectors; the fifth step is to implement adaptation options; and the sixth step is to monitor and re-evaluate implemented adaptation options. This preliminary exercise on adaptation planning that is presented in Table 4.1 as a linear progression is a cyclical, iterative process incorporating at least six steps (Fig. 4.1). This tool stimulates discussion and investigation, and allows social actors and stakeholders to make connections at different spatial scales between and among eco-infrastructures, the ecosystem services they provide, the local impacts, vulnerability, and responses to adaptation and mitigation. Due to time and space limitations, we will only be dealing here with steps 1, 2 and 3.

4.4.1 The Increase in GHG Concentration and Atmospheric Warming

In this first step (first and second column), we begin to identify current and future climate changes relevant to the territory under consideration. The warming of the climate system and the rise in global temperatures is already affecting Latin America's climate and its cities. Temperatures in Latin America increased by about 1°C during the twentieth century, while sea level rise reached 2–3 mm/yr since the 1980s. Changes in precipitation patterns have also been observed. Extreme weather events have become more common in several parts of the region, including more periods of intense rainfall and consecutive dry days.

4.4.2 The Impacts of Climate Change on Eco-infrastructures

Relying on climate projections one must next identify eco-infrastructure vulnerabilities (the third column), determine which risks are likeliest and which could have the biggest impacts, develop adaptation strategies, and prioritize them. As indicated above, some climate projections entail large uncertainties, and sorting matters out can be very hard. An example is whether to construct storm surge barriers in the city harbour to protect the city from flooding. This calls for caution against erecting hard infrastructures. The storm-surge barrier question raises the need to look at both the impacts of climate change not just on the environment, but to the environment as eco-infrastructure for responding and providing solutions to climate change adaptation. Once an inventory of these eco-infrastructures has been completed, the second step the assessment of vulnerability focuses on gaining an understanding of how climate change will impact the goods and services (column 4) provided by natural resources and eco-infrastructures (column 3); human-built infrastructure (column 5) and coastal communities (column 6) at different spatial scales. Vulnerability assessment for climate change in a coastal city-region also considers two other factors: the

Table 4.1 Impacts of climate change on eco-infrastructures at different spatial scales

Relating climate change to eco-infrastructures ecosystem services and impacts		Eco-infrastructures (Large Scale LA/Caribe)	Affected assets/ associated ecosystem services regional/local scale	Localized/sector impacts local scale	Hotspots of vulnerability	Strategic responses to reduce vulnerability to climate change
		Melting andean glaciers/paramos	Paramos: store/provision water for use downstream; energy	*Lower water availability* for irrigation, industry, energy, cities *Flooding; mudslides*	Mountains, rivers; cities	*Reduce exposure to hazard/strategic protections:*
Climate change		*Bleaching of coral reefs (Caribbean)*	Food; Protection of shorelines from storms	Lower food availability, lower protection shorelines Fisheries, tourism	Small islands	• Repair eco-infrastructures • Create autonomous urban spaces
		Damage Coastal wetlands-mangroves	Regulation of hydrological regime; protection from flood/storm; habitats; livelihoods	Destruction productive ecosystem: shrimp, oyster, fish production	Coastal cities	• Create networks of zero carbon towns
Increases in ghg con-centration	Consequences Sea level rise Temperature increases	*Sea level rise*	Ecosystem wetlands/mangrove Natural buffer against flooding; high winds, erosion	*Coastal cities inundation and erosion* *Increased flooding* Mangrove forest Agriculture Migration to cities Beachfront Tourism transport	Low lying delta	*Reduce sensitivity to effects* of cc: • Increase livelihood assets and • Increase opportunities *Increase adaptative capacity:* • Flex multi-actor institutions • Disseminate knowledge needed to deal with uncertain future events and
Atmospheric warming	Change in precipitation patterns	Rainfall patterns weather-related natural disasters Hurricanes Flood mudslides storms	Ecoinfrastructure Engineered infrastructures Protection coastal cities	Tourism Beachfront Transport Health	Drylands	• Empower people in planning and decision-making about adaptation
	Extreme vents	*Biodiversity* Decay of the Amazon rainforest		Loss of cultural bio-diversity heritage		

Source: Compiled by author

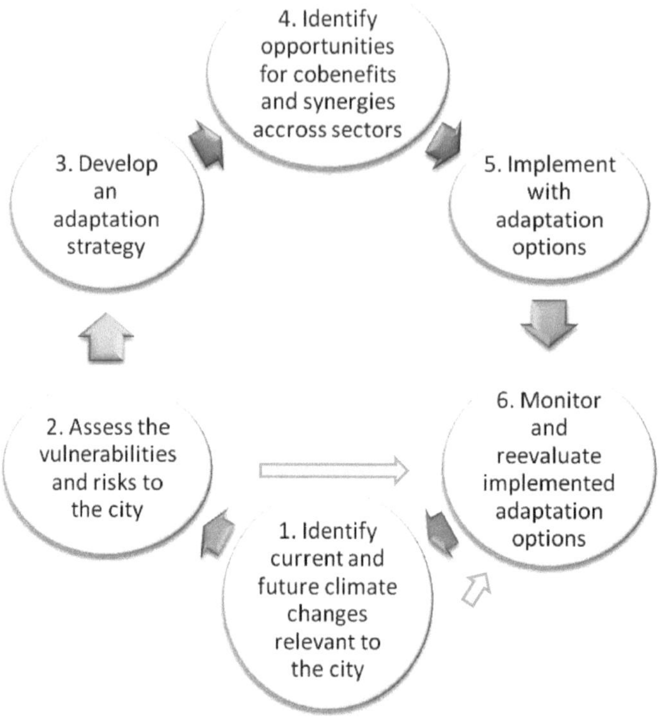

Fig. 4.1 Six-step adaptation planning as a cyclical, iterative process. *Source*: Compiled by author

natural assets and the human capital that will be exposed to and impacted by climate change (and that combined define hotspots of vulnerability (column 6); and starts to evaluate the capacity of communities and ecosystems to adapt to and cope with climate impacts (column 7).

Climate change will impact the health, function and productivity of ecosystems, thus impacting the health and welfare of communities and the people that depend on these natural resources. The main Latin American eco-systems that are already suffering negative effects and impacts from ongoing climate change are outlined in Section 4.2 above

4.4.3 Eco-infrastructures, Eco-system Services and the Affected Assets and Functions

High mountain eco-infrastructures including glaciers and high altitude wetlands (*paramos*) provide numerous and valuable environmental goods and services. They are eco-infrastructures that store and release water for use by downstream populations working in agriculture and living in cities. Large numbers of people in Latin America are dependent on glacier water. The fast melting of the Andean glaciers would deny major cities water supplies and put populations and food supplies at risk in Colombia, Peru, Chile, Venezuela, Ecuador, Argentina and Bolivia. Large cities in the region depend on glacial runoffs for their water supply. Quito, Ecuador's capital city, for example, draws 50% of its water supply from the glacial basin, and Bolivia's capital, La Paz draws 30% of their water supply from the Chacaltaya glacier which is expected to completely melt within 15 years if present trends continue. In Bogota, Colombia, 70% of the city water supply comes from an alpine paramo (a fragile sponge of soil and vegetation), which could dry up under higher temperatures. The volume of the lost glacier surfaces of Peru is equivalent to about 10 years of water supply for Lima (Bradley et al. 2006, Environment News Service 2008, Kaser et al. 2003).

The drastic melt forces people to farm at higher altitudes to grow their crops, adding to deforestation, which in turn undermines water sources and leads to soil erosion and putting the survival of Andean cultures at risk (NEF 2006). The entire range of the tropical Andes, and host to the vital global biodiversity, will be affected. Without this natural storage, more construction of dams and reservoirs would be needed. Power supplies also will be affected as most countries in the Andes are dependent on hydroelectric power generation (Bradley et al. 2006, Francou et al. 2003). Accelerated urban growth, increasing poverty and low investment in water supply will contribute to water shortages in many cities, to high percentages of urban population without access to sanitation services, to an absence of treatment plants, high groundwater pollution and lack of urban drainage systems (IPCC 2007).

Coral reefs eco-infrastructures are an integral part of the Caribbean fabric, threading along thousands of kilometres of coastline. Rich in life and beauty, they serve a multitude of purposes to the Caribbean people. Their fisheries provide food for millions of people, their structure protects shorelines from tropical storm swells; as they bleach, the reefs disintegrate and thus eliminate this protection. Reefs are also a tourism attraction and as they bleach and disintegrate, they lose their aesthetic value (Buddemeier et al. 2008). Coastal wetlands eco-infrastructures provide many environmental services, including the regulation of hydrological regimes, human settlement protection from floods and storms, sustenance for many communities settled along the coast, and habitats for waterfowl and wild life. These wetlands possess the most productive ecosystem and is one of the richest on Earth. Floods, mudslides and hurricanes could have serious impacts on eco-infrastructures and engineered infrastructure along the coastal cities and on the tourism industry, for example, in the form of loss of valuable beachfront. Coastal flooding will negatively affect the ecosystems of mangroves, a plant which is crucial in providing a natural buffer against flooding, high winds, and erosion (IPCC 2007). This may accelerate migration to urban centres (Gleditsch et al. 2007).

4.5 Mapping Potential Hot Spots of Vulnerability at the Regional Scale

On the basis of the first tool – an assessment exercise (Table 4.1), we can elaborate a second tool – a mapping exercise where we start to downscale the information to the regional level and define potential hot spots of vulnerability at the city-region scale and where the impacts of climate change on the eco-infrastructures and their ecosystems may be most dramatic. These hot spots of vulnerability (column 6 of Table 4.1, Fig. 4.2) are locations and places where susceptibility to adverse impacts of climate change is high because of exposure to hazards such as floods and drought or storm surges and because of sensitivity to their effects. These hotspots are the highest priority locations for adaptation, and include:

- The mountains and their rivers where the retreat of glaciers and reduction in the size of snow packs will increase disaster risk and shift the volume and timing of downstream water availability for irrigation, industry and cities. This is the case of La Sierra Nevada de Santa Marta (32 streams of water have disappeared in the last years).
- Small islands where sensitivity to coastal erosion, inundation and salt-water intrusion is high at community levels. This is the case of The Archipelago of San Andrés, Providencia and Rosario islands.
- Lowlying deltas and coastal cities where higher frequency of flooding and coastal inundation will have the most acute impacts. This is the case of the entire Colombian Caribbean coast and of cities such as Cartagena, Barranquilla, and Santa Marta.

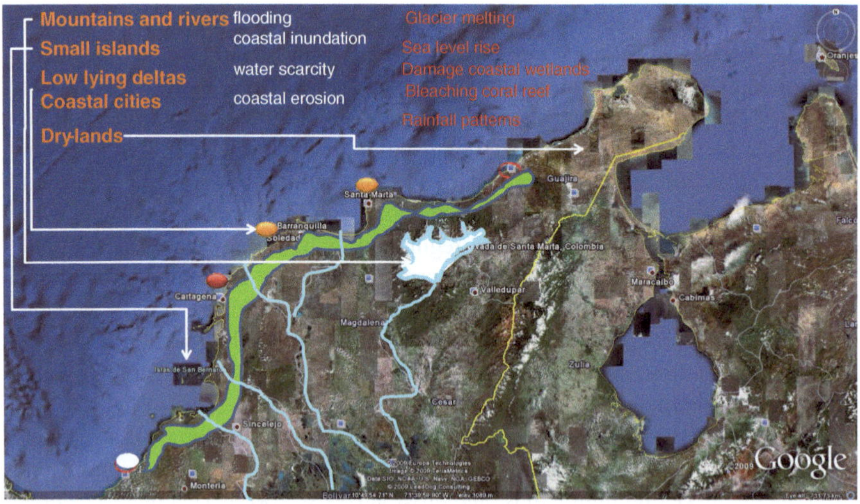

Fig. 4.2 Assessment exercise: defining hot spots of vulnerability at the regional scale. *Source*: Compiled by author http://maps.google.com/maps?t=h&hl=en&ie=UTF8&ll=10.411323,-75.495731&spn=0.027098,0.033002&z=15

- Drylands where susceptibility to more severe or more frequent water scarcity is high because of threats to food security, health and economic development. This is the case of La Guajira, but also of the Monteria-region

As Fig. 4.2 suggests, we find all these hotspots in the Colombian Caribbean coast. Before we move into the formulation of the next question, let us open a parenthesis here and note the following: (1) the regional and the territorial scale of the map is the scale of the ecosystems; (2) it is important to draw climate plans at this territorial scale; (3) it is at this scale where we find the potential to harness revenue streams from regional ecosystem services that could be invested in urban programmes. At this point the critical question to be addressed is: how can vulnerability to the hazards be reduced in the case of each hot-spot on the map? To elaborate on this question requires zooming-in on one of these hotspots where vulnerability is high for the poor and where climate change exacerbates exposure to climatic hazards.[1] This requires to down-scale the information to the city level, and the design of a third tool; a knowledge base for the city and its citizens. This tool is elaborated in the next section.

4.6 Creating a Knowledge Base for the City and Its Citizens

4.6.1 The Localized Impacts of Climate Change: The Case of Colombian Coastal Cities

As allued to earlier, Colombian coastal zones are highly vulnerable to sea-level rise, to coastal erosion, and flooding of low-lying areas. Seven critical zones have been identified: The Archipelago of San Andrés, Providencia y Santa Catalina in the Colombian insular area of the Caribbean; the cities of Cartagena de Indias, Barranquilla, and Santa Marta in the Caribbean continental coast; and the cities of Tumaco and Buenaventura in the Colombian Pacific coast. In the case of Cartagena, neighbourhoods located in the southern border of the *Cienaga de la Virgen* (Fig. 4.3) exhibit high socioeconomic and biophysical vulnerability (Invimar 2005, 2007).

The objective of this more localized assessment of Cartagena de Indias and its ecosystems is to identify the main vulnerable and at risk areas at the city scale. This knowledge is critical for defining priority actions to create secure urbanities. The assessment is not a quantitative tool for ranking cities nor is it intended to be a scientifically rigorous assessment.

Cartagena de Indias is a large seaport, economic hub, as well as a popular tourist destination on the north coast of Colombia. Cartagena faces the Caribbean Sea to the west. To the south is the Bay of Cartagena, which has two entrances: *Bocachica* in the south and *Bocagrande* in the north. The principal water bodies within the urban area are the *Bahía de Cartagena*, *Ciénaga de la Virgen* and *Ciénaga de Juan Polo* that are connected by a complex system of lakes and channels (Alcaldía de Cartagena 2000). Cartagena is also a divided city. It contains a series of antagonisms strong enough to prevent its ecological, social and economic reproduction.

Fig. 4.3 Main antagonisms: Cartagena as a divided city. *Source*: Compiled by author. http://maps.google.com/maps?t=h&hl=en&ie=UTF8&ll=10.411323,-75.495731&spn=0.027098,0.033002&z=15

Three possible antagonisms present themselves: the threat of ecological risks; the inappropriateness of an illegal process of urbanization through which public lands and their ecosystems are privatized; new forms of social exclusion such as new slums and shanty towns (Fig. 4.3). While the threat of ecological risks means that the entire city is in danger of losing everything and of vegetating in an unliveable urban environment, the antagonism between the included and the excluded is a crucial one. Thus, the ethico-political challenge is for all the inhabitants of the city to recognize themselves in this figure of the excluded. In a way, today we are all potentially excluded from nature through climate change impacts, and the only way to avoid actually becoming so is to act preventively through re-design in the form of collective action.

4.6.2 Identifying the City's Eco-infrastructures and Ecosystems and Some of the Forces that Are Degrading Them: The Workbook

The following are some of the most productive and biologically complex ecosystems localized in the Caribbean coastal zone of Cartagena and in its lagoons (Fig. 4.4):

Sandy beaches occupy 90% of the coastal line of the district. The biggest hazard for the preservation of this ecosystem comes from the intense process of urbanization (CIOH 1998). The Rosario islands coral-reefs are endangered by the use of

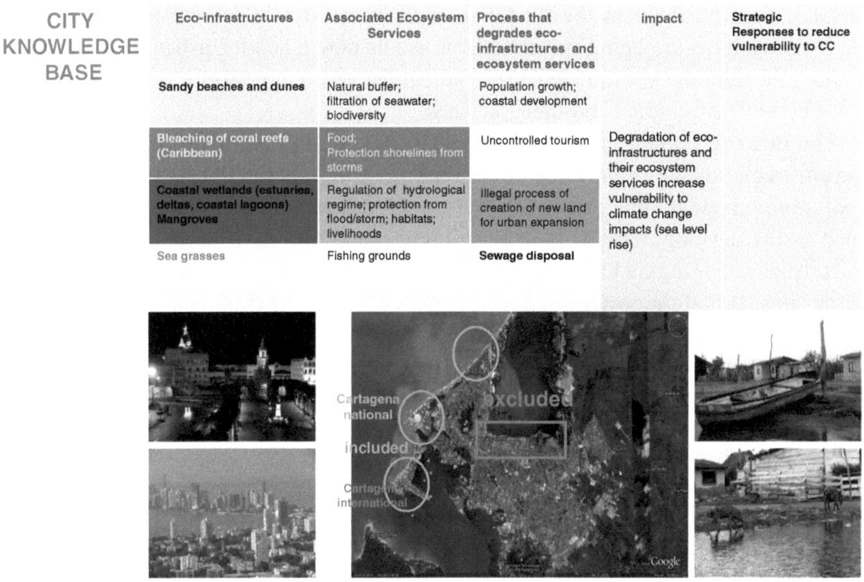

Fig. 4.4 Eco-infrastructures at the city scale. *Source*: compiled by author http://maps.google.com/maps?t=h&hl=en&ie=UTF8&ll=10.411323,-75.495731&spn=0.027098,0.033002&z=15

dynamite as a fishing method, uncontrolled tourists, increase in sea surface temperature and sediment discharge due to dredging of the dike channel (Díaz et al. 2000, Charry et al. 2004). The Cartagena Bay's 76 ha of sea grasses are connected to the open beaches and 58 ha are inside the bay; they are threatened mainly by untreated sewage disposal (Diaz et al. 2003). The biggest coastal lagoon in the area is the *Cienaga de la Virgen*, which has a length of 22.5 km and a mean depth of about 1.5 m. It is separated from the sea by *La Boquilla's bar* and is surrounded by mangrove areas (Alcaldia de Cartagena 2000). The south and west flanks are impacted by urban expansion, as this area is home to several of the city's shanty towns (Niño 2001). The interconnections between the *Cienaga de La Virgen* and Cartagena's Bay are currently interrupted as a result of unplanned urban expansion and garbage accumulation. Unplanned urban and industrial development in the *Cienaga de la Virgen* and Cartagena's Bay has resulted in a lack of sewage treatment that has deteriorated the environment in these areas.

There are several forces that contribute to the process of degradation of urban ecosystem services at the city and regional level; two of them deserve particular mention here: the illegal process of urbanization and coal mining for global markets and their associated logistics. The city of Cartagena is growing and there is scarcity of land. The low-lands around the lagoons of the *Bahía de Cartagena*, *Ciénaga de la Virgen* and *Ciénaga de Juan Polo*, are public lands inhabited by valuable eco-infrastructures such as mangrove forests. These public lands are privatized through an illegal market for urban land: poor people invade illegally the public lands, destroy the urban mangrove forest, fill in the lagoon, create new plots of land, and sell them in the illegal market. These land plots will be later legalized. Ecosystems are being privatized for use as new lands for urban development. Neither the national nor the local government has taken any serious action to protect these eco-infrastructures and their services.

The other important process impacting these eco-infrastructures at the scale of the city-region is the exploration of coal mining for global markets. The logistics of coal mining is degrading the ecosystems and landscapes of the city of Santa Marta such as the development of five coal mining ports (and associated train and truck infrastructure) along 39 km of beautiful beach front (McCausland 2009). These two cases suggest that the synergies and interdependencies of ecological and economic sustainability cannot be assumed as given. This economic model of illegal urbanization negates the ecological component of the synergy-equation (the mangrove forest is destroyed in order to fill in the lagoon and create new land for urban development). These combined process and forces (slum formation and exclusion; coal mining impacts and privatization of public lands through illegal urbanization; etc) have the potential to jeopardize the social reproduction of cities and generate new urban environmental conflicts between different social groups and their values regarding the environment, eco-infrastructures and their ecosystem services; climate change impacts and adaptation responses. Thus, the synergies between the ecological and the economic city still need to be constructed through a process of mediation of these new urban environmental conflicts.[2]

Degradation and destruction of the eco-infrastructures (coastal-wetland and its mangrove forests, Fig. 4.5) and ecosystems leads to loss of these ecosystem services. Of vital importance is the undeniable fact that human well-being can be damaged when these services are degraded, or else costs must be borne to replace or restore the services lost.

To illustrate, exposure to hazards can be reduced through eco-infrastructures and their ecosystem services. The risk of drought can be minimized by eco-infrastructures which store water for use during dry spells in wetlands and groundwater recharge areas, lagoon floodplains, and aquifers. The risk of flooding (Fig. 4.6) can be lessened by floodplains that reduce and control flood by giving water the space needed to dissipate flows. The risk of coastal erosion can be reduced by mangroves, barrier reefs and islands that protect against storm damage, and tidal or storm surges, as witnessed in the Asian tsunami of 2004, where damage from coastal inundation was reduced where mangroves were intact (UNEPWCMC 2006).

Drainage and infilling of wetlands through the process of illegal urbanization means natural water storage is lost and recharge of groundwater reduced, reducing dry season flows and the options available for coping with drought. This process is increasing the city's vulnerability to climate change (Figs. 4.5 and 4.6) by destroying the eco-infrastructures and ecosystem services, needed to reduce exposure to these risks and hazards. The implication is that where ecosystems are lost or degraded, so are the services from them that people use. Without this natural eco-infrastructure, people lose benefits and are exposed to hazards and vulnerabilities they would otherwise be able to avoid or have protection against.

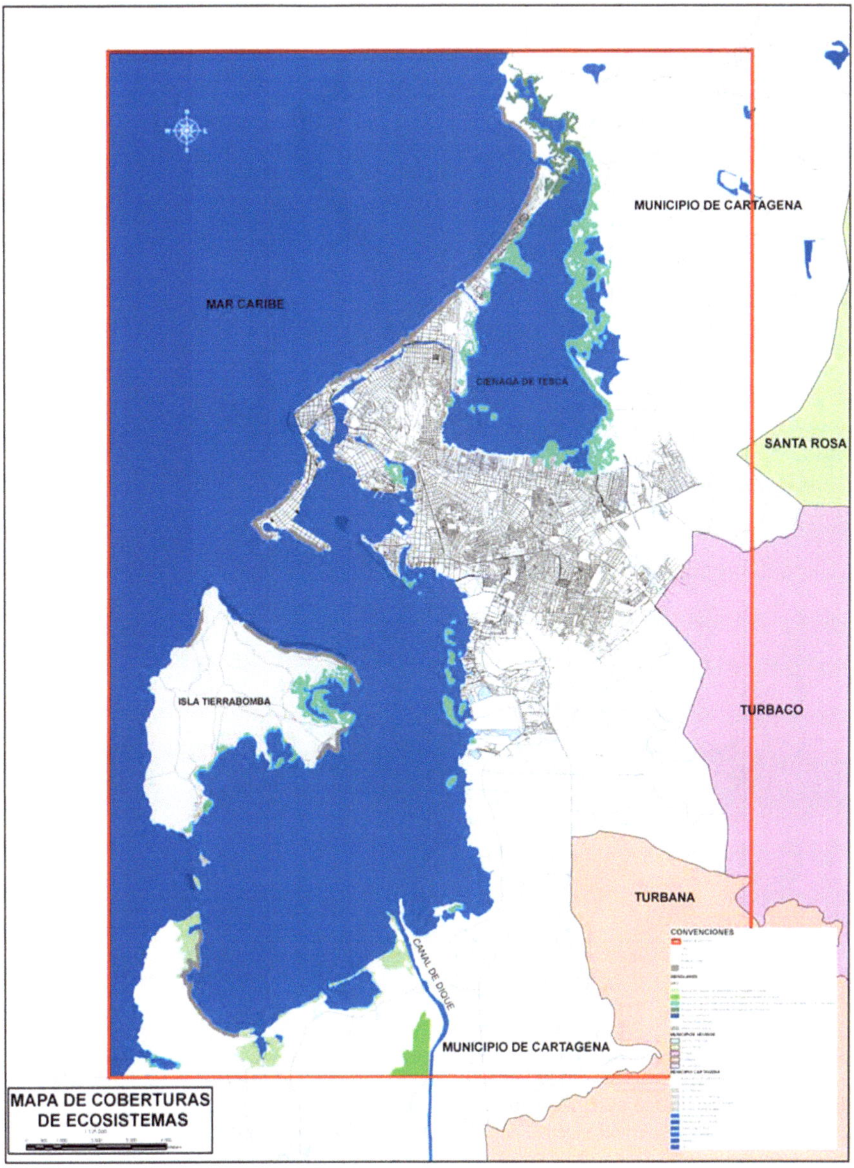

Fig. 4.5 Eco-infrastructures at the city scale: map showing ecosystem land cover in study area. *Source*: INVEMAR-Instituto de Investigaciones Marinas y Costeras "José Benito Vives De Andreis". 2005

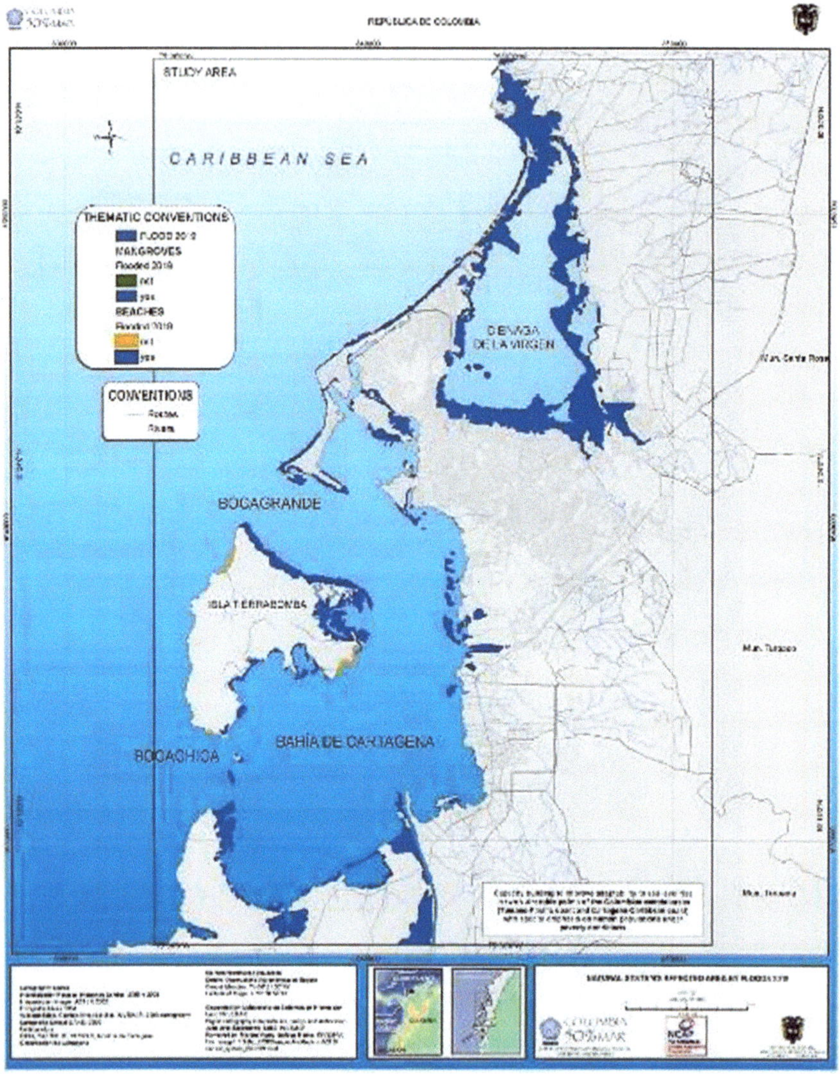

Fig. 4.6 City-knowledge base. Mapping the impacts of climate change (sea level rise scenario). *Source*: Adaptación costera al ascenso del nivel del mar. *Martha P. Vides Ed.* 01/04/2008; Invemar

4.6.3 Conclusion Regarding This Preliminary Assessment Exercise

The preliminary assessment exercise shows that all the ecosystems services and eco-infrastructures in the area needed for adaptation have been seriously threatened and damaged by the mismanagement of the process of urbanization and resource exploration. The natural infrastructure of the lagoons has been damaged, and as a result communities living with these hazards are less able to cope. The adaptive capacity of the ecosystems, eco-infrastructures and communities of Cartagena has become fragile, just when resilience is most needed. Sea level rise will only bring a higher risk of coastal inundation and erosion and lower the resilience[3] of its communities. This coastal city is becoming a hot spot of vulnerability where higher frequency of flooding and coastal inundation will have the most acute impacts.

Put simply, the effects of climate change mean that Cartagena and the *Cienega de la Virgen* will be at greater risk from flooding in future years. Furthermore, many flood risk areas are undergoing development and regeneration, meaning that more people, buildings and infrastructure are likely to be exposed to the risk of flooding in the future. This is the case of *La Cienega de Juan Polo* north of *La Cienega de La Virgen* (Fig. 4.7). Eco-infrastructures are thus needed to reduce the vulnerability of the city to climate change. They need to be integral to portfolios of adaptation measures and strategies. If eco-infrastructures are overlooked, opportunities to reduce vulnerability and increase resilience will be missed. The combination of all these eco-infrastructures and their ecosystem services could reduce exposure to climatic hazards. The focus on reducing vulnerabilities brought by climate change requires that there is new explicit recognition given to the role of eco-infrastructures. This is what our next tool proposes to do.

Fig. 4.7 Cienega de Juan Polo. *Source:* Compiled by author http://maps.google.com/maps?t=h&hl=en&ie=UTF8&ll=10.411323,-75.495731&spn=0.027098,0.033002&z=15

4.7 Securing Urban Space Through Eco-infrastructures: Self-Enclosed Spaces and Coastal Networks of Zero Carbon Settlements

Latin American countries and their cities have a "comparative advantage" in pursuing a low-carbon growth path and mitigating climate change. These cities could play an important role in developing responses to the challenges of climate change and resource constraints. To realize this potential requires developing new styles of eco-infrastructures around the protection of the city, redesign of self-enclosed urban spaces, and the creation of networks of zero carbon settlements. In this way they will be integrating adaptation, mitigation measures and non-regret options and thus reduce the city's emissions at low cost, while at the same time reaping sizable development co-benefits (Tol and Yohe 2006, Lal 2004, Landell-Mills 2002, Vardy 2008, Baker 2008). This is a *prudent approach* based on *precaution and re-design* as a form of collective action (Schneider et al. 2007, Yamin et al. 2006).

4.7.1 First Set of Strategies for Adaptation Planning: A Tapestry of Eco-infrastructures

Based on this approach, a style of spatial planning of cities that considers climate change impacts as vital components of urban development, that requires cities to act cross-sectorally in a holistic rather than sectoral engagement in climate change is proposed. This planning requires in turn, the concept of a multidimensional system of infrastructures that weave together at least six strands of infrastructures: the first layer of blue eco-infrastructures (the flood control function; the sustainable urban drainage system); the second layer of urban forest (mangrove) eco-infrastructures; the third layer of green eco-infrastructures (linked greenways and habitats); the fourth layer of grey infrastructures (the engineering infrastructure and sustainable engineering systems); the fifth layer of human-habitats (the built systems, hard-scapes and regulatory systems); and, the sixth layer of renewable energy infrastructures (solar, wind, biomass, etc) (Fig. 4.8). This web of infrastructures and habitats is the first step of a progressive infrastructure redesign where adaptation planning recognizes the ecosystem services and the climate change adaptation function of these eco-infrastructures.

4.7.1.1 Blue Eco-infrastructure: The Creation of Room for the Water

Coping with floods, drought, storms and sea-level rise will depend on water storage, flood control and coastal defence. In response to climate change, many countries and cities are likely to invest in even more grey-infrastructures for coastal defenses and flood control to reduce the vulnerability of human settlements to climate change. However, providing these functions simply by building grey infrastructures – such as dams, reservoirs, dikes and sea walls – may not be adequate (Palmer et al. 2008). It is here where the blue eco-infrastructures have a critical role to play (Fig. 4.9).

Natural ecosystems can reduce vulnerability to natural hazards and extreme climatic events and complement, or substitute for, more expensive infrastructure investments to protect coastal and riverine settlements. Exposure and the risk to flood is reduced by restoring the function of the floodplains in combination with sound land-use planning, parks and other open public spaces that could function as safety reservoirs in case of floods, and also barrier islands and wetlands. By given space back to the water, the goal is to restore the lagoon's eco-infrastructures

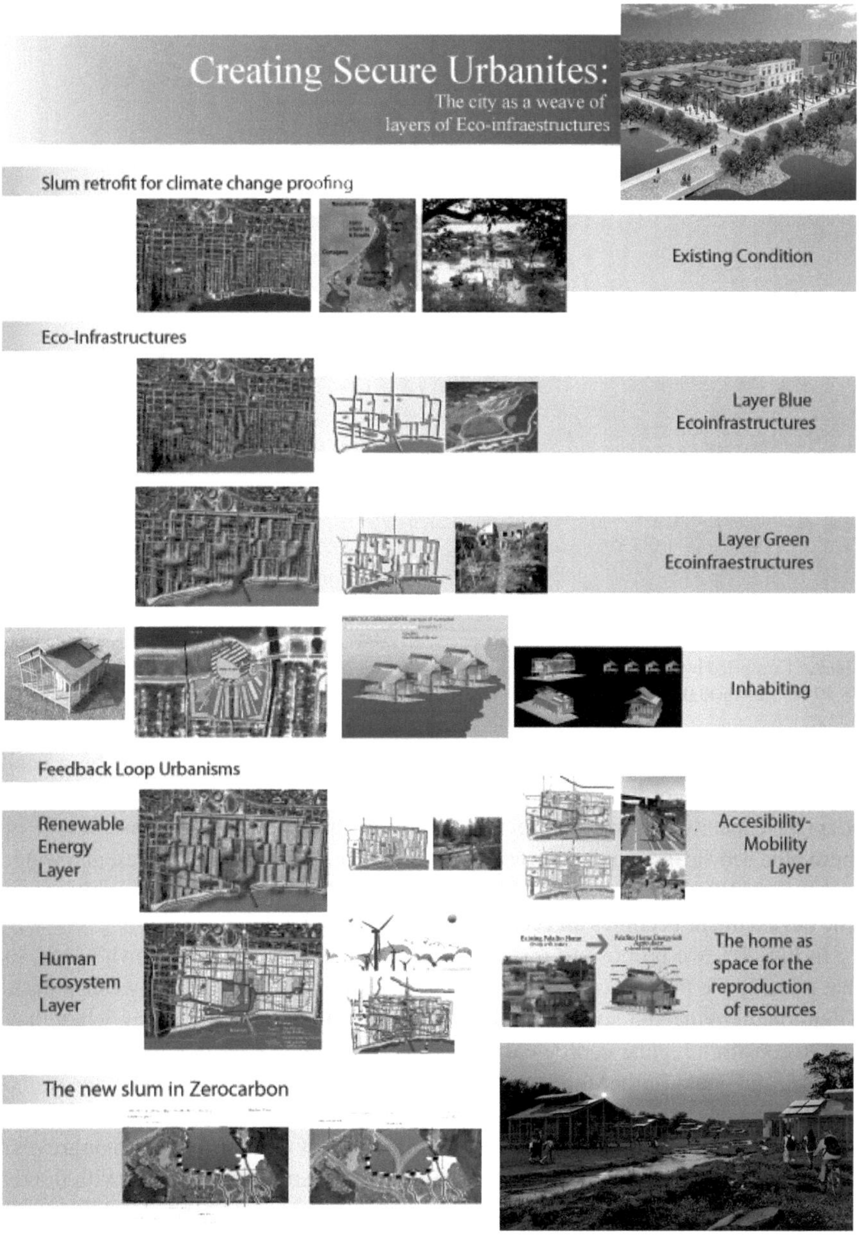

Fig. 4.8 Secure urbanities: strategic protection through eco-infrastructures. *Source*: Compiled by author

Fig. 4.9 Eco-infrastructure: reuse public space for rainwater storage connected through canals. *Source*: Compiled by author http://maps.google.com/maps?t=h&hl=en&ie=UTF8&ll=10.411323,-75.495731&spn=0.027098,0.033002&z=15

as a source of adaptive capacity and renewed resilience. The layer of blue eco-infrastructures incorporates flood risk into urban (re)development and increases adaptive capacity towards future flood impacts. Investing in the conservation of these blue eco-infrastructures provides storm protection, coastal defenses, and water recharge and storage that act as safety and control barriers against natural hazards. The environment becomes an eco-infrastructure for adaptation.

In Fig. 4.8, we propose to restructure and reconstruct a "shanty town" area, so that more space may be created for storage of excess rainfall through water plazas. Traditional engineered solutions often work against nature, particularly when they aim to constrain regular ecological cycles, such as annual river flooding and coastal erosion, and could further threaten ecosystem services if creation of dams, sea walls, and flood canals leads to habitat loss. The idea is to design a flood control project that utilizes the natural storage and recharge properties of critical forests (mangroves) and wetlands (the lagoons) by integrating them into a strategy of "living with floods in water plazas" that incorporate forest protected areas and riparian corridors and protect both communities and natural capital (Fig. 4.10).

4.7.1.2 Green Eco-infrastructure: Restore the Mangrove Urban Forest

The risk of coastal erosion can be reduced by protecting mangroves (Danielsen et al. 2005, Kathiresan and Rajendran 2005). The strategy is to use the potential for mitigation of the urban mangrove forest to reduce emissions at a low cost through

Fig. 4.10 Living around a water plaza. *Source*: Compiled by author

afforestation and reforestation (A/R, REDD Web Platform 2010). The restoration of the mangrove swamp ecosystems can be successful, provided that the hydrological requirements are taken into account, which means that the best results are often gained at locations where mangroves previously existed; which is the case in Cartagena and her *Cienegas*.

The restoration of mangroves can also offer increased protection of coastal areas to sea level rise and extreme weather events such as storms while safeguarding important nursery grounds for local fisheries. These reforestation activities could generate carbon credits for the voluntary market that will be used to finance sustainable livelihood activities in the area, such as fruit tree gardens (see below, green eco-infrastructures), aiming at increasing urban farmers' income, while at the same time reducing pressures on native forests. The opportunity to earn future carbon finance payments can increase the value of the informal and squatter settlement and its marginal lands (Lal 2004, Landell-Mills 2002, Harris et al. 2008, Betancourth 2009a). This will amount to the transformation of the shanty town into a *new "extractive protected area"* (Allegreti 1994), that will reduce emissions from deforestation and degradation of native forests in the city and the region. This mangrove urban forest eco-infrastructure will be a regional park of interconnected networks of natural areas and other open spaces that conserves natural ecosystem values and functions, and sustains clean air and water (Fig. 4.11).

Fig. 4.11 Network of zero carbon settlements within a regional park (the mangrove green-belt). *Source*: Compiled by author http://maps.google.com/maps?t=h&hl=en&ie=UTF8&ll=10.411323,-75.495731&spn=0.027098,0.033002&z=15

The park will enable the urban area of the informal settlements to flourish as a natural habitat for a wide range of wildlife, and deliver a wide array of benefits to people and the natural world alike, such as providing a linked habitat across the urban landscape that permits bird and animal species to move freely. In addition, this urban forest eco-infrastructure can also provide the following services: cleaner air; a reduction in heat-island effect in the urban area; a moderation in the impact of climate change; increased energy efficiency; and the protection of sources of water. In Cartagena we are proposing to re-create and reconstruct the mangrove forest that once covered the *Cienega de la Virgen* plain under a new park concept (Fig. 4.8). The idea is to give the city of Cartagena a big protective mangrove peri-urban forest that can function as a bio-shield against sea level rise, and climate change. The mangrove greenbelt can also provide significant coastal protection from erosion. The mangrove forest will be connected to a network of urban open space lands to preserve a high quality of life, carbon sink creation, and city beautification. The forest will clear the air and treat the water that runs into the lagoon (*Cienega de la Virgen*), re-naturalize the territory and increase its biodiversity, create a living laboratory of environmental monitoring, provide an area for recreation, revitalize the historic/natural memory and strengthen the city identity.

Introducing eco-tourism has the additional benefit of making the forest accessible to citizens, promoting goodwill among the people, and demonstrating the importance of maintaining and improving the forest. It will thus be the community who will begin planting the trees. As part of this urban forestry proposal, all major roads in the area will be provided with green medians and above all, green corridors (Betancourth 2007). The distributed greenery ensures that the roads have high CO_2

absorption capacity in close range of the emission source. The roadside greenery aids in reducing the heat island effect and atmospheric pollution. The urban forest can help mitigate and adapt for temperature changes due to climate change.

4.7.1.3 Green Eco-infrastructure: Urban Agriculture

By re-creating, improving and rehabilitating the ecological connectivity of the immediate environment, the green-infrastructure turns human intervention in the landscape from a negative into a positive. It reverses the fragmentation of natural habitats and encourages increases in biodiversity to restore functioning ecosystems while providing the fabric for sustainable living, and safeguarding and enhancing natural features. Urban forestry and urban agriculture strategies for climate change mitigation are integrated into this green eco-infrastructure This new connectivity of the landscape with the built form (see orange and grey infrastructures) can be both horizontal and vertical (Figs. 4.12 and 4.13).

4.7.1.4 Orange Infrastructure

This layer represents the human community, its built environment (buildings, houses, hardscapes and regulatory systems such as laws, regulations, ethics, etc). Homes are clustered around blue and green eco-infrastructures. The design proposal

Fig. 4.12 Green eco-infrastructure-ecological corridors. *Source*: Compiled by author http://maps.google.com/maps?t=h&hl=en&ie=UTF8&ll=10.411323,-75.495731&spn=0.027098,0.033002&z=15

Fig. 4.13 Orange eco-infrastructure: diversity of homes around a diversity of eco-infrastructures. *Source*: Compiled by author http://maps.google.com/maps?t=h&hl=en&ie=UTF8&ll=10.411323,-75.495731&spn=0.027098,0.033002&z=15

for the individual urban-home/farm extends the ecological corridor (around which homes are clustered (Fig. 4.14) vertically from the ground up to the green gardens on the living roof tops. Thus the blue and green infrastructure network can be used to define the hierarchy and form of the habitats and natural green spaces within a community (see living around a water plaza, Fig. 4.15).

4.7.1.5 Grey Infrastructure

The grey infrastructure is the usual urban engineering infrastructure such as roads, drains, sewerage, water reticulation, telecommunications, energy and electric power distribution systems. This is also the infrastructure of mobility and accessibility. These mobility systems should integrate with the green and blue infrastructures rather than vice versa, and should be designed as sustainable accessibility systems (Fig. 4.16).

4.7.1.6 Renewable Energy Infrastructures

Finally, it is the layer for renewable energies (Fig. 4.17). These last three layers of infrastructures bring us into the second main set of responses and strategies, namely the redesign of feed-back loop urbanisms (a cycle of behavior in which two or more

Fig. 4.14 Water plaza. *Source*: Compiled by author http://maps.google.com/maps?t=h&hl=en&ie=UTF8&ll=10.411323,-75.495731&spn=0.027098,0.033002&z=15

Fig. 4.15 Living around a water plaza. *Source*: Compiled by author

Fig. 4.16 Grey infrastructure: multimodality and accessibility. *Source*: Compiled by author http://maps.google.com/maps?t=h&hl=en&ie=UTF8&ll=10.411323,-75.495731&spn=0.027098,0.033002&z=15

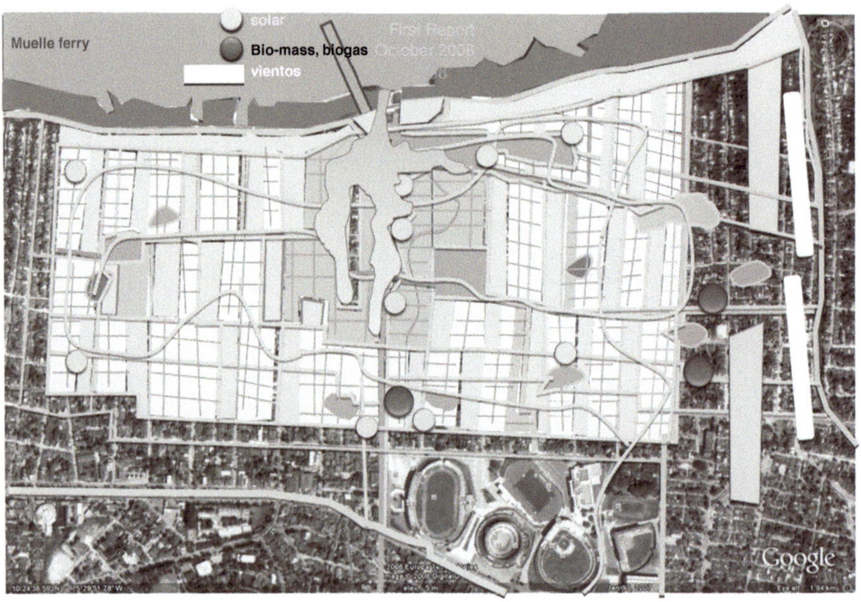

Fig. 4.17 Renewable energy infrastructures. *Source*: Compiled by author http://maps.google.com/maps?t=h&hl=en&ie=UTF8&ll=10.411323,-75.495731&spn=0.027098,0.033002&z=15

infrastructures act to reinforce the other's action) and self-enclosed spaces. But let us first look at the layer of the home.

4.7.1.7 The Home

As it is the case in Cartagena, the urban poor are typically at the highest risk in the event of natural disasters due to the location of low-income settlements. Ensuring that cities continue to drive growth in a sustainable manner is fundamental to development and poverty eradication. An important adaptation strategy for local governments is to provide new shelter options for the poor to avoid the creation of new settlements and slums on marginal land. But, population retreat, a most workable strategy against highly risk areas, generates strong cultural resistance. Despite natural phenomena like earthquakes, subsidence and tsunamis threats, people will not leave their "informal settlements" to start paying for public basic services on a safer house. It is in this regard that the housing tradition of the Pacific coast is relevant. Pacific coast meso-macro tidal regime is subject to a medium to low wave regime associated to wind's influence. Tidal amplitude reaches up to 5 m in some areas, which is 10 times greater than the Caribbean (Invimar 2005, 2007). This natural condition has allowed the development of *palafitic* housing, a dwelling built on a platform over the sea, an autonomous adaptation strategy towards sea level changes. This proven ancient adaptation strategy can be used in areas where rising temperatures due to climate change are becoming a problem. The idea is to transfer this technology from the Pacific coast to the Caribbean and implement this solution for the case of housing around the *Cienega of La Virgen* (Fig. 4.18).

4.7.2 Second Set of Strategies for Adaptation Planning: Nested Closed Urbanism and Decoupling from National Infrastructure and Building Enclosed Self-Sufficient Cities

The last three layers of infrastructures above (mobility, multiple land uses and renewable energies) will be weaved together to conform with nested feedback loop urbanisms. Nested feed-back loop urbanisms are urban developments that can be created to deal with their own infrastructure needs on site, including water supply, storm-water control, sewage treatment, thermal demand for (heating and) cooling and electrical demands. Creating these nested systems will buffer the demand on centralized infrastructure and add system robustness and resilience; all necessary in a world with increased uncertainty in climate effects on infrastructure.

Cities usually seek out resources from locations ever more distant and connected through networks. Developing responses to climate change requires challenging this traditional approach and build greater self-sufficiency by a dual strategy of both decoupling from external reliance on national and regional infrastructures and building up local and decentralized systems for water and energy supply, waste disposal and mobility systems; that is, by building more "self-sufficient" infrastructures of

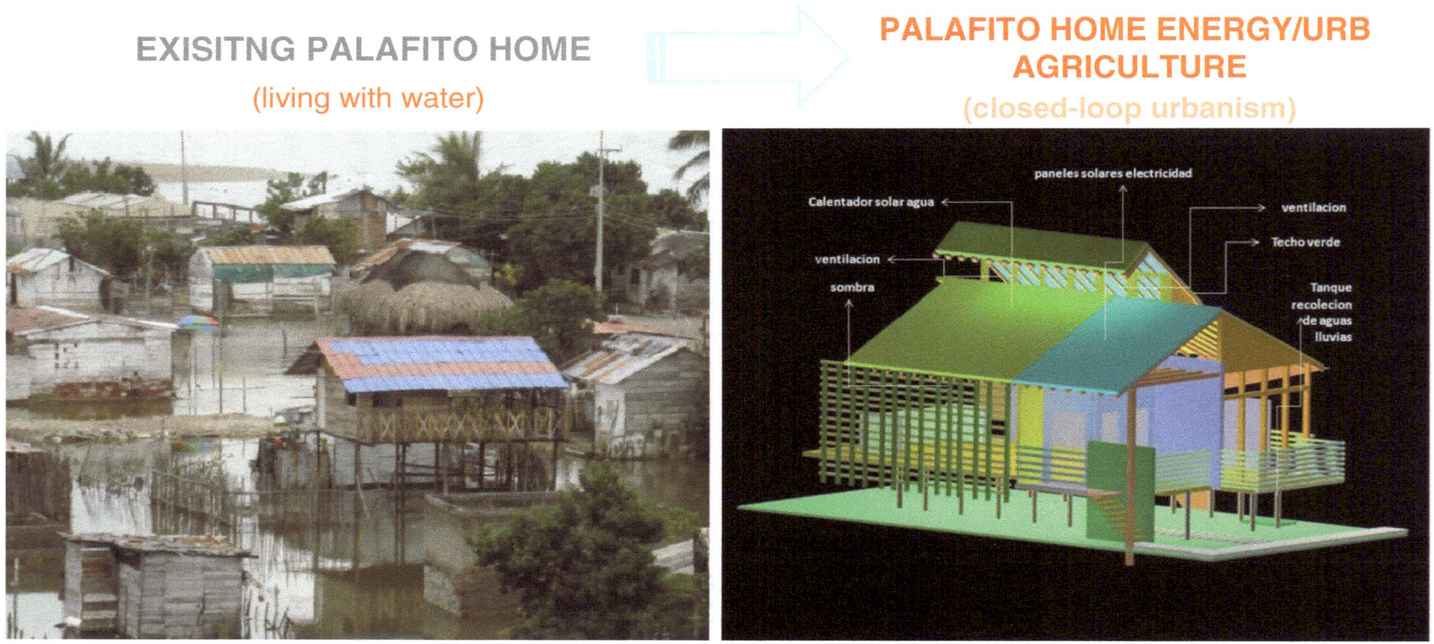

Fig. 4.18 From palafito home to the house as a unit for the production of renewable energy, water conservation and urban agriculture. *Source*: Compiled by author

provision on a city scale. It is important to design a suite of infrastructure strategies for energy, waste and water to minimize the consumption of resources and production of wastes; to consider reuse, develop decentralized energy production and waste treatment technologies; and reduce reliance on external infrastructure to increase the relative self-sufficiency of the city (Fig. 4.18).

4.7.2.1 The Transport Sector

In dealing with the mobility of citizens, the top policy priority in the Latin American region in general and in Colombia in particular, is to slow down the rapidly rising rate of emissions from light vehicles by providing incentives for more efficient cars and for reduced car use. This can only be attained with the integration of mobility services (mobility planning and integrated transport strategies that span across different transportation modes) multiple land uses and development of renewable energies through urban design. There is already in place a mass transit system for Cartagena (Trans-caribe). In order to avoid the pitfalls of the Trans-milenio system in Bogota, we propose to supplement this system with a maritime transport system in the Cienega de la Virgen. Thus, pollution-free buses, and water taxis, powered by fuel-cells or other zero carbon technologies, will run between neighbourhoods. We propose to have only green transport movements along the Cienega de la Virgen's coastline. People will arrive at the coast by boat, traveling along the shore as pedestrians, cyclists, or passengers on sustainable public transport vehicles. What is now a highway will become a trail system along the shore and within the regional mangrove park proposed above (see Fig. 4.11 urban forest).

The mass transit system for Cartagena (Transcaribe) and the city at large will be linked to the Cienega de la Virgen coastline, by a network of pedestrian walkways. The adjacent communities will inhabit and transform these non-regret investments in transport infrastructures through a series of supplementary projects that include: an urban village with multiple uses along the walkway (with a water-canal illuminated by light-emitting diodes (LED)) that connects the terrestrial and maritime system of mass transit (zero carbon vehicles will be allowed only within the walkway such as the already in the area existing *bici-taxis*); a regional commercial node at the intersection of the walkway and the system of mass transit, a Trans-Caribe Station that generates renewable energy (Betancourth 2003); and a market and festival square where the families living in homes that produce urban agriculture and renewable energies along the walkway will trade their products (see below, the home as production system of agriculture and renewable energy products) and thus add and capture value to and from the flow of commuters moving between the terrestrial and maritime system of mobility. Thus the transport system goes beyond being merely a line on a map to rapidly connect two points in the possible shortest way; and, becomes a habitat (Betancourth 2003, 2007) (Fig. 4.19).

High quality densification along the mass transit system would reduce the impact on the environment, while contributing to making Cartagena greener, more sustainable, more livable, and more affordable (Betancourth 2003). The zero carbon

Fig. 4.19 Integration of transport and land use planning along the mass transit corridor. *Source*: Compiled by author http://maps.google.com/maps?t=h&hl=en&ie=UTF8&ll=10.411323,-75.495731&spn=0.027098,0.033002&z=15

settlement proposal for la Cienega de la Virgen (Figs. 4.8 and 4.11) explores increasing density in a variety of contexts: in lower density areas, along transit routes and nodes, and in neighbourhood centres. Of key importance is to support density that is high quality, attractive, energy efficient, and respectful of neighbourhood character, while lowering the city's GHG emissions. The energy efficiency of this transport system will be improved by retrofitting traffic signals and street lights (replacing incandescent fixtures with light-emitting diodes (LED)) as well as by the conversion of outdated lighting to modern, efficient technology in public sector facilities, parking structures, police substations, fire stations, and community centres, resulting in energy savings and in financial savings.

4.7.2.2 Renewable Energy

Wind conditions, high solar radiation levels, geothermal resources, bio-mass, are excellent in many Latin American countries and cities, as well as in Cartagena and La Cienega de la Virgen. Compared to costly grid extensions, off-grid renewable electricity typically is the most cost-effective way of providing power to isolated

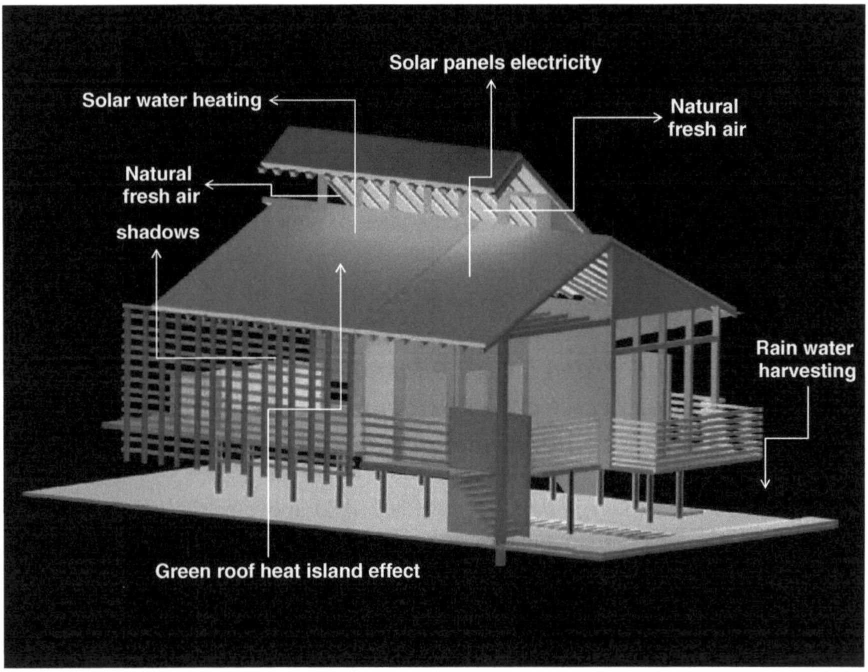

Fig. 4.20 The home as a system for the generation of renewable energy, water collection and urban agriculture. *Source*: Compiled by author

urban, periurban and rural populations (ESMAP 2007). We are proposing the concept of a Zero Carbon settlement for La Cienega de La Virgen (Fig. 4.17). Zero carbon means no net carbon emissions from all energy uses in the home. Key features of a zero carbon development could include technologies such as passive solar energy, thermal solar panels and the conversion of solar energy to electricity in photovoltaic cells. The home is conceived as a system for the generation of renewable energy closely connected to urban agriculture, combining water collection (rainwater harvesting), roof-top living gardens to reduce the impact of urban heat island effect, and recycle building materials (Fig. 4.20).

4.7.2.3 Urban Gardening at Home

Urban agriculture can be defined shortly as the growing of plants and the raising of animals within and around cities (Fig. 4.12). There is an incipient form of urban agriculture already at work in the neighbourhoods around the *Cienega de la Virgen*. It is important to strengthen this initiative by better integrating it with the urban economic and ecological system. This is the most striking feature of urban agriculture, which distinguishes it from rural agriculture. Such linkages include the participation of urban residents as farmers, use of typical urban resources (like organic waste as compost and rain water for irrigation), and direct links with urban consumers. Urban agriculture is an integral part of the urban system: the residential unit allows for collection of rainwater for irrigation (Fig. 4.18); the market in the proposed pedestrian walkway that connects the Cienega de la Virgen with the system of mass transit allows for a direct link with urban consumers (Fig. 4.19). Urban agriculture could help address the problems of food scarcity, unemployment, as well as urban waste and waste water disposal.

4.7.3 Third Set of Strategies for Adaptation Planning: Creating Networks of Zero Carbon Settlements Along the Coast Connected Through Regional Eco-infrastructures

The proposal for La *Cienega de La Virgen* described above is aiming to show that urbanization can be a fundamentally sustainable process, and, that we must rethink the means by which we urbanize. We envision de *Cienega de La Virgen* not as a dormitory town, a single-use housing development, but as an ecologically sustainable, and commercially sustainable zero carbon settlement; a settlement that will run on renewable energy, recycle and re-use waste water, protect the wetlands and mangrove forest by returning land to a wetland state creating a "buffer zone" between the city and the mud flats of La *Cienega de la Virgen*, and protect air quality by creating a system of multimodal mobility integrated with a dynamic layer of multiple land uses; small villages that meet to form a city sub-centre, where all housing is situated within seven minutes' walking distance of terrestrial and maritime public transport. This not only lowers the consumption of energy, but also enables transport to be run on renewable energy to achieve zero carbon emissions. Having compact, efficient, and walkable settlements spread along a landscape of eco-infrastructures, that recognize human relationships with nature and secure their long-term sustainability, is an important mitigation and adaptation measure. This is a settlement as an urban landscape of multifunctional eco-infrastructures where living roofs, large trees and soft landscapes areas absorb rainfall; where a network of street swales and unculverted meadows safely manage large volumes of water.

This is an urban landscape of eco-infrastructures that provides flood protection, that does not waste water but stores and recycles it for irrigation, that saves energy. It is also a landscape where living roofs insulate buildings, and trees shade homes and offices which reduces the need for air conditioning, cleans and cools the air, provides green spaces to encourage exercise and socializing, where you can work or cycle to school or to work through car-free greenways, where meadows run alongside offices and shops, and where you can see food being grown in the park. The idea is to develop this prototype through a demonstration project and then to replicate it along the coast to form a network of such settlements connected through a system of sustainable mobility and of regional eco-infrastructures (see Figs. 4.8 and 4.11,

4.19). This is a vision to help address the climate change challenges that we hope will become a prototype in the implementation phase.

4.7.4 Fourth Set of Strategies for Adaptation Planning: The Impacts of Restoring and Repairing the Eco-infrastructures on Sensitivity and Adaptive Capacity

Expanding livelihood assets and enabling economic development sensitive to climate hazards will assist sustainable management of the blue and green eco-infrastructures proposed above.

Eco-infrastructure governance Adaptive capacity will be built through flexible and coordinated institutions in learning and the dissemination of knowledge needed to empower people in planning and decision-making related to adaptation. Restoring the lagoon's natural eco-infrastructure could become a source of adaptive capacity and renewed resilience.

Community action: participatory and community action for redesigning and restoring the eco-infrastructures can increase resilience to current disasters, for example, by building houses on stilts (*palafito* homes), replanting coastal lowlands (urban mangrove forest), digging and maintaining drainage ditches within the settlement (blue eco-infrastructure). However, city-level commitment is needed for city-wide eco-infrastructures to effectively complete the adaptation for climate change.

4.7.5 The Model to Finance Investments to Repair the Eco-infrastructures

Carbon offsets and carbon credits may be an opportunity for carbon markets to make cities less dependent on national government for financial support. The access of funds through carbon markets could be recognized as an important adaptation initiative (Betancourth 2009a). Bio-rights are also innovative financing mechanisms for reconciling poverty alleviation and environmental conservation. They may offer a novel approach to linking conservation with development. By providing micro-credits for sustainable development, the approach enables local communities to refrain from unsustainable practices and be actively involved in environmental conservation and restoration. Micro-credits are converted into definitive payments upon successful delivery of conservation services at the end of a contracting period. Bio-rights offers an approach in which global stakeholders pay local communities to provide ecosystem services such as carbon sequestration, fresh water supply and biodiversity (Guardian.co.uk 2010b). The approach unites the conservation and development aspirations of NGOs, governments, the private sector and local communities alike. It accomplishes community involvement in the preservation of environmental assets that are of global importance (for example, the mangrove forest).

4.8 Conclusion

This paper has focused on the role of the environment in providing solutions to climate change. There are links to resilience, which accord the environment a critical role in climate change adaptation. We need to recognize the benefits of ecosystem services in strategies for climate change adaptation and improve resilience to climate change impacts on cities through investments in nature's eco-infrastructures. The restoration of the eco-infrastructures of the Cienega de la Virgen lagoon will rebuild ecosystem services that help to reduce exposure to climatic hazards, but especially, it will help to ensure people have more of the assets needed to make urban fishing and farming livelihoods less sensitive to climate change. It will support livelihoods and economic development that reduce sensitivity to hazards, especially for the most vulnerable. Just as important, the learning, flexible institutions and investment that underpin effective management and restoration of the coastland's natural eco-infrastructures provide vital adaptive capacity that is based on resilience.

The case we have presented here demonstrates how adaptation that is based on resilience could reduce exposure to hazards, to impacts and increase in adaptive capacity. In the hot spots of vulnerability along the Colombian Caribbean coast, citizens will cope better with climate change impacts where eco-infrastructures are intact or restored than where they are degraded. Where climate change has led to weakening capacity to cope with shocks and stresses, the key is to increase resilience. With resilience as a goal, the eco-infrastructures, the feedback loop spaces, and the network of zero carbon settlements, must form the heart of effective strategies for climate change adaptation.

The tools drafted above are intended to initiate a learning process for local governments. They look at the issues of climate change, and its potential consequences that can affect ecosystems and cities. The tools recommend a thorough city self-assessment and a comprehensive information base as starting points; they offer strategic responses (eco-infrastructures; enclosed spaces and network of zero carbon settlements) that a city can use as follow-up to building its programs for resilience. The tools aim to generating public awareness and engaging stakeholders as well as to motivate city officials to take actions.

Notes

1. Vulnerability to climate change is high if changes in climate increase the exposure of populations to events such as drought, floods or coastal inundation, because of higher frequency or severity where the ability of people to cope is limited. Capacity to cope is most limited, and thus sensitivity is highest where livelihoods and the economy are based on a narrow range of assets that are easily damaged by climate hazards, with few alternate options or means of managing risk. Vulnerability is therefore especially high for the poor in those "hot spots" where climate change exacerbates exposure to climatic hazards.
2. The analytical framework proposed by the Eco2Cities program tends to assume these synergies as given. Therefore this program needs to be supplemented with strategies to mediate these new urban conflicts (Betancourth 2008a, b). New tools such as the mapping of the social tensions,

their impacts on eco-infrastructures, the construction of consensus starting from those impacts, need to be added to this framework (Launch: Ecocities2. World Bank 2009).

3. Resilience is the amount of disturbance that can be withstood before a system changes its structure and behaviour – before, for example, it breaks down (Folke et al. 2004).

References

Alcaldía de Cartagena (2000). *Plan de Ordenamiento Territorial – Componente General. Diagnóstico General. Despacho del Cartagena de Indias*. Alcaldia Municipal de Cartagena.

Allegretti, M. H. (1994). Reservas extrativistas: parâmetros para o desenvolvimento sustentável na Amazônia. In R. Arnt (Ed.), *O Destino da floresta: reservas extrativistas e desenvolvimento sustentável na Amazônia* (pp. 17–47). Rio de Janeiro: Relume-Dumará.

Arnell, N. W. (2004). Climate change and global water resources: SRES scenarios emissions and socio-economic scenarios. *Global Environmental Change*, 14: 31–52.

Avissar, R. & Werth, D. (2005). How many realizations are needed to detect a significant change in simulations of the global climate? *American Geophysical Union*, Fall Meeting 2005.

Baker, J. L. (2008). *Urban poverty: a global view. Urban Paper Series (UP-5)*. Washington, DC: World Bank.

Betancourth, C. H. (2003). *A competition entry for the design of a BRT system for Cartagena, October–November (Cartagena de Indias-Colombia)*. Bogota, Colombia: OPA International.

Betancourth, C. H. (2007). *Creating civic highways. Celebrating civitas, community and public value in mobility. The case of NM516 Highway*. City of Aztec: Azec, NM.

Betancourth, C. H. (2008a). *Conceptual framework to, balance through re-design the demands of infrastructure mega-projects and the demands of conservation*. Bogota, Colombia: Peace University.

Betancourth, C. H. (2008b). Co-generating multiple stakeholder reception of renewable energy projects. Paper presented at the 2008 behavior, energy and climate change conference, November 16–19 at Sacramento, California. Washington, DC: Marstel-day.

Betancourth, C. H. (2009a). *Creando asentamientos zero-carbon: una triple solución al problema de los desplazados en Colombia: Policy Brief, Accion Social*. Bogota, Colombia: Peace University.

Betancourth, C. H. (2009b). *Developing an urban design and transportation master plan and a design code for the town centre of the City of Bloomfield*. Bloomfield, NM: New Mexico State Department of Finance.

Betancourth, C. H. (2009c). *Developing an urban design and transportation master plan and a design code for the City of Aztec*. Aztec, NM: State Department of Finance Aztec, NM.

Blanco, J. T. & Hernández, D. (2009). The Costs of climate change in tropical vector-borne diseases—a case study of malaria and dengue in Colombia. In W. Vergara (Ed.), *Assessing the consequences of climate destabilization in Latin America* (pp. 69–87). Sustainable Development Working Paper 32. Washington, DC: World Bank.

Bradley, R., Vuille, M., Diaz, H. & Vergara, W. (2006). Threats to water supplies in the tropical Andes. *Science*, 312: 1755.

Buddemeier, R. W., Jokiel, P. L., Zimmerman, K. M., Lane, D. R., Carey, J. M. & Bohling, G. C. (2008). A modeling tool to evaluate regional coral reef responses to changes in climate and ocean chemistry. *Limnology and Oceanography Methods*, 6: 395–411.

C40 Cities Organization (2010) http://www.c40cities.org/climatechange.jsp. Accessed 10 August 2010.

Charry, H. E., Alvarado, M. & Sánchez, J. A. (2004). Annual skeletal extension of two reef-building corals from the Colombian Caribbean Sea. *Boletín de Investigaciones Marinas y Costeras*, 33: 209–222.

Cioh, C. (1998). *Caracterización y diagnostico integral de la zona costera comprendida entre Galerazamba y Bahía Barbacoas. Rep. Tomo II, CIOH – CARDIQUE*. Colombia: Cartagena.

Confalonieri, U., Menne, B., Akhtar, R., Ebi, K. L., Hauengue, M., Kovats, R. S., Revich, B. & Woodward, A. (2007). Human health. Climate change: impacts, adaptation and vulnerability. In M. L. Parry, O. F. Canziani, J. P. Palutikof, P. J. van der Linden & C. E. Hanson (Eds.), *Contribution of working group II to the fourth assessment report of the intergovernmental panel on climate change* (pp. 391–431). Cambridge: Cambridge University Press.

Coundrain, A., Francou, B. & Kundzewicz, Z. W. (2005). Glacier shrinkage in the Andes and consequences for water resources- Editorial. *Hydrological Sciences Journal*, 50(6): 925–932.

Cox, P. M., Betts, R. A., Collins, M., Harris, P. P., Huntingford, C. & Jones, C. D. (2004). Amazonian forest dieback under climate carbon cycle projections for the 21st century. *Theoretical and Applied Climatology*, 78: 137–156.

Cox, P. M., Harris, P., Huntingford, C., Betts, R. A., Collins, M., Jones, C. D., Jupp, T. E., Marengo, J. & Nobre, C. (2008). Increasing risk of Amazonian drought due to decreasing aerosol pollution. *Nature*, 453: 212–216.

Curry, J., Jelinek, M., Foskey, B., Suzuki, A. & Webster, P. (2009). Economic impacts of hurricanes in México, Central America, and the Caribbean ca. 2020–2025. In W. Vergara (Ed.), *Assessing the consequences of climate destabilization in Latin America*. Sustainable Development Working Paper, LCSSD. Washington, DC: World Bank.

Danielsen, F., Sorensen, M. K., Olwig, M. F., Selvam, V., Parish, F., Burgess, N. D., Hiraishi, T., Karunagaran, V. M., Rasmussen, M. S., Hansen, L. B., Quarto, A. & Suryadiputra, N. (2005, 28 October). The Asian tsunami: a protective role for coastal vegetation. *Science*, 310(5748): 643.

Dasgupta, S., Laplante, B., Meisner, C., Wheeler, D. & Yan, J. (2007). *The impact of sea level rise on developing countries; a comparative analysis*. World Bank Policy Research Working Paper 4136. Washington, DC: Development Research Group, World Bank.

De La Torre, A., Nash, J. & Fajnzylber, P. (2009). *Low carbon high growth. Latin American responses to climate change*. Washington, DC: World Bank.

Díaz, J. M., Barrios, L. M., Cendales, M. H., Garzón-Ferreira, J., Geister, J., López-Victoria, M., Ospina, G. H., Parra-Velandia, F., Pinzón, J., Vargas-Ángel, B., Zapatay, F. & Zea, S. (2000). Áreas coralinas de Colombia. INVEMAR, Serie de Publicaciones Especiales Nº 5. Santa Marta, 176 p.

Díaz, J. M., Barrios, L. M. & Gomez, D. I. (Eds.). (2003). Las praderas de pastos marinos en Colombia: Estructura y distribución de un ecosistema estratégico. INVEMAR, Serie Publicaciones Especiales No. 10, Santa Marta, Colombia. 160 p.

Energy Sector Management Assistance Program (ESMAP) Study (2007). *Latin America and the Caribbean, energy sector retrospective review and challenges*. Washington, DC: World Bank.

Environment News Service (2008). Melting Andean glaciers could leave 30 million high and dry. Washington, DC, April 28. http://www.ens-newswire.com/ens/apr2008/2008-04-28-01.asp. Accessed 02 August 2010.

Folke, C., Carpenter, S., Walker, B., Scheffer, M., Elmqvist, T., Gunderson, L. & Holling, C. S. (2004). Regime shifts, resilience, and biodiversity in ecosystem management. *Annual Reviews of Ecology, Evolution, and Systematics*, 35: 557–581.

Francou, B., Vuille, M., Wagnon, P., Mendoza, J. & Sicart, J. E. (2003). Tropical climate change recorded by a glacier in the central Andes during the last decades of the 20th century: Chacaltaya, Bolivia, 16 °S. *Journal of Geophysical Research*, 108(D5): 4059. DOI: 10.129/2002JD002473.

Giddens, A. (2009). *The politics of climate change*. London: Polity.

Girardet, H. (2008). *Cities, people, planet: urban development and climate change* (2nd ed). Chichester: Wiley.

Gleditsch, N. P., Ragnhild N. & Salehyan, I. (2007). Climate change and conflict: the migration link. Coping with Crisis Working Paper Series. New York, NY: International Peace Academy. www.ipacademy.org/our-work/coping-with-crisis/working-papers

Guardian.co.Uk (2010a). World feeling the heat as 17 countries experience record temperatures. www.guardian.co.uk. Accessed 12 August 2010.

Guardian.co.Uk (2010b). Ecuador signed a $3.6bn deal not to exploit oil-rich Amazon reserve. http://www.guardian.co.uk/environment/2010/aug/04/ecuador-oil-drilling-deal-un. Accessed 11 August 2010.

Harris, N., Grimland, S., Pearson, T. & Brown, S. (2008, October). Climate mitigation opportunities from reducing deforestation across Latin America and the Caribbean. Report to World Bank, Winrock International.

Hodson, M. & Marvin, S. (2009). Urban ecological security: a new urban paradigm? *International Journal of Urban and Regional Research*, 33: 193–215.

Hoyos, C. D., Agudelo, P. A., Webster, P. J. & Curry, J. A. (2006). Deconvolution of the factors contributing to the increase in global hurricane intensity. *Science*, 312: 94–97.

Instituto de Investigaciones Marinas y Costeras José Benito Vives de Andréis (2005). *Capacity building to improve adaptability to sea level rise in two vulnerable points of the Colombian coastal areas (Tumaco-Pacific coast and Cartagena-Caribbean coast) with special emphasis on human populations under poverty conditions. Draft technical report.* Santa Marta, Colombia: Invimar.

Instituto de Investigaciones Marinas y Costeras José Benito Vives de Andréis (2007). *Capacity building to improve adaptability to sea level rise in two vulnerable points of the Colombian coastal areas (Tumaco-Pacific coast and Cartagena-Caribbean coast) with special emphasis on human populations under poverty conditions. Vulnerability Assessment.* Santa Marta, Colombia: Invimar.

Intergovernmental Panel on Climate Change (2007). Climate change 2007: synthesis report. Contribution of Working Groups I, II and III to the Fourth assessment report of the Intergovernmental Panel on Climate Change (Core Writing Team, Pachauri, R.K; Reisinger, A., eds.) Geneva, Switzerland. 104 p. http://www.ipcc.ch/ipccreports/ar4-syr.htm. Accessed 02 August 2010.

Kaser, G., Irmgard, J., Georges, C., Gómez, J. & Tamayo, W. (2003). The impact of glaciers on the runoff and the reconstruction of mass balance history from hydrological data in the tropical Cordillera Blanca, Perú. *Journal of Hydrology*, 282(1–4): 130–144.

Kathiresan, K. & Rajendran, N. (2005). Coastal mangrove forests mitigated tsunami. Estuarine. *Coastal and Shelf Science*, 65: 601–606.

Knight, F. (1921). *Risk, uncertainty and profit*. Boston, MA: Houghton Mifflin.

Lal, R. (2004). Soil carbon sequestration impacts on global climate change and food security. *Science*, 304: 1623–1627.

Landell-Mills, N. (2002). Developing markets for forest environmental services: an opportunity for promoting equity while securing efficiency?. In I. R. Swingland, E. C. Bettelheim, J. Grace, G. T. Prance & L. S. Saunders (Eds.), *Carbon, biodiversity, conservation and income: an analysis of a free-market approach to land-use change and forestry in developing and developed countries* (pp. 1817–1825). London: The Royal Society.

Launch: Ecocities2. World Bank (2009). Fifth urban research symposium, Marseille. June 28–30, 2009.

Magrin, G., Gay García, C., Cruz Choque, D., Giménez, J. C., Moreno, A. R., Nagy, G. J., Nobre, C. & Villamizar, A. (2007). Latin America – Climate change 2007: impacts, adaptation and vulnerability. In M. L. Parry, O. F. Canziani, J. P. Palutikof, P. J. van der Linden & C. E. Hanson (Eds.), *Contribution of working group II to the fourth assessment report of the intergovernmental panel on climate change* (pp. 581–615). Cambridge: Cambridge University Press..

McCausland, E. (2009). *El infierno en el paraiso*. Bogota, Cololmbia: El Tiempo.

Medvedev, D. & van der Mensbrugghe, D. (2008). *Climate change in Latin America: impact and mitigation policy options.* Washington, DC: The World Bank.

Mendelsohn, R. (2008). *Impacts and adaptation to climate change in Latin America.* Washington, DC: World Bank.

Milly, P. C. D., Dunne, K. A. & Vecchia, A. V. (2005). Global pattern of trends in streamflow and water availability in a changing climate. *Nature*, 434: 561–562.

New Economics Foundation (NEF) (2006). *Up in smoke? Latin America and the Caribbean: The threat from climate change to the environment and human development*. London: NEF.
Niño, L. (2001). *Caracterización biofísica, in Cartagena de Indias* (pp. 103–137). Bogotá, Colombia: IDEADE.
Palmer, M. A., Reidy-Liermann, C. A., Nilsson, C., Flörke, M., Alcamo, J., Lake, P. S. & Bond, N. (2008). Climate change and the world's river basins: anticipating management options. *Frontiers in Ecology and the Environment*, 6: 81–89.
Petter, G. N., Nordås, R. & Salehyan, I. (2007). Climate change and conflict: the migration link. Working Paper Series. International Peace Academy. http://www.ipacademy.org/asset/file/169/CWC_Working_Paper_Climate_Change.pdf. Accessed 02 August 2010.
Pirages, D. & Cousins, K. (Eds.). (2005). *From resource scarcity to ecological security*. Cambridge, MA: MIT Press.
Raddatz, C. (2008). *The macroeconomic costs of natural disasters: quantification and policy options*. Washington, DC: World Bank.
Ruiz-Carrascal (2008). *Bi-monthly report to the World Bank on environmental changes in Páramo Ecosystems*. LCSSD. Washington, DC: World Bank.
Schneider, S. H., Semenov, S., Patwardhan, A., Burton, I., Magadza, C. H. D., Oppenheimer, M., Pittock, A. B., Rahman, A., Smith, J. B., Suarez, A. & Yamin, F. (2007). Assessing key vulnerabilities and the risk from climate change. Climate Change 2007: impacts, adaptation and vulnerability. In M. L. Parry, O. F. Canziani, J. P. Palutikof, P. J. van der Linden & C. E. Hanson (Eds.), *Contribution of working group II to the fourth assessment report of the intergovernmental panel on climate change* (pp. 779–810). Cambridge: Cambridge University Press.
Stern, N. (2008). The economics of climate change. *American Economic Review*, 98(2): 1–37.
Thomas, C. D., Cameron, A., Green, R. E., Bakkenes, M., Beaumont, J. L. ,Collingham, Y. C. et al. (2004, 8 January). Extinction risk from climate change. *Nature*, 427: 145–148.
Toba, N. (2009). Economic Impacts of Climate Change on the Caribbean Community. In W. Vergara (Ed.), *Assessing the consequences of climate destabilization in Latin America. Sustainable development working paper*, LCSSD. Washington, DC: World Bank.
Tol, R. S. J. & Yohe, G. W. (2006). A review of the stern review. *World Economics*, 7(3): 233–250.
Trans-caribe system map (2010). http://www.itdp.org/index.php/projects/detail/cartagena_brt/. Accessed 16 August 2010.
UN (2004a). *State of the world's cities 2004/2005 – globalisation and urban culture*. New York, NY: United Nations.
UN (2004b). *World urbanisation prospects: the 2003 revision*. New York, NY: United Nations.
UNCHS (2002). *The state of the world cities report 2001*. New York, NY: United Nations Centre for Human Settlements, United Nations.
UNDP Human Development Report (2007/2008). *Fighting climate change: human solidarity in a divided world*. Washington, DC: UNDP.
UNEP-WCMC (2006). *In the front line: shoreline protection and other ecosystem services from mangroves and coral reefs*. Cambridge: UNEP-WCMC.
UNFCCC (2007). Report on the second workshop on reducing emissions from deforestation in developing countries. Available at: http://unfccc.int/resource/docs/2007/sbsta/eng/03.pdf. Accessed 02 August 2010.
UN-Habitat (2008). *State of the worldś cities 2008/2009- harmonious cities*. Nairobi: UNHabitat Publications.
United Nations (2006). *World urbanization prospects: the 2005 revision*. Department of Economic and Social Affairs, Population Division. New York, NY: United Nations.
United Nations Environment Programme (UNEP) (2007). *Global environment outlook 4 (GEO-4): environment for development*. Nairobi: UNEP.
United Nations Environment Programme (UNDP) (2009). *Charting a new low-carbon route to development*. New York, NY: UNDP.

United Nations Framework Convention on Climate Change (UNFCCC) (2006). Background paper—impacts, vulnerability and adaptation to climate change in Latin America. UNFCCC Secretariat. Bonn, Germany: Available at: http://unfccc.int/files/adaptation/adverse_effects_and_response_measures_art_48/application/pdf/200609_background_latin_american_wkshp.pdf. Accessed 02 August 2010.

Van Lieshout, M., Kovats, R. S., Livermore, M. T. J. & Martens, P. (2004). Climate change and malaria: analysis of the SRES climate and socio-economic scenarios. *Global Environmental Change*, 14: 87–99.

Vardy, F. (2008). *Preventing international crises: a global public goods perspective*. Washington, DC: World Bank.

Webster, P. J., Holland, G. J., Curry, J. A. & Chang, H.-R. (2005). Changes in tropical cyclone number, duration, and intensity in a warming environment. *Science*, 309(5742): 1844–1846.

While, A. (2008). Climate change and planning: carbon control and spatial regulation. *Viewpoint in Town Planning Review*, 78(6), vii–xiii.

Yamin, F., Smith, J. B., & Burton, I. (2006). Perspectives on dangerous anthropogenic interference; or how to operationalize. Article 2 o f the UN framework convention on climate change. In H. Schellnhuber (Ed.), *Avoiding dangerous climate change* (pp. 82–91). Cambridge: Cambridge University Press.

Chapter 5
The Relationship of Sustainable Tourism and the Eco-city Concept

Scott Dunn and Walter Jamieson

Abstract Asia currently has more than 100 cities with populations over one million. By 2015, Asia will account for 12 of the world's largest cities. Many of these cities are doubling in population every 15–20 years. Alongside this significant urban growth tourism numbers have also significantly grown over the past 10 years in most major urban centres in Asia. It is within this context of the absolute growth of urban areas and the growing levels of tourism activity that this chapter examines the concept of eco-cities from a tourism perspective. Most eco-city concepts have been developed to deal specifically with resident needs and activities and protecting environmental values. However, developing the co-city concept becomes much more complex when many cities are faced with the challenge of meeting the needs and aspirations of tourists which introduces a number of new stakeholders to be involved in the overall planning and development process. This chapter will first look at the nature of tourism from an urban perspective and the challenges facing planners as they attempt to achieve the principles and goals of the eco-cities concept. The nature of eco-cities as they relate to that definition of tourism is then analyzed with the article concluding with a series of recommendations for innovative sustainable tourism destination creation within the overall objectives of the concept of eco-cities.

5.1 Introduction

The sheer scale of the number and level of growth of Asian metropolitan areas highlights the vital need for developing sound planning and management techniques and approaches for Asia since these cities have a concentration of wealth and economic power of not only their countries but also the region. For example, Mumbai generates one-sixth of the GDP of India. The GNP of Tokyo is twice that of Brazil;

S. Dunn (✉)
AECOM Technology Corporation, Singapore
e-mail: scott.dunn@aecom.com

T.-C. Wong, B. Yuen (eds.), *Eco-city Planning*, DOI 10.1007/978-94-007-0383-4_5,

the GNP of Kansai in Japan is larger than the GNP of Spain. These major centres are expanding rapidly as the need for housing and space for industry and commerce expands. Bangkok, for example, grew from 67 km^2 during the late 1950s to 426 km^2 by the mid-1990s.

5.2 Urban and Metropolitan Tourism

The concept of eco-cities is likely familiar to many of the readers and has been covered in other parts of this book volume. It is less likely that many urban and regional planners understand the nature of tourism and its impact on cities and regions. We shall therefore begin by briefly looking at some of the key challenges and issues related to tourism and more specifically urban and metropolitan tourism.

It has only recently become accepted in some jurisdictions that tourism is an essential part of the overall process of urban planning and governance. Often tourism has been seen as a private sector activity with little impact on the overall governance, management and design of cities. The literature documenting issues of tourism planning and governance is still inadequate. However, there are growing signs that many urban authorities are now recognizing the importance to plan sustainably for tourism in order to achieve the benefits of tourism as part of meeting their overall social, cultural, economic and environmental goals. At the same time many national ministries and departments of tourism have recognized the important contribution of tourism to the GDP and employment which has raised its profile as a tool for economic and community development.

As history has shown tourism has the potential to bring about significant environmental and social positive impacts if properly managed and to be a positive force for development. On the other hand either neglected or improperly managed tourism has been shown to bring about significant negative impacts and considerably contribute to the metropolitan challenge. The United Nations World Tourism Organization (UNWTO) has recognized the responsibility and role of tourism in the larger urban and metropolitan management process. Towards that end it sponsored three international conferences in Kobe, Shanghai and Tucson on the management of metropolitan tourism and produced a monograph "Managing Metropolitan Tourism: An Asian Perspective" in order to begin to meet the knowledge and practice gap that presently exists (UNWTO 2010).

Tourism is a complex industry that in effect is an industry of industries. Beyond the more visible dimensions of hotels and transportation there exist a large number of public as well as private stakeholders all very much involved in the delivery of tourism services and products to the tourists. As the middle class in many parts of Asia increases we have seen corresponding increases in tourism to many parts of the continent and in particular to the gateway cities as well as other urban areas.

As with any other areas of urban economic and social activity, tourism must work within urban environments – especially in developing economies – that lack money, transparent governance structures, project management experience, structures for enforcement of regulations and standards, and the ability and willingness to work in an integrated fashion. These realities make urban tourism management more of a challenge than in European, North American or developed metropolitan areas in Asia.

Many urban areas now face an increasing population with growing middle class expectations and increasing numbers of interregional and domestic tourists. These urban areas face competition from a wide range of destinations for the tourist's dollar and the investment necessary to develop world-class facilities. The nature of the tourist is changing very quickly making planning and management evermore a challenge. Many of the tourists are in fact first time visitors while others are from a younger generation of travelers with quite different expectations than their parents. There is also the recognition of the need to meet the needs of multiple cultures with their diverse lifestyles, religious beliefs and traditions.

Given the rapid rise of tourism there has been a clear pattern of increased air pollution caused by the intensive use of vehicles for tourism/recreation-related mobility, pollution of water and marine ecosystems due to recreational navigation and peaks in the generation of solid and liquid wastes. It has been documented that tourism facilities are responsible for substantial increases in the consumption of fossil fuels for heating and electricity due to the visitors' rising quality expectations for services and facilities. Poorly planned and managed tourism destinations disturb birds feeding habitats and wildlife, cause land erosion and damage to vegetation which leads to erosion in ecologically sensitive areas. In summary while tourism can significantly increase the quality of life for its residents it also brings about significant disruption.

As mentioned earlier many countries position tourism as a pillar of their economic growth and development at local, regional and national levels. For example, China is strongly promoting and driving tourism development across all regions by working proactively and vigorously to promote tourism development for both international and domestic markets. On November 25, 2009, the State Council Executive Meeting chaired by Chinese Premier Wen Jiabao released a statement on "Accelerating the Tourism Industry Development" which emphasizes tourism as a strategic pillar industry in the national economy.

While strong direction for the promotion and development of tourism numbers from national governments such as China are becoming increasingly common unfortunately in many countries and destinations tourism development occurs with an inadequate understanding and planning over the impact of tourism on the quality of life in cities. Many tourism ministries and departments are working in isolation without taking into account regional and urban strategies and plans. Moreover many urban authorities also fail to work effectively with a wide range of tourism stakeholders. There can be no doubt that tourism must be seen as an essential urban planning and management activity and an integral part of eco-city planning and management.

Our introductory discussion on tourism closes with an examination of how the tourism community has responded overall to the challenge of achieving sustainable and responsible tourism. There have been numerous charters and agreements that

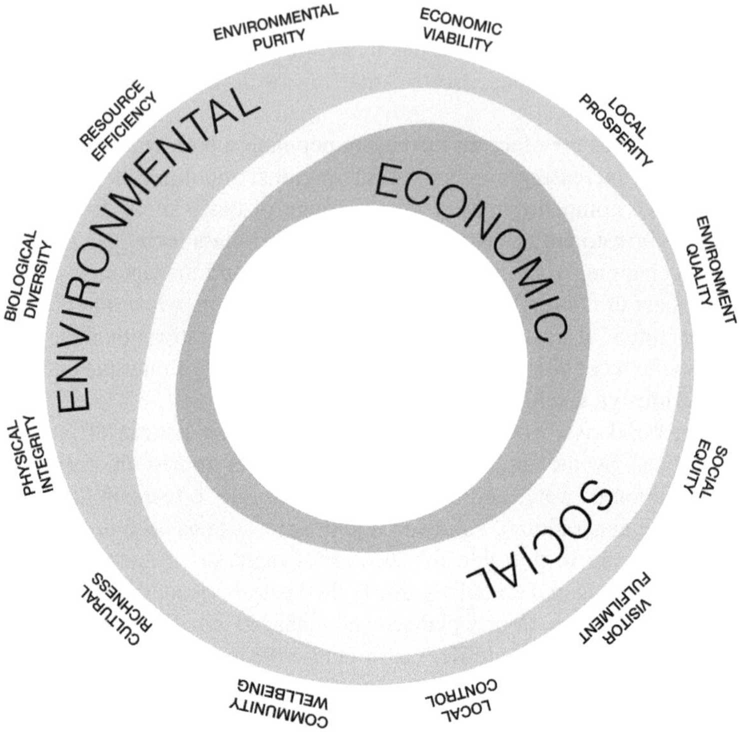

Fig. 5.1 UNWTO sustainable objectives. *Source*: United Nations World Tourism Organization

have helped to ensure that tourism development in many destinations is increasingly respectful to the environment, local cultures and values, cultural traditions and the ways of life of the local people. The UNWTO has identified with 12 different objectives which are now seen to be the guiding principles in many parts of the world (see Fig. 5.1).

Tourism has moved from a process of mass movement of people to people traveling with a wide range of motivations. The public and private sector response has taken many forms, but heritage and its many dimensions and nature-based activities are still seen as the primary reasons why people travel.

One important tourism niche market is the ecotourism. In many areas of activity the trend has been to apply the prefix "eco" to a number of activities which often are by no means sustainable in any significant dimension. The same is true of tourism where many stakeholders have now identified nature-based activities as ecotourism. However, ecotourism has a very specific meaning. The International Ecotourism Society defines ecotourism as a "responsible travel to natural areas that conserves the environment and improves the well-being of local people" (TIES 1990). The Society states that ecotourism is about "uniting conservation, communities, and sustainable travel". This means that those who implement and participate in ecotourism activities should adhere to the following ecotourism principles: minimize impact, build environmental and cultural awareness and respect, provide positive experiences for both visitors and hosts, provide direct financial benefits for conservation, provide financial benefits and empowerment for local people and raise sensitivity to host countries' political, environmental, and social climate. While obviously ecotourism espouses many of the same principles as responsible and sustainable tourism it is about tourism to natural areas. It can be seen as a subset of the larger field of nature-based tourism. Nature-based tourism can be seen as leisure travel undertaken largely or solely for the purpose of enjoying natural attractions and engaging in a variety of outdoor activities. Bird watching, hiking, fishing, and beachcombing are all examples of nature-based tourism.

While clearly some cities can offer nature-based experiences very few have sufficient natural areas to offer a true ecotourism experience. We have taken this time to explore this concept given the misunderstanding that eco-cities surrounding ecotourism.

The process of sustainable destination management which has been developed to deal with any negative externalities in many ways echoes many of the eco-city concepts. The challenge is how urban planning and management can help to create sustainable competitive tourism destinations. The eco-city concept holds such promise and can offer a useful model to examine the management of tourism and urban areas.

While many planners and people concerned with ecological balance and priorities would rather not have large numbers of tourists visiting their cities, given the strength of the industry it is unlikely that many urban areas will escape accommodating a large number of visitors in the future. Those supporting the eco-city concept need to better understand tourism and how it can not only meet the needs of the residents, protect the environment but also sustainably meet the needs and impacts of tourists.

5.3 The Eco-city Concept

While there has been a great deal written on eco-cities the authors felt that it was appropriate to start by sharing their understanding of the eco-city concept. The eco-city is an umbrella concept that encompasses a wide range of approaches that aim to make existing cities and urban development more ecologically sound and livable (Jabaroon 2006). These approaches introduce a number of environmental, social and institutional policies that are directed towards sustainable solutions. The concept mainly promotes the ecological agenda and emphasizes environmental management through a set of institutional and policy tools. According to Register (2002), eco-city zoning is a tool for polycentric restructuring of car-dependent cities by increasing the density around centres and recovering natural and agricultural landscape in the interspaces. It strongly focuses on a scale of ecosystems and habitats dedicated to the minimization of inputs of energy, water and food, and waste output. Generally, an eco-development strives for a carbon-neutral footprint where the human habitat is designed as a closed system. May (2008: 1) suggests that:

> In an eco-city, human habitat is designed with the recognition that the city, as the earth, is a closed system. When a thing ends its life cycle in a place in which it is treated as waste, it is polluting a closed system that will eventually become too full of detritus to support life.

Most definitions of an eco-city or sustainable urban community underscore the environment, economy and society (or quality of life) of a place (Kline 2000). Eco-developments are tackling environmental issues on a broader scale rather than in a piece-meal fashion. Creating the eco-city, therefore, requires several mechanisms including careful management of local resources, long-term planning, establishment of an ecologically sound set of institutions, and different land uses, environmental, social and economic policies (Robinson and Tinker 1998). As noted earlier the tourism community led by the UNWTO has been striving for many of the same goals. For the past 10 years many international organizations, national governments, destinations and private sector groups are increasingly looking at how tourism can contribute positively to the growth of tourism destinations based on principles very similar to that of eco-cities.

Many scholars have attempted to identify the major characteristics of the eco-city. From the planning perspective, Kline (2000), for example, highlights four attributes of eco-city: ecological integrity, economic security, quality of life, and empowerment. Her underlying objective is to use these attributes as a measurement tool or sustainability indicators that can influence the development decisions, track progress and evaluate the results. Gaffron and colleagues (2005) define five elements of the EU-funded ECOCITY project: urban structure, transport, energy and material flows, and socio-economy. More specifically, Kenworthy (2006) proposes a conceptual model of the eco-city based on the core issue of urban transport systems. He discusses ten critical eco-city dimensions: compact, mixed-use urban form; protection of the city's natural areas and food-producing capacity; priority to the development of superior public transport systems and conditions for non-motorized modes; extensive use of environmental technologies for water, energy and waste management; human-oriented centres; high-quality public realm; legible, permeable, robust, varied and visually appropriate physical structure and urban design; maximized economic performance of the city and employment creation; and a visionary process of the city planning. These are very similar to tourism sustainable management principles which were discussed earlier in this article.

In addition to the planning dimension, Jabaroon (2006) describes the eco-city in terms of its relationships to sustainable urban forms. The eco-city might be viewed as a "formless" city or an "eco-amorphous" city (ibid.). Drawn from several design approaches (e.g., the Ecovillage, Sola Village, Environmental City, Green City and Sustainable City), he points out that the distinctive concepts of the eco-city are greening and passive solar design. Many tourism facilities are now being built to the highest possible green standards and increasingly resorts and in fact destinations are adopting internationally accepted standards for sustainability. As will be seen later goals of the eco-city are very much in keeping with responsible and sustainable tourism practices, planning and design.

5.4 Case Examples of Responsible and Sustainable Tourism Development

Many cities have taken innovative approaches to integrating sustainability in their overall tourism planning and management. In order to better understand the relationship between eco-cities and tourism, specific examples are presented here.

5.4.1 Suzhou

In Suzhou, the design of 51 km^2 of green and public domain around Jinji Lake has created a new modern icon for the city. This necklace of projects in Suzhou is a complement to the old city's famed gardens and the city's first viable urban park. Ten years ago the lake was surrounded by farmland and fishing villages. The award-wining design has been an important part of the city's efforts to brand itself as a fitting home for foreign investment just as the park itself restored ecosystems around the Jinji Lake.

The area around the lake has now been developed with business and commercial areas, residential districts, beautiful parks, and a 9-mile walkway around the entire lake. According to a glowing review of Jinji Lake in *The New York Times*, it is a place "where progress is a walk in the park." The open space around the lake has become a major draw for visitors.

There are eight unique neighbourhoods with diverse water and landscape expressions encircling Jinji Lake. Neighbourhoods on the western and northern shores, closer to the city of Suzhou, feature broad promenades that attract residents and workers to the water's edge. Waterfront parks are adjacent to international shopping, entertainment, and cultural destinations. On the eastern and southern shores, farther from Suzhou, lie lakefront destinations for more passive recreation and environmental education.

Along with the restored water system of the lake there are now many water-based activities that bring people onto the lake including a main stage for nightly cultural shows that draw thousands of visitors and residents. The show incorporates the lake water as a main element which would have been impossible 10 years ago due to high levels of pollution.

Figures 5.2 and 5.3 provide an example of the landscape design plan as well as an image of the eventual product.

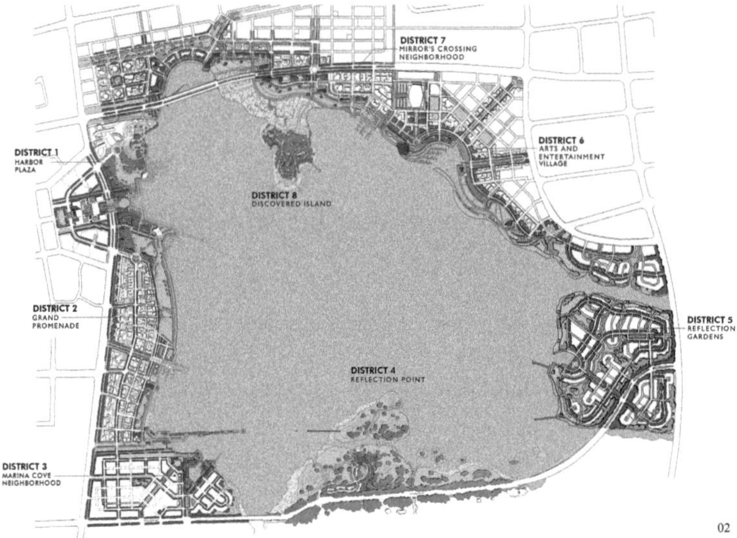

Fig. 5.2 Site plan for Jinji Lake Suzhou. *Source*: Plan Courteous of AECOM

Fig. 5.3 Illustration of landscape quality at Jinji Lake Suzhou. *Source*: Photo courteous of AECOM

5.4.2 Busan, Korea

The Gadeokdo Modalopolis Island project provides a unique perspective into a new paradigm of integrated development. Gadeokdo is located near Busan in South Korea at the end of the Baekdudaegan (Baekdu Great Mountain Chain) and has great potential to become a dynamic tourist destination. As an island along the southern coast of South Korea the area is at the crossroads between the start of the Trans-Siberian railway, the four river inner water transport route, a free economic zone with a major port facility and a proposed new Southeast International Airport.

As the centre of connectivity the Gadeokdo master plan introduces Modalopolis as a concept which integrates nature and human habitation, environment and development, business and tourism at a local and global level. The master plan is a way to connect people into an integrated destination that captures new international tourism trends.

The heart of the new destination is an integrated tourism development that captures the unique environmental factors that showcase the ocean, the land and the air. Multi-linked layers of tourism programmers and attractions in Gadeokdo offer tourism opportunities for both short- and long-term visitors. A spaceport hub connects North Asia to the other major geographies which allow opportunities for fuel efficient trans-ocean flights, thereby reducing travel times and resource consumption.

The island is designated as a no visa environment which allows foreign visitors and airport transit travellers easy access to the island and various activities and cultural experiences. Attractions on the island have been designed to accommodate future sea level rise with floating pods, buffering and controlled flooding. An image of the presentation concept can be found in Fig. 5.4.

5.4.3 Seoul, Korea

A fascinating and innovative example of sustainable urban regeneration is the Cheonggyecheon project in Seoul. (The original name of the Cheonggyecheon (Stream) is "Gaecheon" meaning "Open Stream".) A 6.8 km rivulet with a riverbank area which cuts through the heart of the city, this green lung is not just an ecological attraction. It is a recreational and cultural place with sculptures, fountains, historic bridges and waterfront decks dotting various stretches. It was not always this way. For nearly half a century until 2003, the stream was covered by a four-lane, two-way highway used daily by 170,000 vehicles. In 2002, then-mayor Lee Myung Bak announced the highway would be eliminated, the river would be restored and a 400 ha park created beside it. The project cost an estimated US$386 million and was as ambitious as it was meticulous.

Demolition started in July 2003. Diamond-wire and wheel saws – the most advanced technology available – were used to methodically slice up the highway and minimize noise, dust and other pollution for commercial and residential buildings in the area. The highway was dismantled a year later.

Fig. 5.4 Gadeokdo Modalopolis Island. *Source*: Master Plan Presentation Board, Courteous of AECOM

The restoration was a marriage of technology and creativity. Embankments were built to withstand the worst flood conditions. Sculptures, fountains and murals now dot the riverbanks. Long-buried bridges and foundation stones were restored and reinstated. Fish and birds started migrating to this sanctuary, thanks to the biotopes (spaces with uniform environmental conditions) introduced throughout the city, and credited with reducing the temperature of the surrounding area by between 2 and 3°C.

The value of nearby land and apartments reportedly increased by over 40%. Cafes, restaurants and other lifestyle businesses mushroomed. In the first 16 months after restoration, more than 40 million people visited the river, drawn by various attractions, e.g. 22 historical bridges, nine fountains, Sky Water Site and the Willow Swamp. There is even a Cheonggyecheon Museum which chronicles the history of the river. At Cheonggye Plaza – where the Cheonggyecheon begins – crowds throng cheek by jowl to see the tri-coloured fountain and a beautifully lit waterfall cascading four metres.

This is an excellent example of turning an eyesore into a tourism attraction and amenity for residents. Metropolitan areas will have to continue to invest in environmental and urban improvement and think strategically about increasing the quality of life for residents and tourists if they are to remain competitive.

Before and after images can be found in Figs. 5.5 and 5.6.

Fig. 5.5 The Cheonggye expressway prior to restoration. *Source*: The Preservation Institute Web Site www.preservenet.com/.../FreewaysCheonggye.html

Fig. 5.6 The Cheonggye River after restoration. *Source*: The Preservation Institute Web Site www.preservenet.com/.../FreewaysCheonggye.html

5.4.4 Summary of Case Examples

The case examples demonstrate that it is possible to meet social and environmental objectives while producing important tourism attractions and infrastructure. Others demonstrate that there is an important role for new technologies to meet the needs of urban areas and the tourism industry. What is important is that the examples demonstrate innovative and out-of-the-box approaches to dealing with both the urban condition as well as tourism. They also provide proof that it is possible to combine tourism and sustainable development.

5.5 Recommendations for Sustainable Tourism Within an Eco-city Context

There are a number of possibilities for ensuring that the eco-city concept and tourism development can not only coexist but in fact reinforce each other.

But first must be recognized that the eco-city concept provides significant opportunities for adopting the necessary regulatory and physical structures that will allow for sustainable urban governance and development, and the delivery of sustainable tourism experiences. As noted earlier, tourism is a multifaceted field and by its very nature a multidisciplinary area of activity. Economists, historians, archaeologists, city planners, tourism policy and planning experts, marketers, poverty reduction specialist, environment conservation planners, architects, urban designers, landscape experts, are some examples of the diversity of the knowledge and skills necessary to manage urban tourism destinations in a sustainable and responsible manner. Many if not all of these same stakeholders are an essential part of ensuring that an urban area can be developed within the framework of eco-city principles

Managing tourism within an ecosystem context is about a number of dimensions including creating cityscapes and landscapes that are worth the tourist's long journey; sites full of natural and cultural qualities which provide memorable experiences, business practices that protect the environment and contribute to the social, economic and cultural development of the host communities, respect for the environment, protection of local traditions and lifestyles and finally an increased appreciation of a community's history and traditions. This rather formidable list of objectives exemplifies the complexities of sustainability within a tourism environment. Ultimately the objective is to ensure the overall success of the tourism destination within an increasingly competitive visitor landscape.

5.6 Ensuring that Tourism Is an Important Part of the Eco-city Concept

In a slideshow entitled "Sino-Singapore Tianjin Eco-city" there is a call for a holistic master plan (the slideshow was downloaded from the Internet where there were no reliable reference sources). The PowerPoint presentations calls for a comprehensive

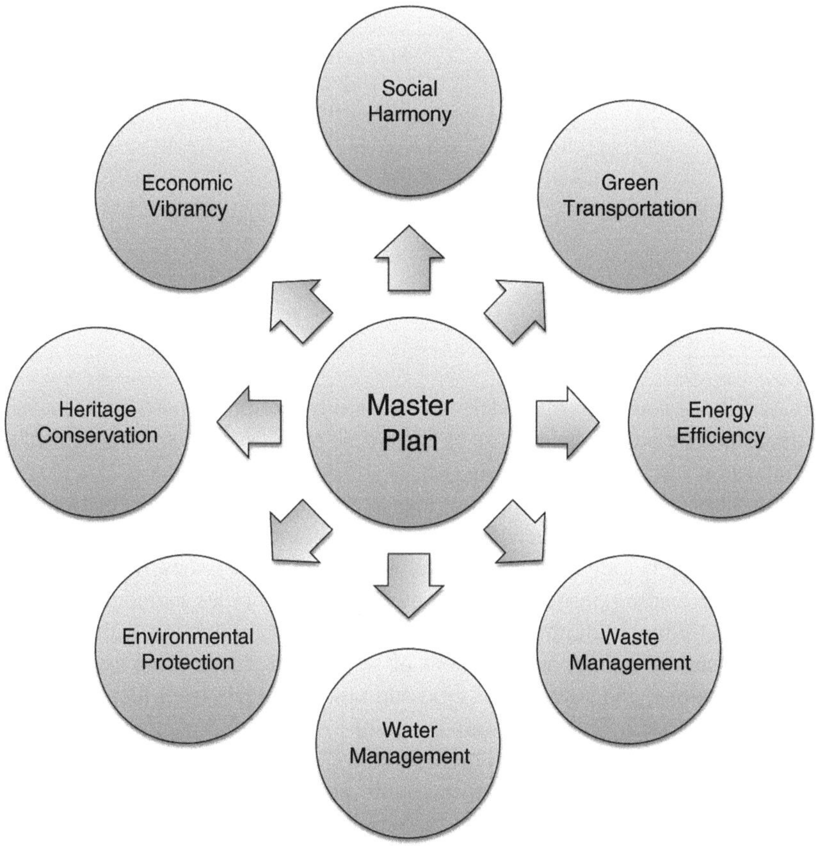

Fig. 5.7 Sino-Singapore Tianjin eco-city master plan approach. *Source*: Sino-Singapore Tianjin Eco-city – Internet

planning approach incorporating best ideas from China and Singapore. The ideas for the master plan are captured in Fig. 5.7.

It is interesting to note that while there is mention of a number of dimensions tourism is not identified as part of the overall master plan process. This is true in many jurisdictions and presently the United Nations World Tourism Organization is working with many governments in ensuring that tourism is seen as an integral element of economic, social and community development. While many mayors recognize that tourism is an incredible creator of jobs and will be for a considerable time many have been unable or unwilling to make the connection between tourism and achieving a sustainable form of urban development and activities.

5.6.1 Adopting an Integrated Approach

For the principles of sustainable tourism to be adopted within the eco-city context it is essential that a solid framework for tourism growth and development be adopted.

Fig. 5.8 Key dimensions of an integrated destination approach. *Source*: Compiled by authors

There have been a number of models put forward by various experts including one that developed by one of the authors of this article (Jamieson 2006). It looks at four key dimensions as can be seen in Fig. 5.8.

The authors recognize that there are many other approaches to destination management. Whatever the approach there can be no argument that well thought out and integrated approaches are necessary to deal with the range of issues that are essential in creating the leading destinations of tomorrow. These issues include community involvement, training, community development, poverty reduction, cultural sensibility, job creation, equity, advancement of women, greening of hotels and other tourism operations, and sourcing of food and other products from local communities. In effect these are integral components of a sustainable planning and design ethos.

The integrated approach needs to take into account that tourism is but one of many functions and concerns within metropolitan areas and it is important that tourism planning and management effectively integrate its concerns and methodologies with those of larger urban management and government structures. Without this integration metropolitan tourism development will not be seen as one of the key strategies in metropolitan growth management and development.

5.6.2 Marketing and Product Development

In order to ensure a fit between tourism demand and supply of products, destinations need to become more sophisticated in how they develop products in a sustainable way while understanding the realities of the tourism market. Equally important are planners and designers is to use marketing strategies that are sustainable and accurately reflect what a destination has to offer.

5.6.3 Stakeholder Involvement

As noted earlier given the complexity of the industry there is an important need to maintain effective stakeholder organizations and processes. Cities and destinations are becoming more sophisticated in their understanding of how to manage stakeholders but it is especially important in the context of eco-cities to look at equitable and far-reaching participation in order to ensure that all segments of the society share equally the benefits of tourism development.

5.6.4 Provision of Services

The provision of basic services and infrastructure in many developing metropolitan areas is still in its infancy. While clean water, transportation infrastructure, solid waste management and pollution control are well accepted as essential elements of urban areas in developed economies they are still in the very early stages of development in many Asian metropolitan areas. The tourists now expect a high level of sanitation and services as they travel. Coincidentally the eco-city movement equally values the need for these basic goods and services to be delivered to all the residents of an urban area. In the case examples presented above it becomes clear that innovative approaches can help to achieve both tourism objectives as well as ensuring high quality of life for the residents.

5.6.5 Developing Appropriate Urban Forms

As congestion continues to grow in many destinations managing the visitor experience becomes increasingly important. Careful thought needs to be given to sustainably facilitating movement within tourism destinations. It is no longer acceptable to have large numbers of buses increasing congestion and pollution around many key tourism sites. It is only with sophisticated visitor management techniques that destinations can begin to meet sustainable tourism principles.

5.6.6 The Promotion of Appropriate Technologies

There are now many appropriate technologies that are designed to promote sustainable tourism development within metropolitan areas. These range from low tech water conservation approaches to the development of entire environmental management systems concerned with minimizing the impact of tourism activity and the use of scarce resources. The challenge, as with many other areas of urban management, is in the implementation and adoption of these technologies. Industry is becoming increasingly aware of the advantages of appropriate technologies and there is a need for incentives and support to ensure a much more widespread use of these technological innovations. There are unique opportunities in the development of new tourism destinations and resorts to use eco-city dimensions as the governing principles for development. These technologies can also become attractions and experiences in themselves and showcase new and innovative approaches especially for the ever-growing number of tourists with a strong concern for sustainability and greening.

5.6.7 Monitoring and Knowledge Management

As noted earlier in this article there is precious little material on urban and metropolitan tourism management especially within an eco-city frame of analysis and development. There is an urgent need to continue to identify examples of good practice, document them in ways that are usable to practitioners and developers and disseminate them in an effective way. There is also the need for sustainable monitoring techniques especially at the destination level. The notion of dashboards is quickly gaining recognition and should be considered as a way of providing comparable data from destination to destination. The case for such an approach has been proposed for Hawaii in a recent academic article (Park and Jamieson 2009).

5.7 Conclusion: A View of the Future

Within an eco-city approach the following tourism planning and management dimensions must be considered if destinations are to be sustainable and competitive:

- All the stakeholders in tourism development should safeguard the natural environment with a view to achieving sound, continuous and sustainable economic growth geared to equitably satisfying the needs and aspirations of present and future generations.
- All forms of tourism development that are conducive to saving rare and precious resources, in particular water and energy, as well as avoiding so far as possible waste production, should be given priority and encouraged by national, regional and local public authorities.
- The staggering in time and space of tourist and visitor flows, particularly those resulting from paid leave and school holidays, and a more even distribution of holidays should be sought so as to reduce the pressure of tourism activity on the environment and enhance its beneficial impact on the tourism industry and the local economy.
- Tourism infrastructure should be designed and tourism activities programmed in such a way as to protect the natural heritage composed of ecosystems and biodiversity and to preserve endangered species of wildlife. Stakeholders in tourism development, and especially professionals, should agree to the imposition of limitations or constraints on their activities when these are exercised in particularly sensitive areas: desert, polar or high mountain regions, coastal areas, tropical forests or wetlands, nature reserves or protected areas;
- Nature tourism and ecotourism are recognized as being particularly conducive to enriching and enhancing the reputation of tourism, provided they respect

the natural heritage and local populations and are in keeping with the carrying capacity of the sites.

It is hoped that this discussion has helped to introduce another element into the debate and implementation of eco-city principles within the larger process of urban and metropolitan planning and management with a special focus on tourism. This debate is especially important given the growth of tourism in many urban areas not prepared for tourism activity.

References

Gaffron, P., Huismans, G. & Skala, F. (2005). *Ecocity book 1: A better place to live*. Hamburg, Vienna: Facultas Verlags- und Buchhasdels AG.

Jabaroon, Y. R. (2006). Sustainable urban forms: their typologies, models, and concepts. *Journal of Planning Education and Research*, 26: 38–52.

Jamieson, W. (Ed.). (2006). *Community destination management in developing economies*. Binghamton, NY: Haworth Press.

Kenworthy, J. R. (2006). The eco-city: ten key transport and planning dimensions for sustainable city development. *Environment and Urbanization*, 18(1): 67–85.

Kline, E. (2000). Planning and creating eco-cities: indicators as a tool for shaping development and measuring progress. *Local Environment*, 5(3): 343–350.

May, S. (2008). Ecological crisis and eco-villages in China. *Counterpunch*, November issue, 21–23.

Park, S.-Y. & Jamieson, W. (2009, 1 March). Developing a tourism destination monitoring system: a case of the Hawaii tourism dashboard. *Asia Pacific Journal of Tourism Research,* 14: 39–57.

Register, R. (2002). *Ecocities: building cities in balance with nature*. Berkeley, CA: Berkeley Hills Books.

Robinson, J. & Tinker, J. (1998). Reconciling ecological, economic, and social imperatives. In J. Schnurr & S. Holtz (Eds.), *The cornerstone of development: integrating environmental, social and economic policies* (pp. 9–43). Ottawa: IDRC-International Development Research Centre and Lewis Publishers.

The International Ecotourism Society (1990). http://www.ecotourism.org/site/c.orLQKXPCLmF/b.4832143/k.CF7C/The_International_Ecotourism_Society__Uniting_Conservation_Communities_and_Sustainable_Travel.htm

United Nations World Tourism Organization (2010). *Managing metropolitan tourism: an Asian perspective*. Madrid: UNWTO.

Part II
Implementation and Practice

Chapter 6
Down with ECO-towns! Up with ECO-communities. Or Is There a Need for Model Eco-towns? A Review of the 2009–2010 Eco-town Proposals in Britain

Eleanor Smith Morris

Abstract The recent Labour Government proposed in England that ten new green clean "eco-towns" should be built by 2020. How did this government programme begin? What are the objectives? Is the British Government creating fabulous models for the future or is it bull-dozing through a programme that will create the slums of the future? The discussion examines the origins of the eco-town programme, and the pros and cons of the proposals. The English eco-towns appeared to be in danger, despite concerns about the under provision of housing. Has the economic crunch paid to the creation of eco-towns? When the Labour Government was under siege, the ongoing row over eco-towns added to their troubles. The idea of eco-towns is valuable as a source of housing but the execution has left a lot to be desired. Many of the original proposals are in the wrong location or are reincarnations of schemes that have already been deemed unsuitable. The new Coalition Government of Conservatives and Liberal Democrats, to the surprise of everyone, announced that they will only keep four of the proposed eco-towns, and at the same time bring back the focus onto brownfield land and urban extensions. Many consider that eco-towns can only make sense of where they are in relation to existing centres of population, transport, infrastructure and employment. Some cities prefer a number of eco-communities or urban extensions in brownfield locations instead of a few free standing eco-towns. The eco-town proposals are compared with the New Urbanism proposals in the United States which burst upon the anti-suburban scene in the 1980s. The principles and concepts of New Urbanism are reviewed with examples where it has been most successful. The proposed new town, Tornagrain, by Inverness, for 10,000 people on a green field site where Andreas Dulany, one of the creators of New Urbanism has prepared a master plan, is examined. In summary, the proposed eco-towns, unlike New Urbanism, offer important opportunities to bring together models of environmental, economic and social sustainability. They will provide testbeds for different methods of delivering, for example: (a) zero carbon building development, (b) offering 30% affordable housing, (c) creating 40% green

---

E.S. Morris (✉)
Commonwealth Human Ecology Council, London, UK
e-mail: emorrischec@yahoo.co.uk

T.-C. Wong, B. Yuen (eds.), *Eco-city Planning*, DOI 10.1007/978-94-007-0383-4_6,

infrastructure; and (d) looking after waste. Some would say that establishing models of development from which others can learn is their most important result and not the provision of 50,000 homes, a small portion of the proposed 3 million homes required for the United Kingdom.

6.1 Introduction: Evolution of New Towns to Eco-towns in Britain

The Eco-Towns, proposed in 2007–2009, are the first revival of the New Town Movement in Britain for 40 years. Previously Britain has had a superb record of creating New Towns from the nineteenth century Utopian, Model New Towns and Garden City New Towns to the magnificent achievement of the first, second and third generation New Towns following the Second World War into the 1970s. In the nineteenth century, Utopian New Towns, such as Buckingham's "Victoria" and Pemberton's "Happy Colony" were envisaged to overcome the squalor, overcrowding and disease of the industrial slum. The principal Utopian New Town to be built in 1817 was New Lanark near Glasgow, Scotland by the industrialist Robert Owen for a manufacturing village of 1,500 persons (Morris 1997).

Model New Towns followed the Utopian communities of which one of the most ambitious was Saltaire, a model industrial town near Bradford, England, built by Sir Titus Salt (1848–1863). It provided vastly improved housing accommodation, lessening the cramped conditions of the city to a newly built town in the countryside. Bourneville, built by the Cadbury Brothers in 1894, further improved the provision of open space, sunlight and environmental conditions. Bourneville was followed by Port Sunlight, built by the Lever Brothers in 1888, again with the emphasis on good housing and generous amenities. The final model town was Earswick, built by Sir Joseph Rowntree in 1905 (Morris 1997).

The success of a handful of benefactors in providing better conditions for their workers could not overcome the extensive slum problem and a more radical approach was required. The public health reformers, like Chadwick, who brought in the 1870 By-Laws to improve workers' housing, made a greater impact on the slum problem than the individual new towns. Thus the reform movement with the greatest positive physical effect on British town planning was the Garden City movement, based on the ideas of Ebenezer Howard as published in Garden Cities of Tomorrow (Howard 1899, 1902). Howard was able to see his proposals realised in the Garden Cities of Letchworth (1903), Welwyn (1919) and Hampstead Garden Suburb (1915). Particularly Letchworth and Welwyn Garden Cities fulfilled Howard's idea with: (a) a wide range of industries and local employment; (b) a spirited community life; (c) houses with gardens and large open spaces; (d) a green belt; and (e) single ownership with excess profit for the benefit of the town. The Garden City concepts formed the basis of the New Town movement after the Second World War until the Futurist City of the linear town planners overturned this approach in the mid-twentieth century with new towns like Cumbernauld and Runcorn (Morris 1997).

Although Letchworth and Welwyn Garden Cities provided tangible evidence that New Towns could achieve the proposals for which they were created, no further practical work occurred until the devastation of the Second World War was felt. The Greater London Plan 1944 proposed eight new towns beyond the Green Belt and the County area (Abercrombie 1945). This spurred the 1946 New Towns Act, one of the most extraordinary phenomena of the post-World War II period, a brilliant feat of creating over 30 New Towns. Internationally Britain achieved a spectacular standard, which other countries including China, Israel and the United States, continue to imitate. Between 1946 and 1950, 14 New Towns, the so-called first generation New Towns were designated; including the most famous Harlow, Stevenage and Crawley. Cumbernauld, Scotland, built with a futurist shopping mega-structure in 1956 was the only New Town of its kind to implement housing and community services focused on a sole centralised structure, unlike Harlow and Crawley with their organic neighbourhoods arranged around Garden City green belts and open space. Then, in a sudden reversal of government policy in 1962, there was a return to the designation of first generation type new towns and five more new towns were created. Finally the concept of Regional cities prompted the creation of Third Generation New Towns, including the most innovative Runcorn and Milton Keynes (Morris 1997).

The 1960s and the 1970s were an exciting period for town planning opportunities. New Towns were built; dispersion and decentralization policies gave many people new opportunities and a new way of life. But it was not to last. By 1979, with the Conservative Prime Minister Thatcher coming to power for 15 years, statutory Structure Plans were installed and any revolutionary new idealistic plans were but a memory of an age based on principles and ideals.

From then on, planning took the form of ad hoc principles, alternative strategies and specific local area objectives. The golden age of planning principles had come to an end (Morris 1997). In the 1980s and early 1990s, the Conservative Government was more interested in Inner City Regeneration, Science Parks and Business Parks than in creating new towns. But to give the Conservatives their due, privately financed "villages" were promoted. In the 1990s over 200 "planned" new villages with an architectural vernacular approach of 4,000–5,000 people were built as the Conservatives favoured new villages to relieve the pressure on the old villages and towns, preventing them from being destroyed by garish new housing estates. The original New Town concept of a "balanced community", which provides local jobs for people living in the town) cannot be fulfilled by small villages. Further the recession of the 1990s also hindered New Town development (Morris 1997). Hence it is intriguing that towards the end of the 1990s with Labour again in power that a mini version of New Towns, the Eco-town should be promoted.

6.2 Background to the Creation of Eco-towns

Considering that strong action was needed to provide inexpensive affordable housing, the Labour Government produced a Housing Green Paper (DCLG 2007). The Housing Green Paper advocated the construction of 240,000 dwellings every year to meet an overall goal of 2 million housing units by 2016 and 3 million housing units by 2020. These figures included 650,000 houses in 29 specified growth areas and 100,000 extra houses in 45 towns and cities which constituted 29 "new growth points" as follows (Lock 2007):

(a) 200,000 new homes to be built on surplus public sector land by 2016 using 340 sites owned by British Rail; 130 sites owned by the Highway Agency and 50 sites by the Ministry of Defence;
(b) 60,000 new homes on brownfield sites to provide affordable rented homes; and
(c) 50,000 new homes to be located in 5 new eco-towns to become new growth points with the towns to achieve zero carbon development standards.

Under the plan, some cities could have access to a £300 million Community Infrastructure Fund earmarked for growth areas, new growth points, and particularly "eco-towns". These new eco-towns were described as "communities with renewable energy sources, high energy efficiency, low carbon emissions, water efficiency, and waste minimalization" (DCLG 2007). The original real purpose of the eco-towns was to help attain the national goal of a 24–36% reduction in carbon emissions by 2020.

Already in May 2007, the then Prime Minister Gordon Brown recommended a series of eco-towns, new free-standing settlements between 5,000 and 20,000 units "intended to exploit the potential to create new settlements to achieve zero carbon development and more sustainable living using the best design and architecture" (Shaw 2007). Yet the programme could not be delivered by the central government but had to be built by private house-builders, housing associations and/or by new types of local housing companies. Long ago during the 1960s and 1970s, local governments each built hundreds of houses per year. What has changed is that the government is now heavily dependent on the private sector to meet the targets. All the talk about roof taxes and planning gain supplement is predicated on the developers' profit margins. But the private sector has to depend on business opportunities in the open housing market which had collapsed since these proposals were made. The growth points initiative that the Government previously in 2005 launched to invite the local authorities to bid on 29 growth points as the location of the eco-towns (Office of the Deputy Prime Minister 2003) faced problems of implementation.

6.2.1 Initial Eco-town Site Proposals

Among the proposals for 57 potential sites for eco-towns submitted, 15 potential sites were nominated in March 2008. The purpose of these eco-towns remained the same: zero carbon development, promoting sustainable living and providing 30–50% affordable homes. In addition, there were to be underground systems for waste recycling, free public transport with car journeys curtailed by a 15 mph limit and green routes to school. Bath water would be recycled and fed to communal flower beds. Each home would pump excess power generated by its solar panels and turbines back into the National Electricity grid. These eco-towns were to

Table 6.1 First 15 Eco-town schemes short-listed for final selection

Site number	Region and town	Number of homes
1	Leeds City region – Selby	Not yet known
2	Nottinghamshire, Rushcliffe	Not yet known
3	Leicestershire, Penn bury (proposed by the Co-op)	12,000–15,000 homes, including 4,000 affordable homes
4	Cornwall, St. Austell. Primary aim is to create jobs affected by the closure of clay pits	5,000 homes
5	Staffordshire, Corborough	5,000 homes
6	Warwickshire, Middle Quinton – (site of old Royal Engineers depot)	6,000 homes
7	East Hampshire, Borden and Whitehill (East Hampshire District Council) – Ministry of Defence sites	5,500 – with 2,000 affordable homes
8	Ford	5,000 homes
9	Oxfordshire, Weston Otmoor	10,000–15,000 homes
10	Bedfordshire, Marston Vale	15,000 homes
11	Northeast Elsenam	5,600 homes including 1,800 affordable homes
12	Cambridgeshire, Hanley Grange (Developed by Tesco)	8,000 homes including 3,000 affordable homes
13	Lincolnshire, Manby (East Lindsay District Council)	5,000 homes
14	Norfolk, Coltishall – An RAF airfield supported by the Dept of Communities & Local Government Rackheath desired by Norfolk DC as part of the planning process	5,000 homes
15	Rossington	15,000 homes
16 (already created)	Cambridge, Northstowe (first official eco-town)	9,500 homes
Total proposed homes		111,600–119,600 (including 10,800 affordable homes)

Source: Collated from various sources

count towards District Housing Targets, in order to make them preferential to urban extensions (Table 6.1).

The Conservatives claimed that the Labour Government chose locations in Tory constituencies, as only 3 of the 15 are in Labour areas, including Rossington. Eventually the Manly, Lincolnshire proposal, the Corborough Consortium and New Marston Gallager Estate proposals were all dropped (Fig. 6.1).

Hanley Grove initially increased its housing numbers from 8,000 to 12,000 to be developed by Jarrow Investments. But by September 2008, Tesco withdrew its 8,000 homes and later decided to re-apply with a modified application through the then Regional Spatial Strategy (RSS) procedure, which has now been abandoned

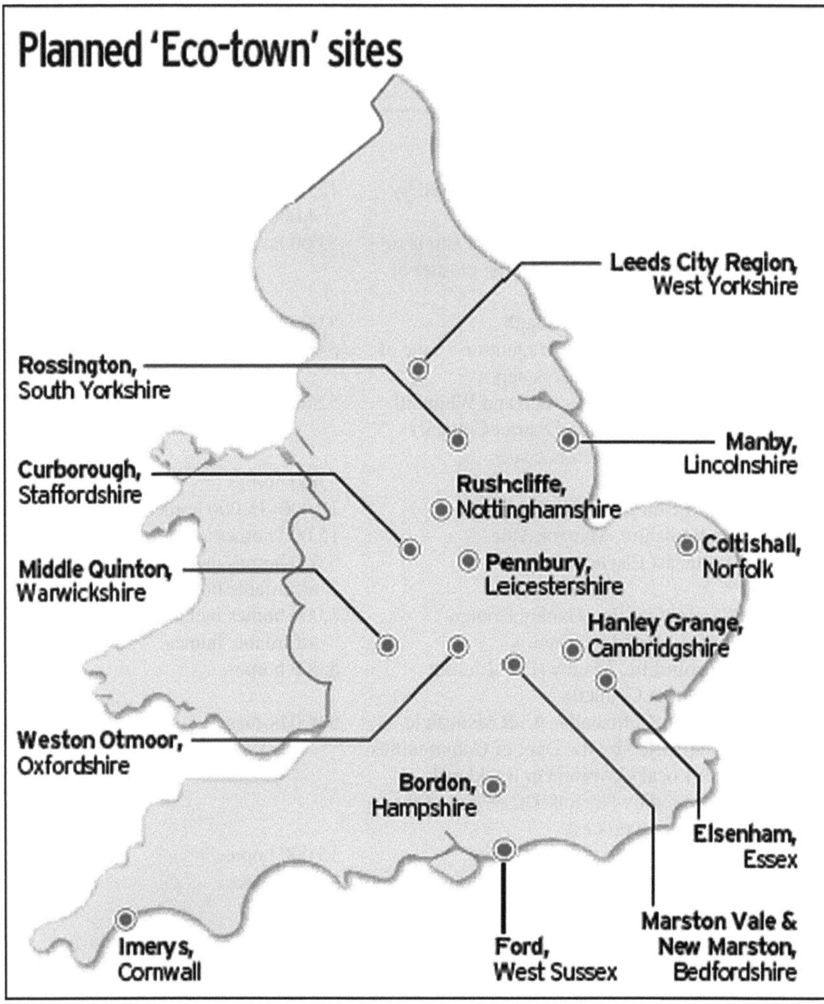

Fig. 6.1 The 15 potential eco-town sites nominated in March 2008. *Source*: Brooksbank-geographyyrl3, Eco-towns in the UK, http://brooksbankgeographyyrl3.wikispaces.com/Case+Study+-+Ecotowns, accessed 28 March 2011

by the Coalition Government. Hazel Blears, then Labour Housing Minister, blocked Multiplex's plan for 5,000 homes in Mereham, Cambridge. Blears was also concerned that the Cambridgeshire Councils could not handle three applications on such a large scale. This left the Northstowe project as the principal eco-town in Cambridgeshire.

In June 2009, Arun District Council challenged the Government Office for the South East (GOSE) for stating it was going to "facilitate" proposals for the eco-town when it should merely "test". The Government had to back track and agree that the eco-town proposal will be subject to full planning procedures.

6.2.2 Choosing the Eco-towns

By June 2008, the 15 chosen towns became 13, which the Department of Communities and Local Government (DCLG) then stated would be whittled down to 10 towns. According to David Lock (2008b), "the term "Eco-Town" turned out to be a powerful pairing of words, much stronger than "urban village" and approaching "garden city" for its ability to stimulate a wide range of people to pool their ideas". As opposition to the eco-towns started appearing against the Labour Government, the Tory Shadow Government announced that there would be no new eco-towns at all when they achieved office.

Contrary to the common public perception, the planning of the eco-towns has complied with the planning process. In order for an eco-town to obtain an outline planning permission, the application will have to include approval in the following aspects:

(a) an environmental appraisal;
(b) a transport assessment;
(c) a sustainability appraisal; and
(d) a community involvement statement.

It is expected that the outline planning application would be "called in" for decision by the Secretary of State, who would hold a public inquiry conducted by an independent inspector. Some people are urging a Special Development Order by the Secretary of State in the manner of the New Town Development Order of the 1981 New Town Development Act. The problem with the outline planning application procedure is that it is painfully slow and allows the huge value on the land to rise, allowing less and less planning gain to provide for the eco-towns. Since the planning gains have to be high, only the best sites will likely survive against the anti-housing lobby.

6.2.3 The Anti-Eco-town Lobby

Throughout 2008 and 2009, the anti-eco-town lobby protested vigorously. Some of the eco-town proposals had to come under the wider Regional Spatial Strategy (RSS) review in early 2010. However, after the May 2010 election Regional Spatial Organizations have been disbanded by the new Coalition Government. Such a case is Middle Quinton in Warwickshire which was to be considered through the West Midlands Regional Spatial Strategy whose review will not be considered at all now. Against the town proposal was the Better Accessible Responsible Development (BARD) who went to the High Court to halt the development without success. They appealed against the High Court decision by saying there was no proper consultation on the Housing Green Paper but they lost that appeal (Fig. 6.2).

Fig. 6.2 Anti-eco-town protestors at Long Marston, Warwickshire. *Source*: Sunday Telegraph 2008

Opponents to Weston Otmoor also fought the eco-town proposal but both groups were over-ruled by the High Court Judge who said the procedure had been adequate. The villagers of Ford, the former location of the RAF Ford Battle of Britain airfield, formed a campaign action group called CAFÉ (Communities against Ford Eco-town). Their objections were based on the lack of transport structure to support communities of up to 20,000 people, the lack of jobs and that the new eco-towns rather than creating local employment would overwhelm the existing prospects.

The campaigners promoted instead for redeveloping the 617,000 vacant properties in England including those in the neglected suburbs, by creating a green template for carbon-neutral neighbourhoods. They were against the Government's commitment to build 3 million new homes by 2020, and the Government's jargon exclaimed by Labour Minister Caroline Flint was "we will revolutionize how people live" (Sunday Telegraph 2009).

The Campaign to Protect Rural England (PPRE) supported rejuvenation of the area. However, the Ford Eco-town proposal could not demonstrate how to incorporate the needs of the local communities, the area's environmental limits and the nature of the infrastructure in the proposal, it was defeated. Meanwhile the Tory Shadow Planning Minister, Bob Neilly, warned the Chairman of the proposed Infrastructure Planning Commission (IPC) that the Tories would scrap any such Infrastructure Planning Commission on decision-making on national infrastructure. This was expected to have a knock-on effect on eco-town development in the United Kingdom (Planning Journal May 2009).

By May 2009, the then Housing Minister, Margaret Beckett, announced that she hoped to approve up to 10 schemes, but she added that the proposals all needed additional work to meet the green standards set by the government. Beckett argued that eco-towns are a good way to set a high bench mark for other housing developments. If Margaret Beckett, a former Foreign Minister, had been able to stay as Housing

Minister, the eco-towns might have had a fair chance. But Beckett had to resign as Housing Minister in the Prime Minister's reshuffle over the MP's expenses scandal. Indeed the turnover of Housing Ministers (Cooper, Flint, Blears, Beckett and Healey) in the past year and a half has been so numerous that it resembled Alice's Tea Party!

In the event the Coalition Government of Conservatives and Liberal Democrats won the May 2010 election, the prospect that eco-towns being scrapped would be high.

6.2.4 New Communities

There is an opposing point of view that the money for new towns should go to new communities as part of urban extensions. The Leeds City-Region Partnership wants to develop a number of eco-communities in place of a single free-standing eco-town. They have located four brownfield locations including the Aire valley and the Bradford canal corridor as being more suitable to meet regeneration and affordable housing demand. A judicial review has caused the Government to admit that alternative approaches to affordable housing may be possible (Fig. 6.3).

In principle, eco-towns should make sense in that besides having available land where new environmental criteria could be met, they must be developed in relation

Fig. 6.3 Aire Valley site where eco-communities are preferred to solitary towns. *Source*: Planning Journal (2008)

to existing centres of population, transport infrastructure and employment. Size does matter. It has been noted that eco-towns of 5,000–10,000 people will not justify public transport unless they are attached to existing cities as urban extensions. They will also struggle to provide diversity of employment unless attached to existing urban areas. It has been suggested that EIA assessments should be paralleled with sustainability assessments in the early stages of choosing sites. People need to be able to walk or cycle or take bus to their activities; otherwise living, working, health and education would become so divorced that the car dominates daily life (Fig. 6.4).

Hence, the Conservatives will opt for regeneration of existing towns with urban extensions and accuse Labour of simply wanting a financial bonanza. Others suggest linking new settlements in a joined up process within the great urban areas. This is sensible as there is less need for high-level self containment; there is the possibility of the connecting thread of transportation, there can be networked local economic

Fig. 6.4 Eco-towns in isolation may not provide the transport or diversity of employment to create thriving towns. *Source*: Country Life 2009

development with accessibility provided by communications technology. There can still be high environmental and carbon dioxide emissions standards within these urban extensions (Shaw 2007).

6.3 New Urbanism

The New Towns of the New Urbanism movement are the newest models for the eco-towns. The architects, Andres Duany and Elizabeth Plater-Zyberk (DPZ), first achieved national fame during the 1980s by creating Seaside, a resort town in the Florida panhandle. It has remained their most famous New Urbanism creation but is still an isolated resort town and not a complete community. In 1988, they created Kentlands, Maryland, the first application of their traditional neighbourhood development principles for a year round working community (Duany and Plater-Zyberk 1991; see Fig. 6.5 below).

The Modernism of the first half of the twentieth century was opposed by the anti-Modernists who were then in turn challenged by the new movement, the New Urbanism. In 1993, Duany and others founded the Congress for the New Urbanism (CNU) which was a deliberate attempt to counteract the 1930s modernist movement, *Congrès International d'Architecture Moderne* (CIAM). The New Urbanism Congress also cleverly allowed them to spread the word not only amongst architects but also amongst public agencies, developers and consumers, something that the older Congress, CIAM never did. In 1966, they created their Bible, the Charter of the New Urbanism, which showed how their approach could be extended beyond neighbourhood and small resorts to suburbia and urban extensions (Leccese and McCormick 2000). The New Urbanism includes the following elements:

Fig. 6.5 Middle Quinton – a British example of New Urbanism. *Source*: Planning Journal (2009)

(a) Interconnected streets, friendly to pedestrians and cyclists in modified grid patterns (no *cul-de- sacs*);
(b) Mixed land uses;
(c) Careful placement of garages and parking spaces to avoid auto-dominated landscapes;
(d) Transit-oriented development;
(e) Well-designed and sited civic buildings and public spaces;
(f) Use of street and building typologies to create coherent urban form;
(g) High-quality parks and conservation lands used to define and connect neighbourhoods and districts; and
(h) Architectural design that shows respect for local history and regional character.

With these key goals, they devised the tool of a zoning code. In the case of Seaside and Kentlands, the DPZ New Urbanism firm devised individual design codes that control the architectural elements and maintain a clear division between private, semi-public and public spaces. Builders and homeowners had to abide by the Code which specifies such details as front porches and white picket fences to promote neighbourliness. The result is that in Kentlands each residential block is a unique ensemble, characterised by varieties of house types as well as fully grown trees and lots of greenery on the periphery.

6.3.1 Kentlands, Maryland, USA

Kentlands was planned for a 356-acre site, surrounded by conventional suburban development, as a community for 5,000 residents and 1,600 dwelling units. By 2001 it was virtually complete. The gross density is low at 14 persons/acre, but higher than the normal density of conventional American suburbs (Dutton 2000). Unlike the cul-de-sacs of normal suburbs or the garden city, Kentlands' streets are based on grids, which are interconnected and adapted to the gently rolling topography, with easy access to the primary schools and the shopping centre. Kentlands has a well organised street hierarchy of residential streets and alleys and boulevards which gather the traffic from the streets and connect to the regional motorways. The residential streets (50 foot right of ways) are narrower than most suburban streets of 70 feet.

One of the New Urbanism principles is the mixture of land uses and the requirement that the neighbourhood plan should contain a variety of housing types and land uses. The different housing types (single family, town houses, multi-family condominiums and multi-family flats) are co-mingled within the same blocks whereas other New Towns build whole blocks of the same type of housing, a process known as cookie-cutter housing. The co-mingling of housing types and the great variety of housing type and lot size are special successful features of New Urbanism. The proportion of single family houses in Kentlands is 31% and the variety of styles is the result of using several different builders in a small area.

One particular feature of the housing units is their tiny gardens or no gardens at all. The housing units are accessible from both the street and the alley, which alleys are unique with all the garages tucked away in the alleys out of sight. They serve as a kind of buffered play area and semi-public social space. Since there are hardly any private gardens, the children tend to play in the service alleys, often making the alley entrance more important than the street entrance.

Kentlands also has squares, like European cities, which are open to the streets. Retail and office facilities are correctly relegated to the edge of the neighbourhood but the shops and supermarkets are big warehouse boxes surrounded by unattractive parking lots. There is nothing to be learnt. The parks are located on an average of 400 ft away from the housing and thus within walking distance. The park system consists of 100 acres or 28% of the total land use and the open spaces vary in size. Greenways and the lake are towards the middle of the site.

6.3.2 Summary of New Urbanism Principles

(1) New Urbanism focuses on vernacular architecture- commonplace buildings of the past, embodying folk wisdom about design and construction, while at the same time giving the interiors light, openness and mechanical convenience expected in houses today. The design of the housing at Tornagrain is based on the vernacular style (see Fig. 6.6);

Fig. 6.6 Housing design at Tornagrain, Scotland, based on the vernacular style. *Source*: Planning Journal (2009)

(2) New Urbanism promotes neighbourliness and a friendly social atmosphere with detailed design features with an emphasis on front porches, picket fences, mews, and garages in the alleys and tight street elevations, all of which provide considerable social interaction;
(3) Although New Urbanism stipulates that neo-traditional designs reduce the number of vehicle trips and trip distances, it is actually the mixed arrangement of the land uses, the densities and the greater number of route choices that reduce the vehicular traffic;
(4) New Urbanism would like transit use. Although commuter rail stations exist in the Washington DC. area, they are not yet connected to Kentlands; and
(5) Financially one pays 12% more for a New Urbanism dwelling, as there are still some builders who think that mixed use is financially risky. However the quality is high that many people are prepared to pay more.

6.4 Summary of the Current Position on Eco-towns

To summarize the position of eco-towns we need to examine: (i) the eco-town and the planning process, and (ii) the criteria for eco-towns, as outlined below.

6.4.1 Eco-towns and the Planning Process

There are many who consider that the eco-town programme should be initiated through the statutory development plan system. This is the view put forward by the Campaign to Protect Rural England (CPRE), the Local Government Association, and naturally the Royal Town Planning Institute (RTPI). But the statutory development plan moves very slowly and it is thought that it might take 7–10 years to prepare the planning application.

The Town and Country Planning Association wishes the Government to shoulder the development risk by means of the existing 1981 New Town Development Act or on a joint venture basis by agreement with the landowners through the participation of an agency like the Homes and Community Agency (Lock 2008a). Using the 1981 Act would still require a full public inquiry in each case. There has also been a draft Planning Policy Statement (PPS) which proposes direct government action through part of the planning system.

6.4.2 The Criteria for the Eco-new town

The eco-town was the Labour government's initiative to deliver new affordable housing needed in England and to demonstrate how to deal with climate change. The only survey as to what people really think of eco-towns has been the YouGov Survey of 2008 which showed that 46% of people in England welcomed the eco-town idea and 34% would not mind seeing one close by where they live (Lock

2008c). The CPRE, RTPI and the Local Government Association are all against the idea. They see a suburban nightmare, car dependent housing estates built on green field sites against the opposition of local people. Building only for 5,000–10,000 people means it has to be car-based and will not be a walking community.

What does the eco-town provide? The main idea of the eco-town is to be a place of experimentation and innovation and to raise standards throughout England. The eco-town's main role therefore is a learning device – the leading edge of the Government's sustainable community's programme. According to Boardman (2007), eco-towns aim to:

(a) Exceed the standards of environmental performance achieved elsewhere in the United Kingdom;
(b) Place emphasis on reaching zero carbon development standards with energy use in housing to be "carbon- neutral";
(c) Provide good facilities and quality infrastructure and deliver new technology particularly in waste management, Combined Heat and Power, district heating, aquifer thermal energy etc;
(d) Provide "affordable" homes as the proponents argue that 50,000 homes is a decent proportion of the 3,000,000 homes required by 2020 with at least 3 in 10 of these should be of low rent;
(e) Provide a green structure in an interconnected network; with the green infrastructure factored into land values; and enhancement of the area's locally distinctive character and to provide multi-functional places, which help adapt the climate process; and
(f) Use brownfield land before green field land, which is not excluded.

In late July 2009, the Department of Communities and Local Government published the Eco-Towns Planning Policy Statement (PPS) as a supplement to Planning Policy Statement 1 and announced that there were to be four approved eco-towns (DCLG 2009) and pledged £60 million over 2 years. Of that sum, £36 million was given to the four eco-towns. After the election, to the surprise of everyone, the Coalition Government accepted the four eco-towns but halved the budget for 2010/11. Despite the 50% cut in eco-town funding, there is still enough start up funding for the projects to proceed. The numbers of homes for these four eco-towns are to be constructed as follows (Matthew 2009):

Eco-towns	Number of homes
Whitehill- Borden, Hampshire	5,500
St. Astell, Cornwall	5,000
Rackheath, Norfolk	6,000
North West Bicester, Oxfordshire	5,000
The total is hoped to be eventually at least	30,000

Specific aspects of these four towns include an emphasis on affordable housing (low rent); improvements to public transport; installing electric car charging points and electric bike charging points, community projects showcasing environmental technologies; developers using up to 30% less carbon than usual; specially designed eco-homes to make them more energy efficient with rainwater re-cycling, low flush toilets, high insulation levels, and environmentally friendly roofs. Development in small stages has begun in all four eco-towns.

The Planning Policy Statement also agreed that projects may be refused if they do not comply with Local Development Frameworks (LDFs). This is a victory for the two-thirds of the local councils which insisted that schemes must fit in with local development frameworks. Already one of these four originally most promising of the eco-towns, St. Austell in Cornwall, is under review. It comprises six eco-settlements achieved by creating villages or expanding existing ones with housing targets. However the location is now considered unsustainable and unsuitable in planning terms for the scale of the development proposed. If it were not for the eco-town initiative the planning system would never have proposed it (Planning Journal 2009).

The surviving four proposals all have the support of their local authorities. As a consequence of all the considerations, an eco-town proposal can now be rejected if it does not comply with the local development framework, which means that future plans must go through the plan-making process. Two of the proposed towns are town extensions and the other one is not on a single site. The Labour planning policy stated that the standards might be adopted by other developers as a way of meeting climate change policy and will ensure that the eco-towns will be "exemplars of good practice and provide a showcase for sustainable living".

6.5 Conclusion

Some of the proposals sound manipulative. One eco-town is to focus on "behaviour change techniques", where residents are to be rewarded by a personal carbon trading scheme if they use low amounts of energy. At other eco-towns, the focus will be on environmental technologies, "green collar" jobs and renewable energy, 40% green open space and high sustainability standards. Some of the standards useful for developing countries keen on the construction of eco-towns include the following (Morad and Plummer 2010):

(a) providing 30% affordable housing (housing for low-income people, particularly local people);
(b) requiring long-term investment into community owned housing rather than private housing which requires a profit;
(c) a zero carbon town which includes public buildings;
(d) providing 40% green open space;
(e) "green housing" using sustainable standards of insulation and thermal efficiency;

(f) providing low carbon homes; and
(g) giving priority to bus and cycling.

6.5.1 What Are the Pitfalls of Eco-towns?

The biggest pitfall is the inability to achieve agglomeration effects with provision of local jobs due to small community sizes; thus basic principle of building a "balanced community" cannot be fulfilled. The jobs are provided in existing towns or satellite business or science parks elsewhere. This makes commuting inevitable. Another pitfall is finding staff that will have the expertise on environmental impact assessment applicable to eco-towns to ensure that any negative environmental impact is timely detected. This means that local authorities will have to increase the extra skills required to deal with the scale and complexity of an eco-town in which exceptionally high standards and technical innovation will be essential.

In conclusion, the fate of the eco-towns remains in the hands of the political process. We would hope that the three eco-towns will survive to set an example to the rest of the country as a new way of life. Already North-East Essex could see eco-town principles applied to major developments after £200,000 was allocated to Haven Gateway (Planning Journal 2010). The money is to be used to conduct studies for using eco-town standards in development planning and to develop master plans. Hopefully the new communities could provide 8,000 homes and many local jobs in eco-towns. With new eco-town standards, the United Kingdom would be able to lead the world in this new way of life which combines affordable housing with green infrastructure.

References

Abercrombie, P. (1945). *Greater London Plan 1944*. London: HMSO.

Boardman, B. (2007). *Home truths: a low–carbon strategy to reduce UK housing emissions by 80% by 2050*. London: Friends of the Earth.

Congress for the New Urbanism & U S Department of Housing and Urban Development (1999). *Principles for inner city neighbourhood design*. San Francisco, CA: Congress for the New Urbanism.

Department of Communities and Local Government (DCLG) (2007, July). *Homes for the future: more affordable, more sustainable. Cmd.7191. Housing Green Paper*. London: HMSO.

Department of Communities and Local Government (DCLG) (2009). *Planning policy statement: eco-towns – a supplement to planning policy statement 1*. London: HMSO.

Duany, A. & Plater-Zyberk, E. (1991). Towns and Town-making Principles. In A. Krieger (Ed.). New York, NY: Rizzoli/Harvard University Graduate School of Design.

Dutton, J. (2000). *New American urbanism*. Milan: Skira.

Howard, E. (1899). *Tomorrow: a peaceful path to real reform*. London: Swan Sonnenschien.

Howard, E. (1902). *Garden cities of tomorrow*. London: Faber.

Leccese, M. & McCormick, K. (Eds.). (2000). *Charter of the new urbanism*. New York, NY: McGraw-Hill.

Lock, D. (2007). Eco-towns helping deliver a step change. *Journal of the Town and Country Planning Association*, 76(8): 238–240.

Lock, D. (2008a). Groans about more reforms. *Journal of the Town and Country Planning Association*, 77(5): 211–212.

Lock, D. (2008b). Eco-towns and planning processes. *Journal of the Town and Country Planning Association*, 77(6): 260–262.

Lock, D. (2008c). Eco-towns push on or push off. *Journal of the Town and Country Planning Association*, 77(10): 398–399.

Matthew, N. (Eds.). (2009). First ecotowns sites get the green light. *Journal of the Town and Country Planning Association*, 78(7/8): 295–296.

Morad, M. & Plummer, M. (2010). Surviving the economic crisis: can eco-towns aid economic development? *Local Economy*, 25: 208–219.

Morris, E. S. (1997). *British town planning and urban design, principles and policies*. Harlow: Longman.

Office of the Deputy Prime Minister (2003). *Sustainable communities: building for the future*. London: HMSO.

Planning Journal, Royal Town Planning Institute. 1st July, 2008, 20 February, 2008, 17 October 2008, 1 May 2009, 22 May 2009, 29 May 2009, 21 August 2009, 9 April 2010.

Shaw, R. (2007). Eco-towns and the next 60 years of planning. *Journal of the Town and Country Planning Association*, 76(8): 1–8, Tomorrow Series Paper.

Sunday Telegraph (2008, 30 March). Anti-eco-town protestors.

Sunday Telegraph (2009, 19 July). Communities against Ford eco-town.

Chapter 7
Eco-cities in China: Pearls in the Sea of Degrading Urban Environments?

Tai-Chee Wong

China's current development is ecologically unsustainable, and the damage will not be reversible once higher GDP has been achieved.

Zhenhua XIE, Minister of State Environmental Protection Agency, China (Arup 2007).

Abstract Economic reforms in China from the 1980s have created substantial material wealth and raised consumption to an unprecedented level. With rising affluence and demand for quality living, densely urbanized zones are increasingly being developed into eco-conscious townships or eco-cities. Whilst commercial entrepreneurship may have adopted norms of eco-city construction in selected sites including coastal areas, major cities and their rapidly extended metropolitan zones have encountered major pollution problems, threatening health and quality of life of ordinary residents. Will eco-cities serve as a normatic model for other Chinese cities to follow towards an improved urban environment? Or are they merely nodal points serving more commercial interests catering to the need of rising middle classes? This chapter investigates the hindrance and potential in developing an environmentally sustainable urban system in a country undergoing a late but rapid urbanization backed up by a huge surplus rural population eager to settle down in the cities. This is followed by analysis of public policy measures in energy saving, promotion of renewable energy, public transport, reforestation, recycling of water and other materials. Finally, the role of ecocities is studied in terms of whether they have the potential to lead a new development path towards a more sustainable urban future in China.

T.-C. Wong (✉)
National Institute of Education, Nanyang Technological University, Singapore
e-mail: taichee.wong@nie.edu.sg

T.-C. Wong, B. Yuen (eds.), *Eco-city Planning*, DOI 10.1007/978-94-007-0383-4_7,

7.1 Introduction

Over the last decade, building ecocities has become a highly fashionable *modus operandi* worldwide. It serves multiple purposes, of which the two most important are to counter the degrading urban environment and, in the process of building it, to create new business opportunities using clean technologies and conservationist measures.

By its most fundamental motivation, at least in theory, an eco-city offers to provide a sustainable lifestyle for both highly interdependent humans and non-human living things. An eco-city aims to provide conditions that enhance the sustainability and productivity of ecosystems, with a broad array of possible life pathways and a capacity to respond to environmental change and undesirable disturbances (see Newman and Jennings 2008: 97 and 102). In other words, such a capacity helps ecosystems to maintain nature's self-regulatory mechanism which could restore the living environment back to operational activities after a disturbance. Accordingly, Newman and Jennings (2008) argue that ecosystems, as long as their resilience and self-renewal ability are not destroyed, should be able to:

> maintain their structure and function under conditions of normal variability. In the face of external or internal disturbance, the structure of the ecosystems may change and functioning may be disrupted, the ecosystem will be able to restore functionality (ibid: 99).

Resilience is defined as:

> the capacity of a system to undergo disturbance and maintain its functions and controls, and may be measured by the magnitude of disturbance the system can tolerate and still persist (Wallington et al. 2005: 4, cited in Newman and Jennings 2008: 99).

Thus, an eco-city has the great potential of being deployed as a technical and pro-environmental instrument in dealing with ecological problems. More specifically, as it tackles urban-sourced environmental issues, its most useful target would be countries currently undergoing high rates of urbanization with haphazard environmental and pollution problems, a typical of which is China under urban reforms.

China is a large country with 1.3 billion people. With a fast growing economic influence and urban population, urban-industrial development over the last 30 years has produced substantial ecological impacts. Adverse effects are expected to spread from population centres to less developed lands in the near future if the deteriorating urban physical environmental conditions are not sufficiently and readily improved. A heavy price has been paid for great emphasis on GDP growth with an outcome of environmental degradation and health hazards. Reportedly in the mid-2000s, 70% of China's lakes and rivers were polluted (Cook 2007: 30).

As a consequence, examining Chinese cities as ecosystems is to look at how the whole urban habitat reacts to the pressing problems, to the ways energy and materials flow and the measures wastes are treated. Realizing the seriousness of the environmental degradation, eco-city development has been one of China's responses to the environmental crisis, with a focus centred at its high density city regions most vulnerable to this threat. Along China's coastal regions and the major river valleys, cities have grown much larger; some growing into forms of urban belts. As material consumption continues to increase rapidly, environmental problems are anticipated to intensify.

This chapter examines China's attempt to remedy the adverse consequences of urban development disassociated largely from the logics of the natural ecological system. In tackling this environmental crisis, eco-city development is concomitantly perceived to be an opportunity for business undertakings where foreign investment and expertise are welcome. The Tianjin Eco-city Joint-Venture between China and Singapore is exemplary of this commercial undertaking. This study will analyze the involvement of Singapore's government-linked companies in constructing an eco-city prototype in the coastal city of Tianjin. First, however, it is crucial to examine why it is an urgency that China needs to manage head-on its run-away urban pollution and environmental degradation.

7.2 Degrading Environments and Demographic Growth of Chinese Cities

Environmental degradation could be traced back to over the last 3,000 years of development history in China's relatively fragile physical environment in feeding a large agriculture-based population. Over this period, its intensive agricultural practice is best mapped by a Han Chinese expansion covering a vast fertile and not so fertile arable lands. This vast movement of population expansion over 20 dynasties went across the central plains in the north, Yangtze Valley in the middle, coastal zones in the east and south, steppes, grasslands in the far north and north-west, and mountains and jungles in the south-west and the west. According to Mark Elvin (2004: 5), this relatively long period of landscape transformation to suit the Chinese permanent and high-density agricultural habitat was characterized by:

> Cutting down most of the trees for clearance, buildings, and fuel, an ever-intensifying garden type of farming and arboriculture, water-control systems both large and small, commercialization, and cities and villages located as near the water's edge as possible.

Deforestation, as in other early civilizations, was necessary to accommodate an expanding population and their activities. Population size was perceived as an important source of collective and individual wealth, and removing mountains was seen as a highly regarded achievement in overcoming barriers imposed by nature. The resulting inherited degraded natural landscape of China which existed at the time of the 1949 Revolution forced new generations to face multiple challenges.

In Mao's China from the early 1950s to the late 1970s, the streets of major Chinese cities were a "world of bicycles". Immediately outside the city built-up area, farmlands dominated the rural landscape and retained many of their pre-1949 features. Anti-urban Marxist doctrine had restricted city size in favour of the core economic sectors of agricultural production and heavy industries. Mobility of peasants to cities was tightly controlled by the *hukou* system, characterized by its rigidity in the transfer of residence permit from a rural to an urban place. From the 1980s, a sharp turn in urbanization trends occurred when Deng Xiaoping championed reforms to transform the economic system in general and, as a result, the urban landscape in particular.

Post-Mao China since the 1980s has not only witnessed urban proliferation and demographic expansion but also a new phase of population relocation from the inner cities to the newly built high-rise apartments in the suburbs. The city centre itself has seen redevelopment to accommodate younger and better qualified couples. Bicycles, though still in large numbers, have given way to city trains and buses serving the large number of commuters. In parallel to this change, highways and other complimentary infrastructure have provided easier links to facilitate the rising mobility of the urban working population. Rising affluence and the concentration of middle classes in major cities have equally produced an increasingly large number of car-dependent commuters, rising consumption of consumer and non-consumer goods have generated high rates of pollutants. In an international assessment of city environment in 2004, China was ranked 100th of 118 countries taking part in the exercise as most polluted. Among the 20 worst polluted cities in the world, China owned 16 of them (Zhai 2009). A key source of pollution has come from the sharp rise in vehicles. From 1980 to 2008, the total number of vehicles rose 28.6 times against a national population growth of merely 34.5% (National Bureau of Statistics of China 2009).

7.2.1 Situation of the Degrading Urban Environment

China's present state of degraded urban environment should be attributed to a fast changing socialist state from Mao's frugal, largely self-reliant and lowly industrialized social organization to an urban-industrial driven economic base supported by a highly successful export-led manufacturing industry. The new scenario is characterized by a changing lifestyle towards an urban-based consumerism and a general lack of practical experience in dealing with complex sources of industrial and transport-related pollution.

Rates of urbanization in the post-1980s till today might be interpreted as a differentiated Chinese "great leap forward" in both physical and demographic scales. In 1980, only 19.4% of the nearly one billion Chinese population was classified as urban against a nearly 800 million peasants (see Table 7.1). By 2008, out of the total 1.328 billion people, the urban population had gone up to 45.7%. If the unregistered floating population of peasant origin who work as migrant workers in the cities are added, the urban proportion would have been even higher to over 50%.

Origins of the urban pollution sources were not attributable to migrant workers but to the expanding industries, and the economic activities and changing lifestyle made possible by rising affluence and consumption levels. Also, fast pace of industrialization and modernization has largely not been met with modern pollution control measures. Large numbers of low- to medium-cost industries were built in the fringe of major cities supported by China's ample supply of low-cost labour

Table 7.1 Urban and rural population change in China during 1980–2008

Year	Total population (1,000)	Urban population (1,000)	Proportion (%)	Rural population (1,000)	Proportion (%)
1980	987,050	191,400	19.4	795,650	80.6
1985	1,058,510	250,940	23.7	807,570	76.3
1990	1,143,330	301,950	26.4	841,380	73.6
1995	1,211,210	351,740	29.0	859,470	71.0
2000	1,267,430	459,060	36.2	808,370	63.8
2005	1,307,560	562,120	43.0	745,440	57.0
2008	1,328,020	606,670	45.7	721,350	54.3

Note: Data for the period 1990–2000 were adjusted using the 2000 National Population census, and the 2008 figure was estimated using the annual national sample surveys on population change
Source: National Bureau of Statistics of China (2009, table 3–1)

available from the rural sector. As Table 7.2 indicates, during the period 1980–2008 growth of civil vehicles was significant, rising from 1.78 million vehicles in 1980 to 16.1 million in 2000, and almost 51 million in 2008. As one can observe from the table, passenger cars saw an out-of-proportion rise from 2000 to 2008, increasing by 4.5 times in a short span of 8 years. Obviously, the rise is mostly in the major cities such as Beijing, Tianjin, Shanghai and Chongqing where the emerging middle and upper middle classes are highly concentrated. Recent trend in vehicular rise in Beijing shows a sharp climb of 31.2% from 2006 to 2008 alone. It is indeed in the major cities where air pollutants are most serious.

Table 7.3 shows that, of the 15 cities studied in terms of particulate matters emission in 2008, their air quality all exceeded the World Health Organization's standard,

Table 7.2 Growth of civil vehicles in China, 1980–2008

Year/city	Total number of vehicles (1,000)	Passenger vehicles (1,000)	Trucks (1,000)	Other vehicles (1,000)
1980	1,782.9	350.8	1,299.0	133.1
1985	3,211.2	794.5	2,232.0	184.7
1990	5,513.6	1,621.9	3,684.8	206.9
1995	10,400.0	4,179.0	5,854.3	366.7
2000	16,089.1	8,537.3	7,163.2	388.6
2005	31,596.6	21,324.6	9,555.6	716.6
2008	50,996.1	38,389.2	11,260.7	1,346.2
Beijing	3,136.8	2,910.2	181.3	45.3
Tianjin	1,084.7	917.1	146.8	20.8
Shanghai	1,321.2	1,107.3	213.9	–
Chongqing	736.4	466.6	254.7	15.1

Note: Figures of Beijing, Tianjin, Shanghai and Chongqing are for 2008
Source: Adjusted from National Bureau of Statistics of China (2009, table 15–26)

Table 7.3 Ambient air quality in major Chinese cities, 2008

City	Particulate matters (PM10)	Sulphur dioxide (SO_2)	Nitrogen dioxide (NO_2)	Days of air quality meeting grade II standards
Beijing	0.123	0.036	0.049	274
Tianjin	0.088	0.061	0.041	332
Taiyuan	0.094	0.073	0.021	303
Shenyang	0.118	0.059	0.037	323
Harbin	0.102	0.043	0.055	308
Shanghai	0.084	0.051	0.056	328
Nanjing	0.098	0.054	0.053	322
Hangzhou	0.110	0.052	0.053	301
Fuzhou	0.071	0.023	0.046	354
Wuhan	0.113	0.051	0.054	294
Guangzhou	0.071	0.046	0.056	345
Chongqing	0.106	0.063	0.043	297
Chengdu	0.111	0.049	0.052	319
Kunming	0.067	0.051	0.039	366
Xi'an	0.113	0.050	0.044	301
WHO Standard[a]	0.020	0.020	0.040	[b]

Source: National Bureau of Statistics of China (2009, tables 11–17, 11–18, 11–19 and 11.24)

Note: All measurements in milligram/cubic metres

[a]World Health Organization 2006. http://libdoc.who.int/hq/2006/WHO_SDE_PHE_OEH_06.02_chi.pdf, accessed May 27, 2009

[b]Using the Air Pollution Index (API) classified as Grade II (50–100), these are the number of days a year where air quality is good enough to allow normal outdoor activities

by 4.2 times (Shanghai) to as high as 6.2 times (Beijing). The emission levels of sulphur dioxide and nitrogen dioxide were not as bad as particulate matters but would still have exceeded WHO's standard up to 3.7 times (Taiyuan) (National Bureau of Statistics of China 2009). As to the number of days per year where air quality was good enough for outdoor activities, Beijing had the lowest of 274 days as against Fuzhou, a coastal city, that enjoyed a high 354 days in 2008.

Solid wastes generated in the large cities are on the rise. Highly polluting heavy industries that used to be located in strategic and key centres such as Chongqing and Kunming in the country's southwestern region are known to have discharged cumulatively huge solid wastes. The percentage of industrial waste treated has improved over the years, yet there is still much room for improvement (Table 7.4). Water is a scarce resource for much of China, notably in the north, northwest and far west of China. Rising consumption since the reforms has seen the decline in storage in both surface and ground levels (Table 7.5). Total water consumption from 2002 to 2008 alone increased by 7.5%. However, living consumption of water to meet daily household needs during the same period rose by 17.9% nation-wide. Much of the waste

Table 7.4 Emission and treatment of industrial solid wastes in major Chinese cities, 2008 (in 10,000 tons)

City	Industrial solid wastes generated	Hazardous wastes	Industrial solid wastes treated	Percentage of industrial solid wastes treated
Beijing	1,157	11.53	835	66.4
Tianjin	1,479	14.79	1,471	98.2
Taiyuan	2,532	3.08	1,202	47.4
Shenyang	479	8.09	461	92.3
Harbin	1,150	1.61	860	74.8
Shanghai	2,347	49.28	2,242	95.5
Nanjing	1,383	18.91	1,282	92.4
Hangzhou	585	7.98	557	95.1
Jinan	1,076	8.87	1,028	94.4
Zhengzhou	1,077	0.24	841	78.1
Wuhan	1,094	1.34	1,007	89.6
Guangzhou	662	16.38	606	91.2
Chongqing	2,311	8.08	1,851	79.1
Chengdu	725	0.79	713	98.3
Kunming	1,989	1.29	790	39.7
Lanzhou	372	17.36	291	78.1

Source: National Bureau of Statistics of China (2009, tables 11–30)

water discharge produced health hazards, turning some rivers into stint waterbodies (Wu et al. 1999, Yangcheng Evening News 2008). The acuteness of environmental harmony has been placed on par with social harmony needed for priority treatment and contemplation (Woo 2007).

7.3 Eco-cities as a Solution to Degrading Environment?

In the face of polluting cities, strengthening environmental governance has been prioritized on the Chinese national agenda for action. In retrospect, economic transformations and growing openness with tightened integration with the market-led advanced capitalist economies have inevitably forced China to change its conventional centrally planned economic style practised during the period 1949–1979. China started to see the urgency to change its *laissez-faire* approach of environmental management which was inefficient and ineffective during this period characterized by low levels of industrialization and pollution.

In 1979, the state Environmental Protection Law was promulgated in China, and in 1984, environmental protection was conceived as a fundamental policy to regulate polluting activities (see Mol and Carter 2007). Subsequently, a series of executive regulations, standards and measures were adopted at four-tier levels (national, provincial, municipal and county) in the 1980s and 1990s after serious references to those of developed economies. Indeed, atmospheric pollution levels would have

Table 7.5 Water and atmospheric conditions in China, 2002–2008

Items	2002	2003	2004	2005	2006	2007	2008
Water available (100 million cu. m)							
Surface water resource	27,243.3	26,250.7	23,126.4	26,982.4	24,358.1	24,242.5	26,377.0
Groundwater resource	8,697.2	8,299.3	7,436.3	8,091.1	7,642.9	7,617.2	8,122.0
Per capital water resource (cu. m)	2,207.2	2,131.3	1,856.3	2,151.8	1,932.1	1,916.3	2,071.1
Water consumption (100 million cu. m)							
Total consumption	5,497.3	5,320.4	5,547.8	5,633.0	5,795.0	5,818.7	5,910.0
Living consumption	618.7	630.9	651.2	675.1	693.8	710.4	729.3
Atmosphere (100 million cu. m)							
Industrial waste air emission	175,257	198,906	237,696	268,988	330,992	–	403,866
Fuel burning	103,776	116,447	139,726	155,238	181,636	–	229,535
Sulphur dioxide emission (1000 tons)	19,270	21,590	22,550	25,490	25,890	–	22,864
Soot emission (1000 tons)	10,130	10,490	10,950	11,830	10,890	–	9,016

Note: All measurements in 100 million cubic metres unless indicated otherwise
Source: Adjusted from National Bureau of Statistics of China (2009, tables 11–17, 11–18, 11–19, 11–24)

been even more spectacular had there not been measures to control the emission of greenhouse gases.

Over the last two decades, actions have comprised greater commitments to international environmental treaties, publicity and education efforts to enhance environmental awareness, more efficient resource use, adoption of newer environment-friendly technologies, cleaner products and closing of heavily polluting factories (Mol and Carter 2007: 2).

Indeed, the Chinese government has taken serious initiatives to develop eco-cities at national, provincial and local levels to counter the adverse effects of environmental degradation. At the central State Council level, the State Environmental Production Agency of China (SEPA) issued in 2003 "The Constructing Indices of Eco-county, Ecocity and Eco-province" which became the general national standard of ecological assessment. Because Chinese cities usually cover within their administrative boundary urban built-up, farming and nature areas, massive rural re-afforestation is often adopted to return the cities to a more natural state to which municipal governments claim that helps them in their efforts to be an eco-city.

During the past 25 years, 390 national demonstration ecopolis have been appraised and named by the Ministry of Environment including prefecture and county level cities such as Yangzhou, Shaoxin, Panjing, Yancheng, Hangzhou, Xuzhou, Guangzhou, Changsha, Haining, Anji, Changsu, Zhangjiagang, Kunshan, Longgang district of Shenzhen, Rizhao and Dujiangyan. Among the many cities being assessed, 32 have passed "environmental model city" appraisal. Furthmore, 13 provinces initiated eco-province development (Hainan, Jilin, Heilongjiang, Fujian, Zhejiang, Shandong, Anhui, Jiangsu, Hebei, Guangxi, Sichuan, Tianjin and Liaoning). 108 experimental cities/counties towards sustainable development covering 29 provinces of China, had been appraised and named by the Ministry of Science and Technology. Big progress had been made in these case studies while some lessons and challenges also emerged such as institutional barrier, behavioural bottleneck and technical malnutrition (Wang et al. 2004, Wang and Xu 2004, Yip 2008).

Despite successes in some aspects, environmental governance has still much to be desired due to the scale of industrial development across the vast country, difficulties in modernizing old and outdated manufacturing plants for fear of job losses, and the creation of new factories at different technological levels. Added to these are a legacy of an older industrial workforce and a fluid social and political environment in the transitional period in which enforcement is a thorny issue. As a strategy at the national level, creating a model city, for the Chinese leaders, is seen as a more workable option considering its potential of demonstration effects.

Eco-cities represent thus a symbolic hope to solving urban degradation problems. According to Rodney White (2002: 3), an eco-city can be defined as "a city that provides an acceptable standard of living for its human occupants without depleting the ecosystems and biogeochemical cycles on which it depends". In energy consumption, the eco-city concept supports a human habitat that, as far as possible, uses non-fossil fuels, and energy saving means to achieve a low aggregate consumption. As a developing country having a relatively low-technology base in pollution control, China has attempted to build up its own standards to guide eco-city planning and development. Adjusting "The Constructing Indices of Eco-county, Eco-city and Eco-province" formulated by "The State Environmental Production Agency of China" in 2003, and using a more sophisticated index classification method covering common characteristics and feature indices,[1] Li Shengsheng and his research partners have worked out a set of criteria that have recognized the different problems faced by different cities. For example, the number of days where air quality is equal to or better than the level 2 standard set by the United Nations fit for outdoor activities is set differently between regions. Li and partners have set 330 days as the minimum acceptable standard for south China but 280 days for the drier north China closer to the arid Inner Mongolia producing often thunder storms sweeping southward during winter. Similarly, disposable income levels would determine the consumption pattern and total personal expenses that would have contributed to total wastes in the cities being compared. They have recommended different limits of personal consumption in monetary terms as an indicator for urban environmental control (Li et al. 2010).

Overall, the low aggregate consumption per capita is translatable into a small and acceptable ecological footprint. As cities are getting larger, and an increasingly large population lives in the cities, actions towards cutting down aggregate consumption have to be concentrated at the local level (the cities). Large cities as nodal points are where consumption of materials and energy is very high on per capita basis. Poor environmental management is bound to lead to a degrading urban environment harmful to different habitats in the urban ecological system including definitely humans. The fundamental concept of eco-cities is to incorporate functions of nature in a miniature manner to serve the interests of human developments. This could be done through "green design" of buildings, infrastructure and integration of nature areas and waterbodies into the urban setting. The resulting lifestyles to be encouraged for citizens to follow would depend essentially on the exploitation of the natural processes (solar radiation, water flows, wind) to achieve desired urban comfort levels, rather than using fossil fuel derived power for heating, lighting and cooling (see Roberts et al. 2009, White 2002).

Taking cities as an ecological system and applying an ecological approach, there is potential that functioning mechanism should fit into a characteristically sustainable urban environment. The development approach is to treat cities ideally as a habitat for animals and plants, and to use as much as possible biological and natural means or resources for the needs of such habitat (Deelstra 1988, Pickett et al. 2001, Mitchell 2004, Hultman 1993, Wheeler and Beatley 2009).

In practice, the conceived approach to achieving eco-city ideals needs to proactively counter industrialist and consumerist lifestyle characterized by, for example, a high intensity of automobile invasion by promoting green energy-based public transit and other environmentally friendly buildings and infrastructure. The aimed outcome by its very nature is to minimize environmental degradation, and to achieve a minimally acceptable quality of life in a sustainable way. This study will focus on eco-cities and their role in transforming the quality of life of residents threatened by a degrading physical environment in a rapidly urbanizing and metropolizing contemporary world (Tibbetts 2002). Ecocities recently developed in China are taken as a case study.

7.4 Eco-city Development in China

Apparently, eco-city development is a brand-new concept in post-Mao China arising from rapid pace of urban proliferation characterized by serious pollution problems. The perception towards eco-city urbanism has inequitably attracted multiple interests, interpretations as well as enquiries from different social spectrums in China.

Critics such as Zhai Ruiming (2009), commented that eco-city projects had attracted at least 100 Chinese cities to bid for fund allocations, and for some interest groups, their primary objective was to use the concept as a pretext to secure land approval in the face of the tightened land control policies. Using Chinese classical and philosophical interpretation, eco-city development is an approach to bring about an integration of "heaven" and "Earth" with the help of high technology and applied ecological principles to achieve an artificial but harmonious urban living environment. Such harmony is achievable via regulating the cyclic mechanism of the ecosystem to meet the standards required of sustainable urban development. Attention is turned to two exemplary eco-city projects being implemented in China.

7.4.1 Dongtan Eco-city

Dongtan covers 8,400 ha and is a small Chongming Island north of Shanghai in the course of Yangtze River. The initiative came in 2005 when the Shanghai Municipal Government instructed its subsidiary "the Shanghai Industrial Investment Corporation" (SIIC) to invite the British consultancy firm, Arup, to design an energy-efficient eco-city. This model city, designed for 500,000 people, would use exclusively sustainable energy and save energy consumption by two-thirds compared to Shanghai. On the technological basis of sustainable development, Dongtan would be designed with the following features (Arup 2007):

- Solar panels, wind turbines and biomass-based fuels to generate energy;
- Buildings to have photovoltaic cell arrays on the roofs. The roofs will have gardens or other greenery to provide insulation and filter rainwater to help reduce energy consumption;
- Design will encourage use of public transport, cycling and walking within a compact city form: 75 dwellings per hectare, with a mixed low-rise and high-density of 3–6 storeys (about 1.2 average plot ratio);
- Distribution of gross floor area: 55% residential, 24% commercial, retail and light industrial, 16% culture, tourism, leisure and hotel, and 5% education and social infrastructure;
- Dongtan's refuse is to be recycled up to 80% (inclusive of organic waste such as rice husks which would be transformed to produce electricity and heat);

- Natural ventilation will be capitalized with adaptation to local microclimatic conditions; and
- Ultimately, the city should achieve an ecological footprint of 2.2 ha per person close to the standard of 1.9 ha per person set by the World Wide Fund for Nature (WWF), but only one-third of the current Shanghai city.

Scheduled to complete the first phase delivery by 2010, Dongtan's implementation nevertheless has been delayed. Critics have been suspicious of Dongtan's impact on the overall Chinese city system accommodating the majority of the urban population who are suffering from the polluting living environment. Some have even described it as a "Potemkin village" (a model unrepresentative of the urban development)!

7.4.2 Tianjin Eco-city Project

The project marks a landmark attempt that China has aimed to build an ecologically sustainable city in northern China known for its aridness in the face of frequent sandstorms from the Mongolian Plateau and Gobi Desert to its northwest.[2] Reportedly, a minimum of 30 billion yuan will be injected into this eco-friendly project situated on a 30 km^2 of coastal marshland, 150 km south-east of Beijing and 40 km from Tianjin. It is a new joint-venture between China and Singapore to build a prototype of "ecological civilization" targeted to achieve "energy-saving, mitigation of pollution and pleasant urban living".

Project management will be undertaken by the Sino-Singapore Tianjin Eco-City Investment and Development Company on a 50–50 basis, represented respectively by a Chinese consortium led by Tianjin TEDA Investment Holding Company and a Singapore group led by the Keppel Group (Quek 2008a, b, People's Net 2009). After the ground breaking ceremony held on 28 September 2008, the eco-city has taken off to construct its Phase 1 covering 4 km^2, and by 2020, it should accommodate 350,000 residents. Like Dongtan, the Tianjin Eco-city will use clean energy, public transport, waste recycling and large tracts of greenery to provide a socially harmonious living style (see Table 7.6).

Similarly, the conceptual framework deployed in Singapore in the early 2000s (integrating work, live and play) has been merged with the Chinese emphasis on harmonious and sustainable development as the planning rationale (Fig. 7.1). By projection, the eco-city will contribute towards the expansion of the Tianjin Municipal Region to become one of the four coastal megacities reaching beyond 10 million in China (see Tibbetts 2002). This gigantic joint project obviously carries certain significance for both Singapore and China.

7.4.3 Significance for Singapore

For Singapore, it matches its objective of regionalization drive and global integration in exporting management expertise and services driven as an economic motivating

Table 7.6 Characteristics of the proposed China-Singapore Tianjin Eco-city project

Item	Characteristics
Total planned area	30 km^2; Phase 1: 3 km^2
Population target	• 2010: 50,000 • 2015: 200,000 • 2020: 350,000
Implementation plan	• Phase 1 to start in June 2008; completed in 3 years • Whole project to be completed in 10–15 years
Targeted indicators	Control targets ensure: • Ecological & environmental health • Socially harmonious living & community growth • Recycling of economically valuable items (total 18) Guidance targets include regional coordinated use in: a. Clean energy b. Public transport capacity c. Water supply & drinking water systems d. Waste recycling e. Urban greenery f. Urban road system g. Community management system h. Culture, education and health research environment
Economic structure	Real estate, business, leisure & recreation, educational training, research & development, cultural innovative development, services outsourcing, modern service & high-end services
Mode of transport	Public transport-oriented concept focused on light rail system, supplemented by bus system, bicycle lanes & pedestrian walkways

Source: Compiled from various sources on Tianjin eco-city websites

Fig. 7.1 The planning concept of Sino-Singapore Tianjin eco-city. *Source*: Keppel Corporation (2008)

force since the early 1990s. It is also a platform to practise "green and sustainable city" ventures in a large scale outside Singapore; such experience acquired would help the city-state to further its overseas businesses. Singapore agencies such as the Housing and Development Board, and the Building and Construction Authority have been working on affordable "green housing" for marketing with their Chinese counterparts (Oon 2009). Further, the project is symbolic of another grand urban planning and business joint venture after the "Suzhou Industrial Park" initiated in the mid-1990s. As always, the eco-city has turned out to be another revenue-generating opportunity.

7.4.4 Significance for China

China has taken the Tianjin project as an experiment with potential to improve urban liveability in a relatively fragile physical environment, especially in its north and western China. The Chinese government's emphasis on harmonious development means not only an important factor in socio-economic development which has seen today the need to narrow widening gaps between rich and poor, but also a common desire to achieve a sustainable physical environment in a rapidly urbanizing state. The venture has served as a lesson to growth-driven enterprises that eco-friendly, energy saving, and for the general public enhancement of civil awareness towards environmental protection, conservation are equally important as part of the development process. Clean environment with economic growth have become a new mandate of governance.

7.5 Discussion and Analysis

In Bossel's systems model of sustainability, as described in Newman and Jennings (2008: Chapter 5), it is highlighted that sustainable ecosystems have to be conditioned by healthy living, zero waste, self-regulating, resilience and self-renewing and flexibility. In meeting energy needs, Bossel's systems model stresses that this is accomplished through green plants, as autotrophs, acting as solar energy collectors. Sunlight is converted to plant biomass. Within an urban setting, eco-city environment facilitates autotrophic system to function in an extent where plants and animals can receive nutrients to live, grow and reproduce. Nutrients include carbon, oxygen, water, nitrogen, phosphorus and sulphur etc taken from the atmosphere itself, waterbodies, rocks and soils. Given that the biosphere is a closed system, nutrients are in fixed supply. Through the respiration processes, organisms produce wastes, and nutrients are cycled continuously between living organisms and air, water and soil in the form of biogeochemical cycles. The cycles produce nutrients and process wastes (see Newman and Jennings 2008: 95–112)

However, all ecosystems are situated within wider ecosystems which are interdependent up to the biosphere scale and covering vast areas or bioregions (ibid: 95).

Undisputably, the two Chinese eco-city projects discussed above have demonstrated a close matching in objectives and action plans but they are merely two nodal points in the sea of a large currently degraded urban environment. In terms of ecosystem coverage, they will have negligible or little impact as the central source of influence. On the contrary, they are vulnerable to adverse effects from the surrounding regions as pollutants do not recognize frontiers, whether national or international. Consequently, the road map for a better solution rests with the spacious urbanized hinterlands. In light of the global warming effects and the signal of melting icebergs in the polar zones, an eco-city's ability should include the adaptation to climate change. Coastal cities, in particular, have to be ready for changing sea-levels. Cities especially those in the tropical zones may experience an intensification of the urban heat island. It is timely now to examine how the green infrastructure is being developed and how the energy consumption of individual buildings is being reduced to mitigate such problems.

Building an eco-city is to build a human habitat towards a sustainable society. However, it is not merely about protecting and enhancing the physical environment. One has to look beyond the environmental aspect to include social and economic aspects of sustainability. In meeting social needs, the eco-city community needs to consolidate the following aspects:

(a) Making the eco-city settlement a "human" scale and form;
(b) Valuing and protecting diversity and local distinctiveness with local cultural identity;
(c) Protecting human health and amenity through safe, clean and pleasant environments;
(d) Ensuring access to quality food, water, housing and fuel at reasonable cost;
(e) Maximizing residents' access to skills and knowledge needed to participate an active social role; and
(f) Empowering the whole resident community to take part in communal decision-making activities including their workplace (see White 2002: 202).

Going beyond social sustainability, the eco-city would have to consider the promotion of economic viability in making the local economy vibrant without damaging the local and regional environment. The Eco-city idea nevertheless has been used hitherto by investors as a business venture at a commercially substantial scale involving a large and varied scope of economic activities whose merits and limitations are now discussed.

7.5.1 Eco-city: A New Form of Business Economics

Using latest ecological concepts, norms and technology to provide an artificial natural state of urban living environment (eco-city) is symbolic of a new knowledge economic sector, known as eco-nomics. Eco-nomics is also an advanced state of capitalism in which green and clean technology is deployed to generate high returns to capital investment, serving at the same time the heatedly pursued environmentally sustainable objective.

The 1992 United Nations Conference on Environment and Development at Rio de Janeiro had affirmed that economic growth and environmental protection were compatible and that resources allocated to counter environmental degradation were justifiable by economic gains (UN 1992). In an international environment characterized by global competition, trade and heated pursuit to sustain high standards of consumption favouring economic growth, a world organization such as the UN had to opt for a material-based developmental stance, at odds with environmentalists who held different views (Clark 1995).

Well integrated into the globalized economic system, Singapore's interest in building eco-cities has become an international business venture. Singapore's expertise in water technology and energy has attracted collaboration from the United Arab Emirates to develop its Masdar City of 6.5 km^2, known as the Masdar Initiative which has an estimated US$22 billion ready for a comprehensive and ambitious undertaking. In developing and commercializing renewable energy technologies, this project has initiated a plan to boast a zero waste and zero carbon footprint. The city has been planned since 2006 by a British consulting firm "Forster and Partners" which designs to use 100% of renewable energies and house 1,500 businesses and 50,000 residents. Irrigation of vegetation and green areas will be solar-powered desalination plant and recycled water (Cheam 2009). Nevertheless, are eco-cities run as profit-oriented ventures free of weaknesses?

7.5.2 Weaknesses of Eco-nomics

Eco-nomics run as private undertakings in particular has limitations in achieving ideal sustainability. It tends to quantify costs and benefits accountable to stakeholders, corporate profits and competitive survival. By its very nature, it fails to see the full intrinsic values of living and non-living things and their interdependency on the Earth. Environmental sustainability, like esthetics, is not a yardstick of financial measurement, and is extremely difficult to quantify, especially in the long-term. Consequently, the operational basis of eco-nomics sees more clearly the profits than the costs that involve destruction of the environment. In our business-led contemplation of environmental protective measures, we may not take the best option of choices in decision-making.

In addressing the consequences of eco-nomics, Al Gore (2006:183–85) rightly points out that capitalistic undertakings do not predict nor are they willing to compute the environmental cost of development. From the same token, international organizations, such as World Bank, International Monetary Fund, regional development banks have unavoidably required recipients to justify their loans and monetary assistance based on economic performance. How in the development process the recipients' environment is destroyed is either beyond the control of the lending authorities or of secondary concern. The business functioning of eco-nomics is unlikely to be successful, at least not in the short- and medium-term in "disempower[ing] the giant corporations immediately, just by not buying their products" (Register 2006: 221). It is difficult for us to imagine shrinking back:

> from the sprawled giants of today with their contradictory internal functions, becoming complex, integrally tuned three-dimensional structure, should produce complexities linked to one another so efficiently as to produce enormous prosperity relative to resources consumed. We may discover that the kind of prosperity of opportunity that enriches life the most is a prosperity of opportunity for untold enjoyment of time, creativity and nature (ibid).

Whether China's market-led eco-city development will meet the above cited challenge by according more priority to ecological benefits will remain to be seen. Although Dongtan's master plan is designed by Arup with ideal sustainability guidelines, critics have questioned the choice of using the Chongming Island, one of China's largest bird reserves for the eco-city project (DAC 2009).

7.6 Conclusion

Within the Earth's own operating system, national boundaries are no barrier to external encroachment of pollutants. Indeed, the scales and impact of ecosystems are so broad in range that they stretch from a local environment to that of the global. Within the complex networks of the global ecosystem, due to its dynamic interactions, a regional or national ecosystem cannot be studied in isolation from the other. Climate change is a case in point. A warmer world is seen as a crisis and a real threat to the common survival of living species (Newman and Jennings 2008: 92–93, White 2002). Much of this crisis associated with global carbon, hydrological and water cycles has a cause-effect with anthropogenic and human activities. Cities must be made part of the natural system, and fully integrated in the ecosystem.

Eco-cities are a response to the contemporary environmental and resource crisis arising mainly from human activities, and climate change. Cities where human groups are in their densest form with most acute problems are where remedial actions are urgently needed. Typically, catching up economically from behind, China's urban growth has witnessed an unprecedented pace accompanied by heavy pollution and environmental degradation since the 1980s. Three decades of Dengism has moved China from leftism to economic development without major turmoil, quadrupling the living standard and laying the foundation for ongoing systemic reforms. Dengism comprises pragmatism and gradualism in favour of material progress via a market-led economy with tight top-down administrative controls to ensure a peaceful transition. Accordingly, the formula that national leaders have to rely on economic growth to protect themselves against recession and inflation has to be compromised with less market-driven economic forces (see Clark 1995: 231).

This Chinese socio-economic transition with little political reforms with a strong emphasis on exponential growth pursuit is being questioned for its lack in compatibility with environmental equilibrium. In technical terms, moreover, eco-city development being used as a business venture may create more wealth and capacity and technological resources to deal with the polluted environment and, in the process, generate more business opportunities in managing the degraded environment. But is this "nodal point effect" a viable treatment towards environmental sustainability and a more lasting ecological health in a vast and populous nation undergoing rapid and seemingly uncontrollable urban sprawl? Or, are eco-cities strategically used to produce a demonstration effect?

In dealing with its specific environmental pollution and degradation, China from a relatively low technological base and heavily GDP-led in approach, has interpreted the eco-city concept somewhat differently from the Western ideas of Newman, Register and White cited earlier in the chapter. In particular, as a late starter, China has to put in practical efforts to change many existing cities into an "eco-city". So long as some basic services are provided to enhance the ecological quality of a working environment, including mining towns, one can call it an eco-town. The peri-urban area of coal mining city of Huaibei in East China where measures of eco-service enhancement and ecosystem restoration have been experimented is one such eco-town. Over the last 50 years, this coal producing area has done much damage to the local surrounding farmland, and caused a high level of pollution and vast patches of subsiding terrain. In early 2009, an action plan was conceived aimed at restoring the original wetland conditions and preparing the eco-town towards a low carbon economy (see Wang et al. 2009). Another case in point is the effort dedicated to building Caofeidian into an eco-city in north China near Tangshan. Handicapped by lack of rainfall and fresh water and threatened by salt water, Caofeidian is to be constructed into an eco-city which is environmentally friendly, conservation-oriented, a high-tech oriented, yet a compact city meeting local norm of high-density living. Besides using wind energy as a key source of power supply, intensive water recycling will be heavily relied upon here, just like other areas in China where water is a scarce resource (Ma 2009, van Dijk 2010).

Having said these, if it is inevitable and necessary to build eco-cities as business operations and as a profit-led strategy, pursuing ecological sustainability objectives has to be more anthropocentric than considering the holistic intrinsic values of nature in the process. To be a green and vital economy may not necessarily rely on the logics of maximum profits and efficiency. Achieving a "compassionate relationship between society and nature", as Richard Register (2006: 214) would have desired, will need, in my view, a multi-pronged approach integrating ecological, social, economic, environmental and technological means. The question lies with where to strike a balance, and the extent set for priority for each. Despite their inherent weaknesses, the Dongtan and Sino-Singapore Tianjin Eco-city Projects are a step forward in building a more livable urban environment. With strong commitments from policy makers to implementers, and positive response from the corporate world and citizens, their demonstration effects will hopefully help create a strong inherent ecological consciousness and psychological base from which a wider spread of eco-friendly and environmental sustainable urban environment will be built.

Acknowledgements I wish to thank Professors Ian Douglas and Pierre Laconte for their invaluable comments on an earlier draft based on which improvements were made. All errors, if any, are mine.

Notes

1. Common indices are those indicators considered to be relevant and suitable for all cities; characteristic indices consider the difficulties of success if certain criteria are used in some cities; and specific indices reflect the unique features of cities (Li et al. 2010).
2. There were four cities initially being considered: Tianjin, Caofeidian Industrial Park north of Tangshan, Baotou (Inner Mongolia and Urumqi (Xinjiang Province). Tianjin was seen as having the greatest economic potential that Singapore emphasized. For China, coastal Tianjin could be designed as a new growth engine in north China (Quek 2007: 2).

References

Arup (2007). Dongtan eco-city, Shanghai. http://www.arup.com/_assets/_download/8CFDEE1A-CC3E-EA1A-25FD80B2315B50FD.pdf. Accessed 20 January 2010.

Cheam, J. (2009). Singapore role in Emirates eco-city. *The Straits Times*, Singapore, January 20, p. B16.

Clark, J. G. (1995). Economic development versus sustainable societies: reflections on the players in a crucial context. *Annual Review of Ecology and Systematics*, 26: 225–248.

Cook, I. (2007). Environmental, health and sustainability in twenty-first century China. In R. Sanders & Y. Chen (Eds.), *China's post-reform economy: achieving harmony, sustainable growth* (pp. 30–43). London: Routledge.

Danish Architecture Centre (DAC) (2009). Sustainable cities – Dongtan: the world's first large-scale eco-city? http://sustainablecities.dk/en/city-projects/cases/dongtan-the-world-s-first-large-scale-eco-city. Accessed 30 May 2010.

Deelstra, T. (1988). An ecological approach to planning. *Town and Country Planning*, 57(4): 105–107.

Elvin, M. (2004). *The retreat of the elephants: an environmental history of China*. New Haven, CT: Yale University Press.

Gore, A. (2006). *Earth in the balance: ecology and the human spirit*. New York, NY: Rodale.

Hultman, J. (1993). Approaches and methods in urban ecology. *Geografiska Annaler*, 75B(1): 41–49.

Keppel Corporation (2008). Three harmonies of the Sino-Singapore Tianjin Eco-city. http://images.google.com.sg/imgres?imgurl. Accessed 26 May 2009.

Li, S.-S., Zhang, Y., Li, Y.-T. & Yang, N.-J. (2010). Research on the eco-city index system based on the city classification. Bioinformatics and biomedical engineering (ICBEE) 2010 4th international conference (pp. 1–4). Chengdu.

Ma, Q. (2009). Eco-city and eco-planning in China: taking an example for Caofeidian eco-city. Proceedings of the 4th international conference forum on urbanism, Amsterdam/Delft, 511–520.

Mitchell, G. (2004). Forecasting urban futures: a systems analytical perspective on the development of sustainable urban regions. In M. Purvis & A. Grainer (Eds.), *Exploring sustainable development: geographical perspectives* (pp. 109–127). London: Earthscan.

Mol, A. P. J. & Carter, N. T. (2007). China's environmental governance in transition. In N. T. Carter & A. P. J. Mol (Eds.), *Environmental governance in China* (pp. 1–22). London: Routledge.

National Bureau of Statistics of China (2007). *China statistical yearbook*. Beijing: China Statistics Press.

National Bureau of Statistics of China (2009). *China statistical yearbook*. Beijing: China Statistics Press.
Newman, P. & Jennings, I. (2008). *Cities as sustainable ecosystems: principles and practice*. Washington, DC: Island Press.
Oon, C. (2009). Mr Eco-city goes the extra mile. *The Straits Times*, January 9, 2009, p. A28.
People's Net (2009). Tianjin Eco-city in the new step forward in the development of green GDP as the main driving force, http://news.022china.com/2009/09-28/158103_0.html. Accessed 20 January 2010.
Pickett, S. T. A., Cadenasso, M. L., Grove, J. M., Nilon, C. H., Pouyat, R. V., Zipperer, W. C. & Costanza, R. (2001). Urban ecological systems: linking terrestrial, ecological, physical, and socioeconomic components of metropolitan areas. *Annual Review of Ecology and Systematics*, 32: 127–158.
Quek, T. (2007). How Tianjin won fight to be an eco-city. *The Straits Times*, November 22, pp. 1–2.
Quek, T. (2008a). Singapore's eco-city can expect keen contest. *The Straits Times*, February 27, p. 9.
Quek, T. (2008b). SM Goh outlines vision for eco-city. *The Straits Times*, September 26, p. A19.
Register, R. (2006). *Ecocities: rebuilding cities in balance with nature*. Gabriola Island, British Columbia: New Society Publishers.
Roberts, P., Ravetz, J. & George, C. (2009). *Environment and the city*. London: Routledge.
Tibetts, J. (2002). Coastal cities living on the edge. *Environmental Health Perspectives*, 110(11): A674–A681.
United Nations (UN) (1992). *UN convention on environment and development*. New York, NY: Agenda 21: The United Nations Programme of Action from Rio.
van Dijk, M. P. (2010). Ecological cities in China, what are we heading for, just more ecological urban water systems? Unpublished paper. http://www.switchurbanwater.eu/outputs/pdfs/PAP_Ecological_cities_in_China.pdf. Accessed 21 September 2010.
Wallington, T., Hobbes, R. & Moore, S. (2005). Implications of current ecological thinking for biodiversity conservation: a review of the salient issues. *Ecology and Society*, 10(1): 15.
Wang, R.-S., Li, F., Yang, W. & Zhang, X.-F. (2009). Eco-service enhaancement in peri-urban area of coal mining city of Huaibei in East China. *Acta Ecologica Sinicia*, 29: 1–6.
Wang, R.-S., Lin, S.-K. & Ouyang, Z.-Y. (2004). *The theory and practice of Hainan eco-province development*. Beijing: Chemical Engineering Press.
Wang, R. S. & Xu, H. (2004). *Methodology of ecopolis planning with a case of Yangzhou*. Beijing: China Science and Technology Press.
Wheeler, S. M. & Beatley, T. (Eds.). (2009). *The sustainable urban development reader* (2nd ed). London: Routledge.
White, R. R. (2002). *Building the ecological city*. Cambridge: Woodhead.
Woo, W.-T. (2007). The origins of China's quest for a harmonious society. In R. Sanders & Y. Chen (Eds.), *China's post-reform economy: achieving harmony, sustainable growth* (pp. 15–29). London: Routledge.
Wu, C.-H., Maurer, C., Wang, Y., Xie, S.-Z. & Davis, D. L. (1999). Water pollution and human health in China. *Environmental Health Perspectives*, 107(4): 251–256.
Yangcheng Evening News (2008). The Shijing River: Why does it remain stint after years of treatment? December 3, p. 1.
Yip, S. C. T.. (2008). Planning for eco-cities in China: visions, approaches and challenges. Paper presented at 44th ISOPCARP Congress 2008, 1–12.
Zhai, R.-M. (2009). Over 100 Chinese cities compete earnestly as eco-cities. Chaijing Times, Noverber 23, http://finance.sina.com.cn/g/20071123/10444209289.shtml. Accessed 29 Jan 2009.

Chapter 8
Green Urbanism: Holistic Pathways to the Rejuvenation of Mature Housing Estates in Singapore

Steffen Lehmann

Abstract Cities play a crucial role in the way out of the environmental crisis. This chapter argues that our fast growing cities need to develop as more compact, polycentric mixed-use urban clusters, strongly inter-connected by public transport and highly mixed-use, towards sustainable "network city" models (Castells, *The rise of the network society*. Oxford: Blackwell, 1996). Cities are systems already under stress; cities are resource-intensive, and can sometimes be messy and chaotic. Not everything in cities can always be planned to last more than 25 or 30 years; mature components, such as housing estates, have to be re-engineered and retrofitted. Today, many mature housing estates, which play such a significant role of Singapore's urban fabric, are over 3 decades old and in need of urgent rejuvenation and retrofitting. Some of them are relatively energy-inefficient and highly air-conditioning dependent – but what could be the most appropriate model for such rejuvenation? It is timely to rethink and re-conceptualize these aged estates and districts of Singapore, in order to future-proof them for a fast approaching low-to-no-carbon society. Eco-city planning and the retrofitting of existing inefficient housing estates involves the introduction of mixed-use programmes and smart densification of the urban form. These concepts go far beyond environmental aspects; they include systems' integration and holistic thinking, rather than piecemeal approach or single-minded "techno-fix" approaches. System-integration and holistic conceptual approaches are necessary to ensure that these rejuvenated estates become part of a larger sustainable ecosystem, in regard to their management of waste, energy, water, public transport, materials and food supply. What is needed is a practical strategy for re-energising tired housing, to undergo radical modernization, to meet the changing aspirations and lifestyles of contemporary Singaporeans. It also requires new typologies for both public and private housing, appropriate to the

S. Lehmann (✉)
Research Centre for Sustainable Design & Behaviour, University of South Australia, Adelaide, SA, Australia
e-mail: steffen.lehmann@unisa.edu.au

T.-C. Wong, B. Yuen (eds.), *Eco-city Planning*, DOI 10.1007/978-94-007-0383-4_8,

tropical climate, with terraced gardens, courtyards, and environment friendly solutions. This study explores the typology and findings of a German case study: the city of Freiburg, where two recently completed eco-districts are analysed, as they could inform urban developments in Singapore. This case study shows that cities need to always find local solutions appropriate to their particular circumstances, and that government is key in driving the outcome. The argument is that good urban governance and governmental leadership is crucial to eco-development. In connection with this, the paper also examines a study conducted by the author at the National University of Singapore: an architecture master class, which was looking at careful neighbourhood re-configuration and the integration of the existing estates, avoiding the negative impact of demolition of these estates, to maintain the social community networks.

8.1 Introduction

As more and more of the Earth submits to urbanization, urban planners and architects are being confronted with a series of design challenges and an urgent need to act on them. Among the most significant environmental challenges of our time is the fossil-fuel dependency of existing cities, districts and buildings, and their growing demand for energy, land water and food security. In this context, retrofitting of the existing building stock has widely been recognized as a matter of urgency (Rees and Wackernagel 1995, Jenks and Burgess 2000, Lehmann 2006, Head 2008).

It is increasingly understood that avoiding mistakes in urban development at the early stages could lead to more sustainable, polycentric and compact cities, avoiding car traffic and therefore releasing less greenhouse-gas emissions. This paper presents research in *Green Urbanism* as a holistic pathway towards the rejuvenation of existing city districts, and introduces concepts for the urban intensification of neighbourhoods, to show how cities can transform from out-dated fossil-fuel based models to models based on renewable energy sources and mixed-use densification.

We can observe now strong moves by the government to establish Singapore as a green building hub for the tropics and as a best practice model of "sustainable city" for the Asia-Pacific region (URA 2009, BCA 2009). Singapore Government's first "zero-energy building" (ZEB) in Braddell Road, a retrofit project launched in 2009, is a good example for such an effort. It is now timely to expand these initiatives and explore emergent forms of urbanism, as well as models of affordability, based on new paradigms that will guide the transformation of the shape of districts and housing estates to come.

As the public housing authority of Singapore, the Housing and Development Board (HDB) has played a key role in meeting the housing needs of Singaporeans since its foundation in the early 1960s. Today, around 85% of Singapore's population live in HDB flats, out of which more than 90% own their flats (this is some of the highest home ownership ratio worldwide). The high degree of home ownership has been an advantage, as home owners take usually better care of their neighbourhood.

Since the 1960s, the HDB has constructed large-scale new towns as housing estates, starting with the Toa Payoh Estate in 1961, still following Le Corbusier's model of the "Unité d'Habitation" modernistic slab typology, isolated residential towers in a garden landscape (as coined by the Swiss architect in 1955).

8.2 Singapore's Urban Transformation and Leadership

In a global context, Singapore has done very well over the last two decades in re-inventing and positioning itself as "global city" and living laboratory for good infrastructure and urban planning (sometimes even called a "First World oasis in a Third World region", aiming to differentiate themselves from the rest of the region). Big cities are always in a global competition with each other. According to the recent *Global Liveable Cities Index* (2010), the city state is ranking on place three as one of the most liveable cities in the world, behind Swiss cities Geneva and Zurich, but well ahead of Hong Kong, Tokyo and Osaka. However, Singapore only ranked 14th out of 64 cities in the area of environmental friendliness and sustainability, one of the criteria used in the index. Today, Singapore is seen as one of the global leaders in the following planning areas:

- Achieving a competitive economy and strong real-estate market
- Developing housing typologies for multi-apartment living
- Implementing efficient, affordable public transport
- Leading in urban water management
- Ensuring the integration of urban greenery into planning.

However, in a situation where almost 90% of all materials and food need to be imported to the city state, the situation is fragile and new thinking about urban agriculture, local food supply ("agricultural villages") and resource recovery has gained strong momentum. There are also research initiatives into intergenerational neighbourhoods and behavioural change, in order to find strategies to reduce consumption without reducing lifestyle. In 2010, National Development Minister Mah Bow Tan flagged the idea of a "Learning Network for Cities", where best practice of sustainable urban development, liveability and green technology is identified on a global platform, led by Singapore. "Singapore is in a great position to lead this emerging global discourse on cities", he said at the World Cities Summit in June 2010. He also acknowledged that Singapore has to identify its own urban solutions based on its unique situation as a city state, and commented in discussion with Dr. Dieter Salomon, mayor of the German city of Freiburg: "Singapore is not Freiburg. So we need to understand what has worked well at a smaller, innovative place like Freiburg and see if there are lessons to be learnt, and if these can be transferred to Singapore's situation" (Mah 2010). Despite its differences in scale and cultural context, the second half of this paper will have a closer look at the sustainable urban development principles that were applied in Freiburg and evaluate which lessons could be learnt from it that might be relevant for Singapore.

With the number of city dwellers in Singapore expected to increase from 4.8 million to around 6.5 million by 2035, accompanied with significant demographical shifts (in-migration, ageing population, increase of single households, reduced fertility rate, etc), it is essential to identify strategies for maintaining the current quality of life in Singapore. While incomes of Singaporeans have significantly gone up, lifestyle adjustments have been lagging behind. Singapore has emerged "as major centre for shipping and transport, as well as a major financial trading centre and hub of investment banking, in a matter of decades" (Girardet 2008). However, Singapore needs now to develop an urban vision that goes beyond the common "City in a Garden" concept, and find new pathways to rejuvenate its mature housing estates without entire demolition of these estates. Every demolition means the loss of community history and damages in terms of social sustainability, as all community ties and active networks in these estates are lost. Once residents have been relocated for demolition of the mature estate, they rarely move back to their former estate's location, but settle in another area of Singapore.

The HDB new towns consist of neighbourhoods and precincts, the latter being the smallest unit of 3–5 ha in size, with around 1,000 families, and plot ratios around 1:5–1:8. Singapore is losing its image as a "place for families", becomes more and more unaffordable to bring up a family, and the question that is now frequently asked: How can we create dense urban spaces that can also accommodate families?

8.2.1 HDB Initiatives: From New Towns as Global Post-WWII Phenomenon to Punggol 21

In 2007, Mr. Tay Kim Poh, former CEO of the Housing and Development Board, announced an eco-demonstration project in the north-eastern part of the Singapore Island: A major milestone in the overall plan to transform the HDB towns and estates was the unveiling of the "Remaking Our Heartland (ROH)" blueprint in August 2007. Mr. Tay said: "The coastal town of Punggol was selected as one of the pilot ROH towns, with new strategies and plans formulated to reinforce and realise the vision of "A Waterfront Town of the twenty-first century", or *Punggol 21*. This is HDB's first demonstration eco-precinct, *Treelodge@Punggol*, launched in March 2007, with the first waterfront housing precinct to be launched in mid-2010. When the town is substantially completed in the near future, *Punggol 21* will set the new benchmark for quality living and environmental sustainability for HDB towns" (HDB 2008).

This paper suggests, what Singapore needs is not only luxury housing developments on greenfield sites in the north, far away from the city centre (which increases the need for residents to commute), but to keep the population close to the centre through practical concepts to achieve affordable retrofitting of existing housing estates. HDB estates are (since the 1970s) dispersed all over the southern part of the island, with many of them still close to the city centre. Pedestrian connectivity is everything, and the right densification of these estates towards a more compact, polycentric Singapore will help to improve the walkability of the city.

"Redevelopment" means usually demolition of the entire existing estate. However, rejuvenation solutions (keeping the existing and integrating it in a *retrofit-master plan*) are most of the time lower both in environmental impact and whole-life costs than comparative redevelopments. Paul Sloman from Arup notes in this regard: "These retrofits can reduce energy use by 20–50% in existing buildings, and pay for themselves over several years through the resulting cost savings on energy bills. The greenest buildings may actually be well-managed, retrofitted existing buildings" (Sloman 2008, Arup 2008).

After the Second World War, a large series of *New Towns* was built all over the globe. These towns were planned from scratch, based on the combined ideologies of the Garden City, CIAM-Modernism and the British neighbourhood principle. From Western Europe to Asia, from Africa to the former communist countries, the original universal model of the *New Town* was only slightly adapted to local cultures, economics and politics (from the "superquadras" in Brasilia, to the neighbourhood-modules in Milton Keynes and Almere New Town). It is surprising to realize that one model could simultaneously lead to Scandinavian cleanliness, Indian visual richness, Singaporean repetitive planning lay-out, and Chinese high density.

Typical for these *New Towns* is that they were designed for a new district or quarter, on a very large scale – which is most likely the reason why they often went wrong. In addition, these *New Towns* failed to take into account the various local traditions. Singapore's particular version of new towns is based on the concept of "Housing in a Park", which sets public housing slab and towers within a scenic park-like environment, where residents can enjoy lush greenery close to home. It complements Singapore's vision of the "City in a Garden" (see Figs. 8.1, 8.2, and 8.3).

8.2.2 The Historical Development of Singapore's Housing Estates

In addition, Mr. Tay Kim Poh (HDB) noted: "Soon after Singapore attained self-government in 1959, one of its key challenges was to ease a severe housing shortage. The Housing and Development Board, which was set up a year later to handle this task, opted to provide small and utilitarian flats, which it was able to build quickly and at low cost to house a fast-growing population. Once the housing shortage eased, the Board's challenge was to keep up with the changing needs and aspirations of the people, who were beginning to seek bigger and better flats, and more comprehensive

Fig. 8.1 (**a**) Typical Singapore tower housing estates – built reality (*left*); (**b**) Typical Singapore tower housing estates – urban model (*right*). Note: The modernistic planning concepts have been a mix of slab and point tower typologies (sometimes also courtyard typologies). How to best transform these mature estates into sustainable models, without "tabula rasa" demolition? The mature estates represent a socially healthy microcosm, occupied by a mixture of multi-national communities. (Photos by S. Lehmann 2009)

Fig. 8.2 (**a**) *Top left*: Model photo of a typical Singapore HDB housing estates. (**b**) *Top right*: Singapore is an example for efficient and affordable public transport. *Note*: As lifestyle of Singaporean people has changed, there is now a need to transform these ones step-by-step and upgrade the spaces between the buildings. Higher densities are appropriate around transit nodes and public transport corridors. (Photos by S. Lehmann 2009)

facilities. This is a critical challenge since living in HDB flats is a way of life for most Singaporeans" (Tay HDB 2008).

Singapore has now 4.8 million population (data 2009. Ethnical mix: 75% are Chinese, 15% Malay origin, 10% of other origins), and the population is targetted to increase to 6.5 million within the next 25 years. The lifestyle of Singaporeans has gone through significant changes. We need to ask:

- How do Singaporeans want to live in the next decade?
- How can we adapt the existing estates to climate change?

Fig. 8.3 (**a**) La Salle Art School courtyard (*left*); (**b**) Roof garden on Vivo City shopping centre (*right*). *Note*: While Singapore is experimenting with new types of "quasi" public spaces, most of these spaces are not truly public/civic, but located on roof tops of shopping centres or semi-internalised spaces, which are privately owned and controlled. (Photos by S. Lehmann 2009)

8.3 Learning from Germany's Policies: Why State Is Key

Most urbanization in the next 20 years will occur in the Asia-Pacific region. With climate change, Asia has to lead with new urban models, and Singapore is well placed to play a key role in this. Singapore Government has recently started using policies, such as the "2nd Green Building Masterplan" as drivers to implement sustainable development, and has set the key target for "at least 80% of the buildings in Singapore to be green by 2030" (BCA, 2009). Germany has been using similar policies and a system of incentives successfully over the last two decades: for instance, one much quoted example is the "electricity feed-in tariff" for renewable energy sources, legislated in 1999 (Herzog 2007).

The German Federal Government has specified in its fifth energy research programme (2005) the goal for all new buildings to reduce the primary energy demand, i.e. the energy demand for heating and cooling, domestic hot water, ventilation, air-conditioning, lighting and auxiliary energy by half – compared to the current state of the art. The long-term goal is net-zero emission buildings. A recent EU-Directive (2009) requires all new buildings in the European Union to be net-zero energy buildings by 2020. These are good examples, how policies can accelerate the required paradigm shift and drive the implementation of sustainability measures.

8.3.1 Good Governance and Governmental Leadership is Key to Eco-development

The German case studies show that good governance is crucial to urban development, especially in the introduction of innovative thinking regarding the development of eco-districts, if we want to transform existing cities into sustainable compact communities.

Government and municipalities have to provide public transport, public space and affordable housing, and without political support change will not happen. City council needs therefore strong management and political support for a strategic direction in order to manage sustainability through coherent combined management and governance approaches (including decision-making and accountability), which include evolutionary and adaptive policies linked to a balanced process of review. Public consultation exercises and grassroots participation are essential to ensuring people-sensitive urban design and to encouraging community participation. Empowering and enabling people to be actively involved in shaping their community and urban environment is one of the hallmarks of a democracy. Therefore, a city that leads and designs holistically, that implements change harmoniously, such as Freiburg has done, and where decision-making and responsibility is shared with the empowered citizenry is a city is on its road to sustainable practices (Boddy and Parkinson 2004).

8.3.2 Applying Best Practice: Freiburg's Inner-City Eco-districts

There are two innovative solar city estates in the City of Freiburg, which display well the current approaches towards eco-district development: The green district Vauban, and the Solarsiedlung am Schlierberg. The city of Freiburg in the south-west of Germany is one of the sunniest places in the country (lat. 48°, longitude 7.5°), with an annual total irradiation of about maximum 1.100 kWh/m^2 (in comparison, Singapore receives over 50% more sun radiation) and an average temperature 10°C. Freiburg is a university town with some 30 years of environmentally sensitive policies and practices, and has often been called the "European Capital of Environmentalism".

The two model projects close to the city centre, on the former area of a French barrack site (brownfield), are smaller compared to most housing estates in Singapore; and they have around half the density of a typical Singapore HDB housing estate. However, the applied concepts are highly replicable and pragmatic. Together with the Hammarby-Sjöstad district in Stockholm, it is probably Vauban and Schlierberg that have set the most replicable benchmarks for eco-districts up until today (see Fig. 8.4).

Both estates were built as pilot projects on an inner-city former barracks area, integrating some existing buildings; they have been an ongoing testing ground for holistic sustainable thinking and ecological construction, e.g. the estates include innovative concepts of water management and eco-mobility.

The Solarsiedlung am Schlierberg estate (built during 1999–2006), is located three kilometres south of the historic centre, bordering directly on Vauban. The architect of this estate is Rolf Disch, a pioneer of "solar architecture", who invented the *plus-energy house*. The solar PV-covered roofs of these houses produce more energy than the building consumes: around 15 kW/m^2 per year surplus.

Fig. 8.4 (**a**) The two solar districts in central Freiburg (South Germany) (*top left*); (**b**) The two solar districts in central Freiburg (South Germany) (*top right*); (**c**) Images showing Vauban and Solar District Schlierberg inner-city densification estates, with solar roofs and a light railway (*bottom*). (Photos by L. Lehmann 2009)

Today, about 170 residents live in the 59 terrace houses at Schlierberg. Nine of the houses are placed on the roof of the so-called Sonnenschiff ("Sun Ship"), a block of offices and shops, acting as noise barrier to the nearby main road. The terrace houses are of different widths and extend over two or three storeys, so that the living areas vary from 75 to 200 m^2. In accordance with classic solar building principles, the living and dining rooms are oriented to the southern (sunny) side, access is via a central core and the service zones are on the northern side, including kitchens, bathrooms and building services.

The larger city district of Vauban is interesting for its strong guiding principles and implementation model for the planning and design phase. Vauban comprises a 38-ha former barracks site that was purchased by the city in 1994 with the goal to convert it into a flagship environmental and social demonstration project (size 38 ha; density 140 persons/ha). It is a mixed-use estate, including 2,200 homes (20% of the 2,200 units are public housing), accommodating around 5,000 people, as well as business units to provide about 600 workplaces. The project was completed in 2006 and is widely seen as one of the most positive examples of environmental thinking in relation to urban design. The concept offers an increased building density, social and functional mixes, flat roof greening, and rainwater disposal within the building boundaries. The reused, renovated barrack buildings offer affordable housing for students and special functions to service the quarter, such as schools, shops and workplaces. The main goal of the project was to implement a green city district in a co-operative, participatory way which met ecological, social, economic and cultural requirements.

8.3.3 A Social Agenda for Better User Participation

The city of Freiburg had bought the area from the Federal Authorities. As owner of the Vauban area, the city was responsible for its development and realized the importance of design thinking in policy. The principle of "learning while planning", which was adopted by the city, allowed flexibility in reacting to the developments and to start an extended citizen participation that went far beyond the legal requirements; it enabled citizens to participate directly in the planning process. The citizen's association "Forum Vauban" (which has NGO-status) became the major driving force for the development of Vauban, with the commitment of the future residents to create a sustainable, flourishing community (it turned out that the project was particularly appealing to academics and the middle-class population segment).

In Vauban's new apartment buildings, innovative plan layouts were applied, that allow for openness and a multitude of uses through flexibility, so that changes in family room type and furnishing composition are possible. There was a strong focus on the public space between the buildings, at different scales, created with an emphasis on public safety and reduced car-traffic. Vauban is a car-reduced neighbourhood both through removing the need for automobiles as well as restrictions to car parking. Tramlines form the backbone of public transportation, linking the new city quarter with the rest of the city, while many amenities and public institutions are located within walking distance.

From the beginning, Vauban has been designed to reduce the need for car use and to cut overall journey distance. Tram and bus stops are placed not more than 200 m from any residential building. Car parking garages are located at the edge of the development and car access is restricted to the main access road. A free bus runs through the district and there is a car speed limit of 30 km/h on the main thoroughfare, while the side access roads inside the estate have a limit of 10 km/h and are no-parking zones, except for set-downs and deliveries.

Many of the environmental measures at Schlierberg and Vauban even exceed the strict German regulations: for instance, all buildings (new and retrofitted) must meet low energy house requirements of an annual heating energy consumption of 65 kWh/m^2 or less.

Most buildings were restricted to a height of four to five floors to ensure good climatic performance and day-lighting of the outdoor spaces. Most buildings are equipped with solar panels, others have green roofs. Buildings consume only 30% of the energy compared with conventional buildings, and 65% of the energy comes from renewable energy sources. About two thirds of Vauban's buildings are served by a combined heat and power (CHP) plant, which is powered by a mix of wood-chips (80%) and natural gas (20%). Most of the new buildings at Vauban and Schlierberg fulfil the German "Passive House" standards: walls and roofs are insulated with 400 mm of mineral wool or polyurethane insulation, windows are triple-glazed (See: www.passiv.de).

8.3.4 The Main Concepts of Freiburg's Eco-districts

In the fields of energy, traffic and mobility, user participation, public spaces and social interaction, a series of new concepts were successfully put into practice. In the Vauban district:

- the project's structure integrates legal, political, social and economic actors from grassroots-level up to the city administration;
- all houses are built at least to an improved low energy standard (max. 65 kWh/m^2 per annum); in addition at least 100 units with "passive house" (15 kWh/m^2 per annum) or "plus energy" standard (houses which produce more energy than they need;
- a highly efficient co-generation plant (combined-heat-power CHP) operating on wood-chips, connected to the district's heating grid (the wood-fired community power plant supplies heating);
- solar collectors and photovoltaics (about 2,000 m^2 installed by 2008) are the common element on the district's roofs; the LED-street lighting is solar-powered;
- an ecological traffic and mobility concept was implemented, with a reduced number of private cars, to be parked in the periphery (about 40% of the households are car-free, or agreed to live without owning a car; car ownership is only 150 cars per 1,000 persons; compared to adjacent Freiburg city centre, with 400 cars per 1,000 persons). There is a good public transport system (free bus loops and light rail), and a convenient car sharing system, where car sharers get a free annual pass for the tram;
- car-reduced streets and other public spaces act as playgrounds for kids and for places for social interaction;
- joint building projects (about 30 groups of building owners, the "Geneva Co-operative" and a self-organized settlement initiative) are the fertile ground for a stable community, raising ecological awareness; and
- there is a far-reaching participation and social network organized by "Forum Vauban", giving a voice to the people's needs, supporting their initiatives, promoting innovative ecological and social concepts, and setting-up a communication structure, including meetings, workshops, a 3-monthly district news magazine, publications on special issues and internet-presentations. Social aspects include a co-operative organic food store and a farmers' market initiative.

More information is available: Freiburg's two model districts have been described to great detail by a series of researchers, including Schroepfer and Hee (2008), as well as Heinze and Voss (2009).

8.4 Reducing Greenhouse Gas Emissions in Fast Growing Asian Cities

For the protection of food security, ecosystems and biodiversity, and to enable sustainable urban development, we need to carry out urgent and large greenhouse gas (GHG) reductions (Brundtland 1987, UN – IPCC 2007). It is now understood that nowhere will the impact of climate change be felt more than in Asian cities, where urban growth will far outstrip other regions, and more than double the population by 2050, with a staggering increase of almost 2 billion people (UN-Habitat 2008). The direct link between urbanization and climate change is widely accepted: In general, cities now cater for 3.4 billion people worldwide, using about 2% of the global land area, with over 1 million people migrating to cities each week (Stern 2007, Arup 2008).

8.4.1 Rapid Urbanization: Asian Cities Are Different

It's important to note that the cities in Asia have an entirely different history and development scenario compared with their US, European, or Australian counterparts. Today, most Asian cities are characterized by the following unsustainable trends (see Lehmann 2010a,b):

- There is a high number of inefficient older districts in need of regeneration, with mature housing estates desperate for rejuvenation;
- The existing building stock is out-dated and not energy-efficient;
- Structural problems, e.g. expansion of large shopping malls, but lack of non-commercial, catalytic, mixed-use, socially sustainable city projects;
- High carbon energy supply, and the need to de-carbonize this supply;
- Inefficient water, waste and transport operations; and
- Population growth, aging population trends, combined with job losses and demographical shifts.

However, these cities also share the more resilient characteristics of:

- Lesser impact on the surrounding land for agricultural/food and waste, compared to cities in the US, Europe, or Australia;
- Closer community ties, with a strong attachment to history and place (which is often in need to be better protected);
- Due to the higher population densities, a high percentage of residents in Asian cities are using efficient public mass transit; and
- Public space is here usually more lively and vibrant than in the US or Australian cities.

In addition, over the last decade, Singapore has emerged as leader in thinking about urban greenery ("skygardens") and the role of plants in the sustainable city, as mitigators for the urban heat island effect (NParks 2009).

8.5 Design Studio: A Master Class on the Neighbourhood Re-configuration of Dawson

The raised concerns in regard to Singapore's new town estates and the question of the most appropriate model for rejuvenation were used as starting point for a master class:

Field studies and a design master class, conducted at the National University of Singapore from August to September 2009, further explored the issue of a typical mature housing estate: Dawson Estate in Queenstown, at Commonwealth Avenue, was chosen as field of exploration. Dawson currently houses around 22,000 people and is in many ways testing ground for the identification of possible future approaches (Low 2006).

The specific aim of the master class, involving a cohort of 30 final year students, was to identify best practice and study holistic urban and architectural design solutions for the intensification, rejuvenation, retrofitting, re-energising, compacting and future-proofing of a typical Singapore's housing estates. The aim was to illustrate approaches to design inquiry, which might inform better policy-making in eco-development.

The starting hypothesis was that the lifestyle of the Singaporean people has gone through significant change over the last 20 years; however, over 80% of Singapore's population (over 3.5 million people) still live in HDB apartments that do not properly reflect this demographical shift or change of lifestyle. Furthermore, around half of the estates are mature building stock that is highly inefficient, inappropriate for natural cross-ventilation and highly air-condition dependent. Much of the building stock fails to deal with the tropical climate and the challenges that emerge from climate change and peak oil, as well as the increasing expectations of comfort by its residents (see Figs. 8.5 and 8.6).

While the outcome of such master class exercises and charrettes are usually limited by nature, they have the potential to contribute quickly with a series of suggestions. The students identified a wide variety of solutions – from practical and unachievable – for the transformation of mature housing estates towards self-sufficient, zero carbon districts. Suggestions included:

- Inventing new programmes of mixed use intensification, which allow working from home, leading to new housing typologies;
- Enhancing social sustainability for "aging in place", including community gardens and amenities for all generations;
- Intensifying the use of rooftops for urban farming (hydroponics) and urban heat island mitigation, as well as improving the space between the buildings.

Fig. 8.5 (**a**) Dawson Estate at Queenstown built in the early 1970s (*top left*); (**b**) Typical wide space between the slabs in older housing estate (*top right*). *Note*: The estate is currently too homogeneous and mono-functional; the buildings themselves are not dealing well with the tropical climate, lack proper balconies and western façade shading devices. Today, there are around 900,000 HDB flats across Singapore, housing over 80% of the population (this is around 3.5 million people). HDB has played a unique and significant role over the last four decades and has been crucial to Singapore's urban growth. However, we are now at a point where we have to rethink these existing typical 1960s–1980s new town housing estates, many of which have issues of energy-ineffectiveness and inappropriate, out-dated design lay-outs for living and working in a global city in the tropics

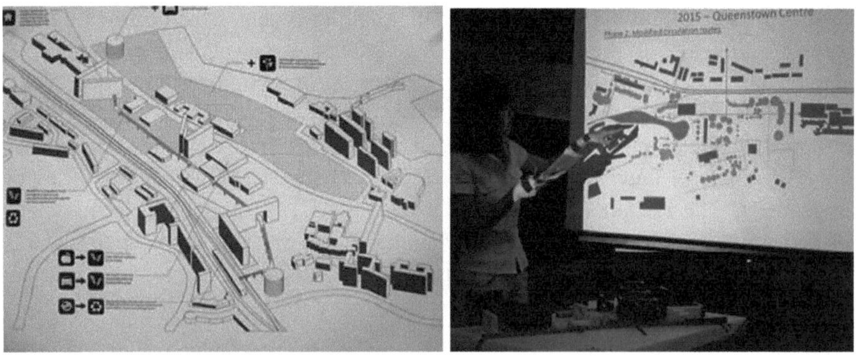

Fig. 8.6 (**a** and **b**) Some images from the final presentation of the students' work, at NUS in September 2009

8.5.1 Re-adaptation Efforts of Singapore New Town Estates

There are now multiple re-adaptation efforts going on in Singapore's housing estates. While the initial planning intention for the adoption of Le Corbusier's compact urban form (predominantly through the *Unité d'Habitation* model, willingly adopted in the 1960s–1970s) was originally meant to save land-take by going high-rise and high-density, this high density model might be seen as environmentally sustainable. However, new policy measures towards eco-development of public housing are required and currently developed. These include the innovative integration of greenery into high-rise buildings, rooftop greenery, concepts of urban farming, increased practice of recycling, water collection and storage, the use of solar energy, ecologically-friendly building materials, and the revitalization of passive design principles.

Wong has extensively researched on indoor thermal comfort and cooling loads of high-density public housing in Singapore. He found that thermal comfort varies between residents living in flats with different sizes and vertical positions (for instance, there are differences in energy consumption caused by urban geometry: if the unit is located in a high point tower, or in a less high slab block or courtyard typology; the unit's orientation and sky view factor of the adjacent street canyons have also an effect on energy consumption). Other findings point out that building design and how an apartment is used are paramount to reducing the environmental impact of the Singaporean home, not so much the walling materials used in construction. Since air-conditioning accounts for a significant portion of energy consumption, passive cooling and natural cross-ventilation are understood as major strategies for reducing energy consumption in tropical housing (Wong et al. 2002, Ng et al. 2006).

Reduced cooling load is often achieved by enhancing air circulation and reducing solar heat gain through façade design and external shading devices. West-facing apartments have in general higher cooling loads, while high point towers offer a better air circulation. Leung points out that in high-density housing clusters in Singapore, where considerable urban obstruction exists, passive cooling potential is also influenced by the geometry of adjacent buildings. Due to their proximity, adjacent buildings modify the amount of sunlight and wind that individual flats are subject to. Therefore, the urban geometry of the housing type becomes an indispensable component in the evaluation of the indoor thermal environment in high-density housing (Wong et al. 2002, Leung and Steemers 2010).

8.5.2 Queenstown: A Resilient Housing Estate in Its Transformation

Queenstown was, when built in 1970–1972, one of the test beds for Singapore's public housing initiatives. Today, life expectancy in Singapore has significantly risen from 64.5 years in 1965, to 79.7 years in 2009. A growing aging population combined with falling birth rates are a serious concern for the city state (Singapore's fertility rate in 2009 was only 1.28).

In many Singaporean estates, such as Dawson Queenstown and Red Hill, there is a need to attract younger families back to these estates (where they grew up, but left). It is also recommendable to generally rethink the role of greenery and landscape, in order to maximize biodiversity and introduce principles of urban farming for local food production. Models of "international best practice" and successful neighbourhood re-configuration were analyzed at the beginning of the studio, and ideas for new types of productive urban landscapes developed, where local food production and improved food security play an essential role.

8.5.3 Growing Population, Changing Lifestyles: Towards a More Resilient Singapore

The main research question, which the students were asked to address, was to identify appropriate and practical solutions for the rejuvenation of mature housing estates, with strategies and concepts suitable to the tropical climate.

The ecological footprint of an estate can easily be calculated, using established methods (such as the "EF" method developed by Rees and Wakernagel (1995). Most of the future energy demand will have to come from on-site renewable energy sources (over 50% as target, from solar and biomass), through the integration of PV-cells into the buildings and infrastructure, and the introduction of innovative solar cooling technology.

8.5.4 What Is Already Happening: The Two HDB Programmes, Remaining Structural Discrepancies?

Building a new estate on greenfield sites from scratch is always easier than dealing with the complexity of existing ones, hence certain reluctance by HDB to change its practice and the preference in the development of entire new estates. HDB has currently the following two different programmes for dealing with mature housing estates:

- First one is the renovation of the existing buildings (*Main Upgrading Programme*, called MUP/HIP, launched 1989); this includes upgrading of access, e.g. putting in elevators, and other pragmatic measurements. However, the main urban problems in the estates remain: for instance, MUP does not improve the lack of mixed-use (the estates remain too homogeneous and mono-functional), and the often unpleasant space between the buildings is not changed.

- The second programme, the *Selective En Bloc Redevelopment Scheme* (called SERS, introduced in 1995), is the demolition of the entire existing estates en bloc, to make way for new redevelopment of the precinct. However, this has significant disadvantages for social sustainability; for instance, that residents have to be resettled and that the existing community ties, which developed and evolved over decades, are destroyed and lost forever.

For a long time, these two models have served Singapore well in meeting the housing needs of its people, while providing them with a quality living environment through provision of adequate social spaces and other amenities. However, in the context of our explorations and from speaking to residents, it became obvious that there is a need for a third way today, with a different emphasis: the reconfiguration of the existing estates, whereby most of the buildings are kept and integrated in an energy and densification master plan.

8.5.5 Identifying a Third Way: Starting Questions for Neighbourhood Re-configuration

The starting point of the design studio was the following three questions:

Q1: How can the entire estate become energy independent, by producing its own energy, cleaning its own water, growing its own food supply?

Q2: How can we attract younger residents, such as young married couples back, to improve the demographic and socio-economic profile mix of the residents (e.g. to live near their aging parents)? How can we maintain the social and historical memory of place?

Q3: There is a high percentage of older residents living in Queenstown; so how will the retrofitted estate better provide for the elderly and cater for all three generations (e.g. with new mixed-use typologies)?

8.6 Concepts for Regenerating the Mature Housing District of Queenstown

Learning from the German examples and in regard to achieving self-sufficiency of mature housing estates, students were asked to address aspects, such as:

- Energy (especially decentralized energy generation, where every citizen can generate energy locally, with small solar units);
- Urban water management (with consequences in regard to roof scapes and landscaping);

- Transport: new concepts of eco-mobility to be introduced into the estate, with a strong focus on walking, cycling, and link to mass transit;
- Material flow and holistic concepts in regard to waste management (McDonough and Braungart 2002);
- Landscape, biodiversity and urban food production, including biomass facilities for organic waste composting;
- Construction systems for retrofitting, with a focus on modular prefabrication of entire building elements, such as add-on balconies or double-skin façade systems;
- A full understanding of the historical and social circumstances in the mature housing estates, including aspects of changing demographics and inter-generational relationships; estates representing a socially healthy microcosm; and
- Not to limit ourselves to the upmarket styling that is bound to come over the existing estates, where many will simply get demolished; but to search for an alternative that maintains the character and network of the existing.

8.6.1 Holistic Approaches for a Pathway to Low-to-Zero Carbon Are Needed

The students were introduced through lectures to the conceptual model of "green urbanism". It became soon obvious to the teams that what is needed is a robust, generic framework for future-proofing the existing, "to achieve an optimal relationship between footprint and population density" (Burton 1997, Hall 2005).

New technologies of decentralized energy generation (energy produced close to the point of consumption, using solar PV, solar thermal, and biomass) are understood as particularly promising concepts, with the potential to achieve a better symbiosis between the urban environment and the precious surrounding garden landscape of Singapore.

Singapore will take on a leadership role for the entire region, by mitigating the environmental impact through:

- Application of international best practice in urban developments and climate-responsive urbanism (introduced, tested and embedded via demonstration and pilot projects);
- Innovation and utilization of key technologies, such as renewable energy technologies, prefabrication and the integration of information technologies;
- Proper incentives and regulations, so that all existing and new housing estates can become carbon-neutral;
- Strong leadership by national and town council leaders, local community groups, planners and academics; and
- Enhanced knowledge transfer, training and awareness of all citizens.

8.6.2 The Conceptual Model of "Green Urbanism" and "Energy Master Planning"

Rather than demolish sections of a city, or build completely new suburbs, more GHG-emissions will be saved through remodelling and densifying the existing districts. Significant environmental, economic and social benefits can be expected in developing more sustainable urban districts and rejuvenating mature housing estates to attract residents of all ages and classes back, to live in these inner-city residential centres closer to their workplaces. More sustainable urban districts will better capitalize on the existing infrastructure of buildings and public transport, and allow population increase using less embodied energy. Highly sustainable city district adaptations will lead to re-energized estates that enable the city's residents to live a high quality of life whilst supporting maximum biodiversity and using minimal natural resources.

Connaughton points out: "New sustainable buildings use more embodied energy than refurbished ones, due to the high embodied energy of constructing new buildings and infrastructure" (Connaughton et al. 2008). However, since it is easier to build new, we find that there is frequently a great reluctance to innovate in the housing sector (JLL 2005).

8.6.3 Green Districts and Exergy Principles: Turning the Estates and City Districts into "Power Stations"

Low-emission energy generation technologies can turn the entire city districts themselves into power stations, where energy is generated close to the point of consumption. Localized energy generation on-site is using renewable energy sources (in Singapore especially solar and biomass), and complemented by distributed cooling systems and solar hot water systems: this has a huge potential to reduce Singapore's built environment's energy demand and emissions. Such decentralized, distributed systems, where every citizen can generate the energy needed, will eliminate transmission losses and transmission costs (which always occur with the large grid and inefficient base-load power stations outside the city) for the local consumer. The *exergy* principles look at capturing and harvesting waste heat and waste water streams, and how the strategic arrangement of programmes within mixed-use urban blocks and estates can lead to unleashing the currently unused energy potential. Currently, Singapore uses only 3.5% of energy from renewable energy sources (data: 2009). However, with a large population and a high number of biomass from greenery, there is a great potential for micro-biogas plants to be integrated in the new districts for local power generation.

These concepts can be considered for both existing and new buildings: Small power generators are positioned within communities to provide electricity for local

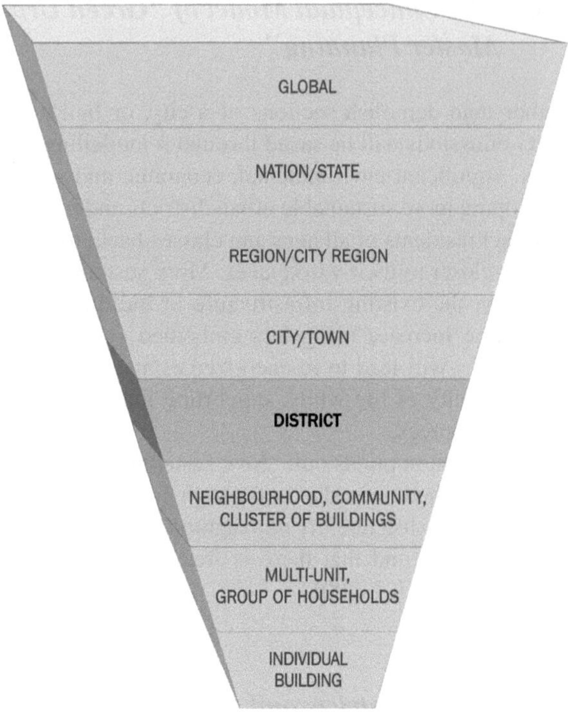

Fig. 8.7 District level energy supply (*left*). Rather than on the building-scale, working on the district-level of energy-effectiveness is most promising. This is highly relevant to the need to retrofit the existing cities and to de-carbonise the energy supply, on a district-scale (Lehmann 2006)

consumption, and the waste heat they produce is captured for co-generation (CHP; or for tri-generation, where the waste heat also produces chilled water for cooling), used for space conditioning via a local district cooling system (see Figs. 8.7, 8.8, 8.9, and 8.10).

8.6.4 Further Issues, the Students Considered

In addition, we asked the students to consider the following issues:

- Increasing the compactness and reconsidering the spaces between the buildings (to achieve a better public space network and stronger connectivity for pedestrians), overall more appropriate to the tropical "outdoor lifestyle", which is less based on air-condition dependency;
- Introducing intensive uses for roof tops, including urban farming and greening, for mitigation of the Urban Heat Island (UHI) effect;
- Modular prefabrication of building elements that are to be inserted or attached to the existing housing slabs, including large balconies, link-ways, break-out spaces, sun-shading devices, solar arrays, planting, and other elements;

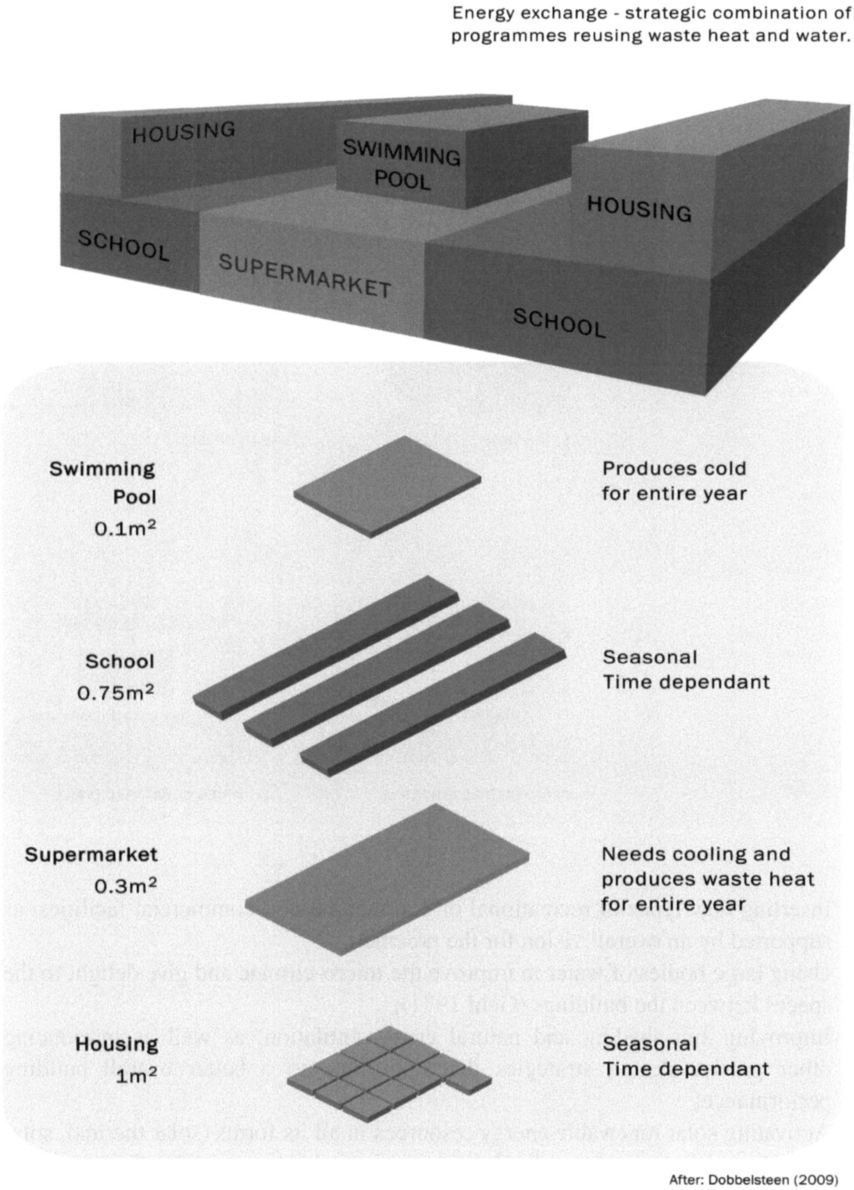

Fig. 8.8 Energy exchange through the strategic combination of programmes reusing waste heat and waste water. Currently, Singapore uses only 3.5% of energy from renewable energy sources; more tri-generation and cascading technologies should be used (Lehmann 2006)

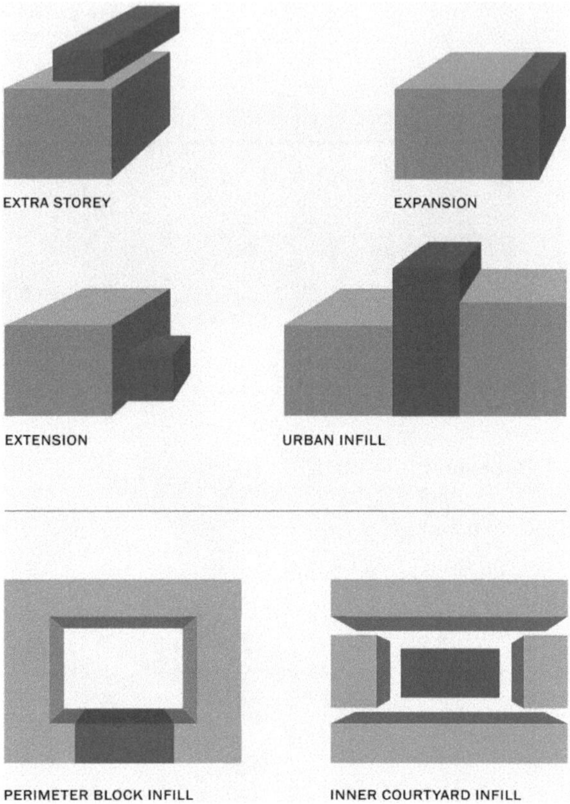

Fig. 8.9 Neighbourhood re-configuration: different arrangements for infill and densification are possible

- Inserting new types of recreational or commercial/non-commercial facilities, as supported by an overall vision for the precinct;
- Using large bodies of water to improve the micro-climate and give delight to the spaces between the buildings (Gehl 1971);
- Improving sun shading and natural cross-ventilation, as well as introducing other passive design strategies that contribute to a better overall building performance;
- Activating solar renewable energy resources in all its forms (solar thermal, solar PV, solar cooling, passive solar design principles, biomass), with a focus on local energy production, to turn the district into a "power station";
- Developing short and long-term strategies for the transformation of the existing district (a plan in 2 or 3 stages); clarifying which densities are required and recommendable; and
- Including other innovative strategies that deal with the particular challenges of Singapore (limited land, resources/materials/food supply), which we will need to develop, in order to future-proof the city against climate change impact (for instance, Singapore currently recycles less than 20% of its waste; this figure is too low and needs to double within the next 10 years).

Fig. 8.10 Diagram shows the conceptual model of "Green Urbanism". The optimum interaction of the three pillars of energy and materials, water and biodiversity, and urban planning and transport improves the environmental and social sustainability of cities. It is a holistic model, which identifies 15 core principles (Lehmann 2006)

Very soon, a couple of challenges for the urban design emerged in the Queenstown study; for instance:

- A focus on local energy generation, urban farming and concepts of waste management started to drive the master planning;
- The poor connectivity between the MRT station and the inner precinct area was recognized as major issue that needed to be rectified; and

- The urban design had to resolve the contrast of two very different sides: noisy Commonwealth Avenue on one side, and quiet, slow Margaret Drive on the other.

8.6.5 Pedagogical Strategies for the Master Class

This workshop was interested to address all these topics in a holistic and integrated way and use it to inform the urban designs. The students were asked to be mainly "strategic thinkers" on the urban scale and to invent new programmes as part of an overall vision, while avoiding to "get stuck in details". Being aware that this is a risky exercise, we were mainly interested in discussing initial concepts that would lead to further individual explorations. Throughout the master class, the Singaporean students were challenged with the thought that architectural "highlights" or spectacular designs contribute very little to the city's urban development in regard to the real issue of climate change.

8.7 Concluding Remarks

The problem of city-making today is as much about making new cities as it is about transforming our existing metropolises, especially housing estates, suburban building stock and edge city developments, which are too mono-functional and which need to become more mixed-use. This understanding is relatively young. We have yet to develop coherent strategies for transforming metropolitan agglomerations into urban configurations that are ecologically, economically, and socially sustainable while creating environments that are memorable and provide architectural delight. Social interaction is best created through intensification of mixed-use programmes and pleasant outdoor spaces, with high quality landscaping between the buildings.

Any vibrant authentic city has grown over years and has buildings which date from different eras. Redevelopment and retrofitting of the existing, mature housing precincts (without demolition of these estates, but integration) includes the increase of densities and other large-scale strategies, which need to be clearly redefined for Singapore's particular condition.

There is a re-affirmation of the following three thoughts:

- *Cities* and urbanization play a mayor role in the battle against climate change;
- *Cities* are resource-intensive and systems already under stress; and
- *Cities* need to be re-engineered to become more sustainable and resilient.

Today, an urban low-emission future is already technically feasible. How will Asian cities adapt, if countries are to meet international obligations such as those outlined in international emission agreements? There is urgency; without incentives, policy directions and updating the building codes, the stationary energy demand across all sectors is projected to increase further. What is needed are some cutting-edge demonstration projects that showcase how these available concepts and technologies can be brought together and set new benchmarks. These practical and achievable solutions for pilot (demonstration) projects would have generic, replicable strategies as outcome, with the potential to be applied to other similar housing estates and rolled-out in large scale, over the next decade.

One of the main arguments is that governmental leadership, good governance and strong guidance by the state is crucial to the development of eco-districts. This became obvious from the German cases. Any city leadership applying best practice for urban governance and sustainable procurement methods will accelerate the transition towards eco-planning. The question is: which networks and skills can be activated and utilized through engaging the local community and key stakeholders, to ensure sustainable outcomes?

8.7.1 Good Urban Governance and Policies Are the Lesson from Freiburg

The German cases illustrate that good urban governance is extremely important if we want to transform existing cities into sustainable compact communities. It has to provide public transport, public space and affordable housing, and without political support change it will not happen. City council needs therefore strong management and political support for their urban visions to be realized. It needs strong support for a strategic direction in order to manage sustainability through coherent combined management and governance approaches, which include evolutionary and adaptive policies linked to a balanced process of review, and public authorities overcoming their own unsustainable consumption practices and changing their methods of urban decision-making. A city that leads and designs holistically, that implements change harmoniously (such as Freiburg), and where decision-making and responsibility is shared with the empowered citizenry is a city that is on the road to sustainable practices. Public consultation exercises and grassroots participation are essential to ensuring people-sensitive urban design and to encouraging community participation. Empowering and enabling people to be actively involved in shaping their community and urban environment is one of the hallmarks of a democracy.

Like in Freiburg, every city leadership needs to ask itself: which are the networks and skills that can be activated and utilized through engaging the local community and key stakeholders, to ensure sustainable outcomes? City councils need strong political support for their strategic direction in order to manage sustainability through coherent combined management and governance approaches, which need to include evolutionary and adaptive policies linked to a balanced process of review. A city that leads and designs holistically, that implements change harmoniously, and where decision-making and responsibility is shared with the empowered citizenry, is a city on the road to sustainable practices. In balancing community needs with development, public consultation exercises and grassroots participation are essential to ensuring people-sensitive urban design and to encouraging community participation. Enabling local residents to be actively involved in shaping their community and urban environment is one of the hallmarks of a democracy.

A good public space network is essential for the liveability of a city. Easy pedestrian connectivity is the backbone of environmental sustainability and open spaces always change to respond to new needs, acting often as catalysts for urban renewal. Cities are a collective responsibility. As far as bureaucratic urban governance and best practice is concerned, authorities could consider many of the following: updating building code and regulations; creating a database of best practice and worldwide policies for eco-cities; revising contracts for construction projects and integrated public management; improving planning participation and policy-making; implementing anti-sprawl land-use policies; legislating for controls in density and supporting high quality densification; implementing environmental emergency management; introducing a programme of incentives, subsidies and tax exemptions for sustainable projects that foster green jobs; eliminating fossil-fuel subsidies; developing mechanisms for incentives to accelerate renewable energy take-up; implementing integrated land-use planning; having a sustainability assessment and certification of urban development projects. Urban design requires multi-disciplinary approaches, where design and engineering are fully integrated with all other disciplines throughout all phases of each project. This concept must be supported; and new policy frameworks should be created, which accelerate behavioural change, waste reduction and the uptake of renewable energy, which increase cultural diversity and economic opportunity.

This case study shows that cities need to always find local solutions appropriate to their particular circumstances, and that government is key in driving the outcome. The argument is that good urban governance and governmental leadership is crucial to eco-development. In summary, we can identify the following essential points for achieving sustainable urban development: Five basic concepts, to transform districts and housing estates towards low-to-no-carbon urbanism.

(a) The battle against climate change must be fought in cities. Sustainable urban design has the potential to deliver significant positive effects. The quality of a city's public transport and waste management services are hereby good indicators of a city's governance;
(b) It is particularly important not to demolish existing buildings, due to their embodied energy and materials. There needs to be a focus on integration, on adaptive reuse and on retrofitting of the existing building stock;
(c) A compact urban form with mixed-use programmes and a strong focus on low-impact public transport will deliver the best outcomes. To de-carbonize the energy supply, we need to install small distributed systems in the estates, based on renewable energy sources;

(d) Stop enlarging the urban footprint and halt sprawl, and protect the precious landscape and agricultural land. Therefore increasing the density of the districts and intensifying uses within the existing city boundary is recommended; and
(e) Change includes a whole range of different initiatives that will deliver significant CO_2-emission reductions – it is not one strategy or measure alone.

8.8 Related Web Sites

8.8.1 German Case Studies

http://www.vauban.de. Accessed 25 March 2010.
http://www.enob.inf. Accessed 25 March 2010.
http://www.oekosiedlungen.de. Accessed 25 March 2010.
http://www.solarsiedlungen.de. Accessed 25 March 2010.

Acknowledgements The author's Master Class at NUS, from August to September 2009, was supported by: Assoc. Prof. Wong Y.C., Mr Cheah Kok Ming, Dr. Nirmal Kishnani. Visiting critics were: Dr Johnny Wong (HDB), Tan See Nin and Sonja Sing (URA), Frven Lim Yew Tiong (Surbana) and Cheong Yew kee (SIA). The author thanks Mrs Cheong Koon Hean and Mr Tay Kim Poh for their insightful comments.

References

Arup: Report (2008). Into an ecological age. London, UK. Based on: Head, Peter (2008). *Entering the ecological age*. Peter Head's talk for The Brunel Lecture Series 2008, London. Resource document. http://www.arup.com. Accessed 25 March 2010.
Boddy, M. & M. Parkinson (Eds.). (2004). *City matters. Competitiveness, cohesion and urban governance*. London: Policy Press.
Brundtland, G. H. (1987). *The Brundtland report: our common future. Report of the World Commission on Environment and Development*. Oslo, Norway/United Nations: Oxford University Press.
Building and Construction Authority (BCA) (2009). *Public presentation by Dr. John Keung, BCA. 2nd green building masterplan*. Singapore: Building and Construction Authority.
Burton, E. (1997/2000). The compact city: just or just compact? A preliminary analysis. *Journal of Urban Studies*, 37(11).
Castells, M. (1996). *The rise of the network society*. Oxford: Blackwell.
Connaughton, J., Rawlinson, S. & Weight, D. (2008). *Embodied carbon assessment: a new carbon-rating scheme for buildings*. Proceedings of world conference SB08. Melbourne, Australia, Conference Proceedings.
Gehl, J. (1971). *Life between buildings, using public space*. Copenhagen: The Danish Architecture Press.
Girardet, H. (2008). *Cities, people, planet: urban development and climate change*. London: Wiley.
Hall, P. (2005). *The sustainable city: a mythical beast. Keynote: L'Enfant Lecture*. Washington, DC: American Planning Association and National Building Museum.
Head, P. (2008). *Entering the ecological age*. The Brunel Lecture 2008, London.
Heinze, M. & Voss, K. (2009). Goal: zero energy building. *Journal of Green Building*, 4(4).

Herzog, T. (Ed.). (2007). *The charter for solar energy in architecture and urban planning. Munic.* Germany: Prestel Publisher.

Housing and Development Board (HDB) (2008). The twin pillars of estate rejuvenation. Presentation by Tay Kim Poh, former CEO of the Housing and Development Board, given at World Cities Summit, June 2008. Singapore, World Cities Summit 2008. Resource document. http://www.worldcities.com.sg. Accessed 05 January 2010.

Intergovernmental Panel for Climate Change (IPCC) Report (2007). *Technical summary. Report: climate change: mitigation.* Contribution Working Group III to the Fourth Assessment.

Jenks, M. & Burgess, R. (Eds.). (2000). *Compact cities: sustainable urban forms for developing countries.* London: Spon Press

Jones Lang LaSalle (JLL). (2005). Building refurbishment – positioning your assets for success. Report published by JLL, London. Resource document. http://www.joneslanglasalle.com/csr/SiteCollectionDocuments/. Accessed 05 January 2010.

Lehmann, S. (2006). Towards a sustainable city centre: integrating ecologically sustainable development principles into urban renewal. *Journal of Green Building*, 1(3): 85–104.

Lehmann, S. (Ed.). (2009). *Back to the city: strategies for informal urban interventions.* Stuttgart/Berlin, Germany: Hatje Cantz Publisher.

Lehmann, S. (2010a). Mature housing estates in Singapore. *Singapore Architect, SIA/MICA,* March issue, 164–171.

Lehmann, S. (2010b). *The principles of green urbanism: transforming the city for sustainability.* London: Earthscan Publisher.

Leung, K. S. & Steemers, K. (2010). *Urban geometry, indoor thermal comfort and cooling load – an empirical study on high-density tropical housing. Proceedings of International Conference SAUD2010.* Amman, Jordan: Conference Proceeding, CSAAR-Press.

Low, C. (2006/2007). *10 stories: Queenstown through the years.* Singapore: National Heritage Board, Education and Outreach Division.

Mah, B. T. (2010). Minister for National Development, Singapore Government. Quote from speech on 30th June 2010 at the 2nd World Cities Summit, Closing Plenary Session. Singapore: World Cities Summit 2010.

McDonough, W. & Braungart, M. (2002). *Cradle to cradle: remaking the way we make things.* New York, NY: North Point Press.

Ng, E., Chan, T., Cheng, V., Wong, N. H. & Han, M. (2006). Tropical sustainable architecture. In J. Bay & B. L. Ong (Eds), *Tropical sustainable architecture – social and environmental dimensions.* Amsterdam, The Netherlands: Elsevier.

NParks, Singapore (2009). Leaf area index of tropical plants: a guidebook on its use in the calculation of green plot ratio. In NParks (Ed.), *Carbon storage and sequestration by urban trees in Singapore.* Resource document. http://www.research.cuge.com.sg/index.php. Accessed 05 January 2010.

Rees, W. & Wackernagel, M. (1995). *Our ecological footprint: reducing human impact on the earth.* Philadelphia, PA: New Society Publishers.

Schroepfer, T. & Hee, L. (2008). Emerging forms of sustainable Urbanism. Case studies. *Journal of Green Building*, 3(2).

Sloman, P. (2008). *Report: existing buildings survival strategies: a toolbox for re-energising tired assets.* Sydney, Australia: Arup. Resource document. http://www.arup.com. Accessed 25 March 2010

Stern, S. N. (2007). *The Stern review: the economics of climate change.* Cambridge: Cambridge University Press. Resource document launched October 2006, published January 2007. http://www.sternreview.org.uk. Accessed 05 January 2010.

UN-Habitat (2008). *State of the world's cities 2008/2009 – harmonious cities.* Nairobi: UN-Habitat. London, UK: Earthscan Publisher.

UN-IPCC (2007). [B. Metz, O.R. Davidson, P. R. Bosch, R. Dave, L.A. Meyer (Eds)]. Geneva/Cambridge, UK: Cambridge University Press. Resource document.

Urban Redevelopment Agency (URA) (2009). Quoted from conversations with Mr. Tan Siong Leng, Deputy CEO of URA, Mrs. Chong Koon Hean, CEO of URA. Sep. 2009 and 2009. Oct. 2009, during IGBC international green building conference, Singapore, 29 October Further information at: http://www.igbc.com.sg.

Wong, N. H., Feriadi, H., Lim, P. Y., Tham, K. W., Sekhar, C. & Cheong, K. H. (2002). Thermal comfort evaluation of naturally ventilated public housing in Singapore. *Journal of Building and Environment*, 37(12), Accessed 05 January 2010.

Chapter 9
Challenges of Sustainable Urban Development: The Case of Umoja 1 Residential Community in Nairobi City, Kenya

Asfaw Kumssa and Isaac K. Mwangi

Abstract Ineffective planning and implementation problems of urban residential plan in Umoja 1 have undermined the development of sustainable and livable urban community in line with the principles of affordable housing for eco-cities. Consequently, ex post measures designed to guide urban planning and implementation in the community have failed. Multi-story apartments are built in the community although these are not provided for in Umoja 1 residential comprehensive development plan. The project has failed to achieve its objective of building sustainable residential community due to several problems. For one, the planned capacity of roads and streets, water supply and sewerage disposal facilities can no longer cope with the new developments and/or those that result from unauthorized alterations of the original semi-detached units. Poor maintenance has degraded the roads and streets while social spaces are allocated and developed into private property. Chronic water shortage and periodic sewerage spills are common malaise in the community. Overstretched water supply and poor sewerage disposal systems have also exacerbated the problem. All these problems have severely altered the physical, ecological and social character of the community. Lack of consultation and participation of affected interest groups in implementation is one of the factors that have undermined sustainable urban development in the community. This chapter examines Umoja 1 residential plan and the challenges of plan implementation process. It focuses on factors that undermine sustainable development from eco-city perspective.

9.1 Introduction

In the late 1970s and throughout the 1980s, the government of Kenya and the Nairobi City Council implemented a strategy of site and service housing schemes.

A. Kumssa (✉)
United Nations Centre for Regional Development (UNCRD) Africa Office, Nairobi, Kenya
e-mail: asfaw.kumssa.uncrd@undp.org

The views expressed here are the authors' own and not necessarily those of the United Nations Centre for Regional Development.

T.-C. Wong, B. Yuen (eds.), *Eco-city Planning*, DOI 10.1007/978-94-007-0383-4_9,

The objective of these schemes was to narrow an ever-widening annual shortfall of houses in Kenyan towns (Kenya 2004, UN-HABITAT 2006). This shortfall has however persisted to the present day. According to the National Housing Corporation (2008), an estimated 150,000 housing units are required annually in the urban areas of Kenya to cater for the backlog of unmet demand. However, only about 30,000 units are built every year. Overall, the gap between supply and demand for housing has been widening for all income and social groups in the last four decades.

The Government of Kenya's urban housing assistance programmes focused on two urban social groups. The first social group, the urban poor, lives in informal settlements such as slums and squatter communities.[1] The second group, the low-income households, is the target group for site and service housing. Consideration for housing assistance for this group of households derives from their being in the formal employment.[2] Income from formal employment is reliable and regular. If well managed, the income would enhance the potential of households to access urban shelter. However, this potential is undermined by the inability to raise sufficient levels of finance to underwrite the prohibitive costs of land, infrastructure and services for conventional housing units. Structural urban land markets largely hinder low-income households from owning houses. Better financial terms would support their entry into forms of transitory tenant purchase site and service housing arrangements whereby housing is provided in different phases of completion before the households move in. The households then construct additional rooms at their own cost and convenience; but in line with the tenant purchase agreement and according to the approved plan and design of the units.

The government then initiated a site and service housing programme designed to overcome the housing problems experienced by these social groups. Mokongeni and Dandora site and service schemes in Thika town and in the city of Nairobi, respectively, are good examples of site and service housing that have addressed the shelter needs of this social group. Umoja I residential scheme is another scheme. The houses are tailor-designed and formulated to address the housing needs of first time low-income homeowners who are expected to complete the construction of the houses upon moving in. Formal employment acts as security in tenant purchase agreement and, together with a commitment by the owners to complete construction of their houses according to the approved design and layout plans, are important criteria for the success of Umoja I project. The employment status of those qualified for houses were assessed, vetted; and registered. Completion of the houses, which involved developing the houses would begin as soon as the new tenants moved into the rooms, which were completed during the ex ante phase of implementation, which preceded the longer term ex post phase.[3]

However, the ex post plan implementation activities and development outcomes of the built houses and social environments defy characterization. The house owners have violated provisions and rules of tenant purchase agreement. Public green spaces in the community were also grabbed and gradually developed into private buildings. These actions have undermined sustainable urban development and hindered promotion of eco-city orientations in Umoja I (Ecocity 2009, Ecocity World

Summit 2009). Structural and physical forms of the built environment and the social character of Umoja I are a complete departure from the type of residential community that was originally designed and developed at the end of the ex ante phase.

Moreover, there is potential for environmental risks from frequent shortage of domestic water supply and the breakdown of the sewerage system in Umoja I (Nairobi Chronicle 2008a, b). These problems have precipitated a sanitation nightmare, further complicating the risks associated with water shortage and health risks in urban communities. Other threats to the environment and health include poor surface drainage that results in the inability to manage surface water run-off that clog sewer drains during rainy seasons. This poses a constant threat to environmental pollution and the ever-present risk of diseases from sewage spills and leaks (Nairobi Chronicle 2009). Also, a large number of households have moved into multi-storey apartments as tenants. This has led to overcrowding. The size of private courts and public green spaces for recreation has been reduced, and their quality degraded. The overuse and overloading of transportation infrastructure and social facilities have eroded their quality, reduced their number; and undermined performance standards. All these are sources of threats to sustainability of the urban environment and hinder Umoja I from evolving into an eco-city community.

This chapter is about challenges to sustainable urban development that Umoja 1 community in the city of Nairobi, Kenya, faces. The community offers relevant lesson for understanding the challenges of planning for sustainable urban development and how these challenges translate into constraints in planning for eco-city communities in Africa. Following the introduction is a discussion of the concept of sustainable development in relation to urban development and the relevance of this to eco-city practices. The next section deals with conceptual and practical issues of site and service housing schemes in relation to them providing a planning and implementation context for Umoja I community. The third section provides a background to Umoja 1 residential community's goal, as well as a concept plan and development principles of the community. The role of households and Nairobi City Council in plan implementation during the ex post phase then follows. The section on ex post plan implementation phase is divided into two parts: the results of a study that was conducive to the appraisal of ex post plan implementation during the first 10 years of Umoja I from 1978 to 1988, are presented and discussed. Finally, the results of data analysis from a field survey recently conducted by the authors to ascertain the tread of development from 1989 to 2008 are discussed before ending the chapter with a conclusion.

9.2 Sustainable Urban Development in Africa

The term "sustainability" has no single definition. Scholars and policymakers define the concept differently. However, it was the World Commission on Environment and Development that gave it a comprehensive and greatly accepted definition (World Commission on Environment and Development 1987) when it defined the term "sustainability" as the need to use and improve the living conditions of the present generation without compromising the ability of the future generations to meet their needs. The word sustainability therefore straddles the economic, social and environmental aspects of development. Others have argued that sustainability should include institutional systems because both public and private institutions, and the decisions they make, affect and determine sustainable development. The issue of sustainability is important in development in general and urban development in particular. This is because the way we use our resources determines both the quality of our lives today as well as that of future generations.

The challenges for Africa in terms of deepening economic and social progress, and sustaining this progress over the next two to three decades, include addressing environmental and ecological resources issues, and mobilizing resources for development. All these are critical in increasing the capacity of African countries for economic acceleration and sustainable growth. Death from disease is clearly linked to poor nutrition as well as to a polluted environment exacerbated by a lack of safe drinking water, poor sanitation and chemical pollution. The environmental threats facing Africa are a combination of the degradation of local and global ecosystems. In Africa and other developing regions, one of the greatest environmental threats is that of water, whose scarcity is increasingly becoming a critical factor in fostering ethnic strife and political tension. Air pollution and deforestation are also some of the major environmental threats in Africa.

Another set of problems with significant social development implications for Africa stems from the social, economic and environmental consequences of urbanization. The rapid growth in Africa's urban population is a direct result of a shift in the balance between the urban and rural economies, as well as due to the natural growth of urban population. Although Africa is the least urbanized continent in the world, it has the highest urbanization rate of 3% per annum. In 2007, the African urban population was 373.4 million. It is projected that 759.4 million Africans will be living in urban areas by the year 2030, and there will be more than 1.2 billion urban dwellers by the year 2050 (UN-HABITAT 2008). Rapid urbanization in Africa is primarily a result of development strategies that stressed urban growth at the expense of agriculture and rural development. A dismal consequence of this scenario is that the rate of increase in the size of the non-agricultural population now exceeds the rate of increase in meaningful non-agricultural employment, leading to what is known as *over-urbanization* (Hope 1997). Table 9.1 depicts the past,

Table 9.1 Proportion of African population residing in urban areas by sub-regions, 1980–2030

Region	1980 %	1990 %	2000 %	2010 %	2020 %	2030 %
Africa	27.9	32.0	35.9	39.9	44.6	50.0
Eastern Africa	14.4	17.7	21.1	24.6	29.0	34.8
Northern Africa	44.4	48.5	51.1	53.5	56.8	61.3
Southern Africa	31.5	36.7	42.1	47.1	52.3	57.9
Western Africa	29.2	33.0	38.4	44.1	50.1	56.1

Source: UN-HABITAT (2008)

current and projected urban population by sub-regions in Africa between 1980 and 2030. The table indicates that Northern and Southern Africa are the most urbanized regions of Africa, while East Africa is the least urbanized, but nevertheless the most rapidly urbanizing region of Africa.

The urban areas in Africa remain the focal point of both the governmental and private sector activities and, as such, are the rational settling place for the population. But as the cities and other urban areas grow, further productive activities tend to concentrate within them. These urban areas generate about 55% of Africa's GDP, and yet 43% of its urban population lives below the poverty line (UN-HABITAT 2008). Although the urban areas are the main catalysts of economic growth in Africa, their economic attraction and the resulting urbanization have been major contributors to urban poverty and environmental degradation. Besides, despite the fact that poverty is more pronounced in the rural areas of Africa, urban poverty is increasing substantially.

As the spectacular demographic upheaval that the continent has experienced in the past 20–30 years has shown, urbanization comes with its own problems. As more rural migrants attempt to escape rural poverty, they flood the cities in search of income-earning opportunities. Thus, urbanization is both a contributor to, and a casualty of, the massive migration of people from rural to urban areas, and of the inordinate demands placed on the scarce resources found in the cities. This influx not only intensifies urbanization but also adds another dimension of urban poverty with all of its attendant consequences for further environmental degradation. The urban poor suffer most from this environmental damage.

Often, urbanization is associated with industrialization and development as the growing cities act as pivotal centres of economic growth, generating goods and services as well as employment for the growing urban population. This is what happened in Europe, North America, Latin America, and most recently in a number of Asian countries where urbanization has led to increased per capita income and improved standards of living. Unfortunately in Africa, urbanization is not accompanied by economic growth or better livelihoods. This is a unique phenomenon, which the World Bank has called "urbanization without growth". This pattern is the result of misguided policies that could not cater for properly managed and planned urban development in Africa. Urbanization in Africa is "a poverty-driven process and not the industrialization induced socio-economic transition it represented in other major world regions" (Kenya 1978, UN-HABITAT 2008: 7).

The state-led and urban-biased development strategies of African countries, combined with various external factors, have weakened the potential of urban regions in Africa to absorb the growing urban populations and provide them with the necessary employment opportunities and services. At the same time, effective formal urban planning is almost non-existent in cities and towns of Africa. Linkages between the process of urban development and planning on the one hand, and rural development on the other, are therefore very weak (UN-HABITAT 2005). As noted earlier, the process of urbanization is not accompanied and supported by industrialization of African economies; and this has translated into a weak urban economic base and high levels of unemployment (UNDP 2005). As a result, in most cities of

Africa, shanty towns and squatter settlements are developing along the periphery of major cities. The poor who live in these areas face tremendous economic and social hardships. They do not have access to basic human requirements such as shelter, land, water, safe cooking fuel and electricity, heating, sanitation, garbage collection, drainage, paved roads, footpaths, street lighting, etc (Suzuki et al. 2009).

Worsening political and ethnic conflicts, the erosion of traditional safety nets, and the deteriorating physical infrastructure and absence of general security in rural areas have further contributed to the problem of rapid urbanization by forcing thousands of people to migrate daily to the relatively safe cities, thereby adding more pressure to the socioeconomic burden of African cities. On the other hand, the Structural Adjustment Programmes (SAPs) implemented in many African countries in the mid-1980s that led to elimination of price controls, reduction of government expenditure on social services, and privatization schemes, resulted in massive lay-offs of the "redundant" labour force. This added and exacerbated the economic and social crisis already plaguing the cities of Africa. These and related problems have stoked rapid urbanization in Africa, leading to not only environmental degradation but also putting undue pressure on urban service delivery (water, housing, solid waste management, road, etc.)

Urbanization has also exerted adverse effects on the environment as industries and cars in urban areas release vast amounts of greenhouse gas emissions, which are in turn responsible for worldwide climatic changes. Undoubtedly, climate change is an emerging threat to humanity. This phenomenon has become a major national, regional and international problem, cutting across developed and developing countries. Marked variations in average annual rainfall, daily temperatures, wind direction and speed are some of the common features of climate change that have emerged since the 1950s. This has been followed by increased incidences of natural disasters such as floods, drought and diseases, both in urban and rural areas. Unfortunately, climate change disproportionately affects the poor and those living in the slums. This is mainly because climate change adversely affects the very things that the poor depend on. It also causes extreme warming of the ocean and rise in sea level, thereby adversely affecting coastal cities and islands.

As mentioned above, East Africa is urbanizing at a very high rate. Kenya, for instance, is projected to be 50% urban by 2030 (OXFAM 2009). It is estimated that half of Kenya's poor will soon be living in urban areas. What is worrisome is that income and social inequality in urban areas continue to increase. Nevertheless, recent moderate economic growth rate and access to educational and health facilities have improved Kenya's Human Development Index (HDI) from 0.520 in 2004 to 0.532 in 2005, pushing Kenya to the medium human development level. Kenya also succeeded in reducing the prevalence of HIV/AIDS from 13.9% in 1999 to 6.7% in 2003 (UNDP 2006). Although these are commendable achievements the country is still entangled in growing inequality, poor governance, ethnic tension, among other social vices. As correctly indicated by OXFAM (2009: 2):

> urban poverty and inequality can have catastrophic social consequences when combined with poor governance and ethnic resentment as the violence in urban informal settlements following the 2007 presidential election made all too clear.

Man-made and naturally induced environmental threats such as unclean and unsafe water, air pollution, and unhygienic habitation cause serious illnesses and reduce life expectancy in Kenyan cities (UNDP 2006). A study conducted by OXFAM (2009) highlights the complex and intertwined and mutually reinforcing challenges of urban development in Kenya. These are:

- Urban residents are almost twice as likely to be infected with HIV as their rural counterparts;
- Children in Nairobi's slums are among the unhealthiest in the country. Over half are likely to suffer acute respiratory infection, and almost half under 5 are stunted; moreover, they are less likely to be immunized than children elsewhere in Kenya, and more prone to diarrhea and fever;
- Population densities can be higher than 1,000 people per hectare in the slums – compared with as low as 4 per hectare in Nairobi's wealthy areas;
- The combination of climate change and unplanned urban growth has led to ever greater numbers of urban houses being severely affected by flooding;
- Social support networks are considered to be weaker in cities than in rural areas and the tendency for ethnicity to be mobilized for destructive ends is on the increase;
- Nearly two-thirds of slum residents interviewed said they did not feel safe inside their settlements; and
- Almost half of Nairobi's population admitted to actively participating in bribery; and almost all thought corruption was endemic in the city (OXFAM 2009: 3–4).

The other major challenge is the rise of crime and violence in Africa's urban areas, including Kenya's. According to the UN-HABITAT (2008: 19), "urban crime and violence is now the first threat to security in African towns". As much as 70% of the urban population in Africa has been victimised by crime (UN-HABITAT 2008). Urban violence and crime are complex socio-economic phenomena that are caused by multi-dimensional problems such as poverty, unemployment, inequity, poor governance, erosion of traditional family values, social and physical segregation, social decay, etc. Strategies to tackle urban violence and crime therefore require an understanding of the social, political and economic issues, which are interconnected, and which often reinforce each other. In this regard, innovative approaches to address this complex and multi-faceted problem are required. Urban violence and crime affect private investment opportunities, reduce public investment in infrastructure, and disrupt effective delivery of services. Most importantly, urban violence and crime affect the majority of the poor, including the elderly, women and the unemployed, who usually live in vulnerable physical conditions and as such cannot afford protection and private security services.

It is suggested that policy response should not halt the urbanization process but rather to develop effective policies for developing basic eco-cities and meeting the housing, infrastructure, and services needs of urban population (White et al. 2008). Specifically, African countries should proactively focus on building the capacity of small- and medium-sized cities so that they can provide effective urban services to the spiraling numbers of poor urban dwellers (UN-HABITAT 2008). Obviously, urban environmental issues should be of main concern to policymakers and urban planners. The issue of sustainability should be addressed within the context of economic, social and environment framework, and also within institutional systems which will have impact on the quality of settlements and life in African urban areas. According to Suzuki et al. (2009), if proper urban policies are not put in place, the current pace of urbanization will be accompanied by unprecedented consumption and loss of natural resources. Most importantly, from the experiences of several countries which have promoted the principles of eco-cities, developing countries in Africa can achieve urban sustainability in their own way, as highlighted by Suzuki et al. (2009: 23–25) recently as follows:

- Many solutions are affordable – even when budgets are limited;
- Significant success can be achieved using existing, well proven technologies and appropriate new technologies;
- Developing countries should take pride in developing home-made solutions; and
- Many solutions can benefit the poor directly and indirectly.

9.3 Site and Service Housing and Urban Development

Norms and practices of planning and implementation in site and service housing programmes are similar throughout Africa, Asia and South America. Srinvas (2009a) outlines three principles that guide implementation of site and service housing units. First, site and services schemes focus on providing the urban poor, who lack finance and power in urban socio-political set up, to access adequate and appropriate housing as a basic need and as a right to a dignified human life. The second principle pertains to the major components of site and service development. This includes land that is available in the form of plots, infrastructure (road and street network and non-motorized movement ways, water supply, electricity and sanitation facilities) and housing structures at various stages of completion.[4] Finance and technical assistance from experts who manage the implementation of different aspects of house and infrastructure construction before target households move in are also part of the second principle. Third, site and services scheme leverage the capacity of the urban poor and low-income households, who would otherwise not own homes, to actively participate in the construction of their own houses at a pace and resource loading they can afford. This absolves the government and other public agencies from solely bearing the burden of urban housing for economically disadvantaged groups in society.

Srinvas (2009b) further points out those housing units are planned, designed; and built at various stages. This gives the low-income households, who own the units under tenant purchase arrangement, the opportunity to easily own affordable houses, with the option of expanding or enhancing the houses repeated at a pace commensurate with their financial capability.

Construction of the house during ex ante phase has to meet five major requirements before owners can move in. First, the structures must meet minimum residential housing space standards. For example, a habitable room must not be less than 3 m^2. Second, construction costs must be reconciled with affordability by target households in the beneficiary social group. Third, land tenure arrangement and level of infrastructure and services provision have to address long-term needs of secure ownership and improved quality of living. Fourth, there is a degree of involvement by the government, municipal agencies and external development assistance agencies, especially in construction of the housing units, infrastructure and services provision at ex ante phase. Fifth, the tenant purchasers of the site and service housing units are the key implementers as they construct the remaining areas of the house during ex post phase. Local Government Councils are charged with the responsibility of providing oversight implementation management over the building activities of the new owners under the tenant purchase arrangements. Housing structures are built to varying levels of completion in site and service communities. Transportation network (roads, streets) and services (water supply, sanitation and drainage network, education, health and community market facilities) are also fully developed.

Two sets of actors drive the implementation activities during ex ante and ex post phases of site and service schemes. Project financiers, government and urban local government authorities drive the ex ante phase. The first rooms of the house, all infrastructure and service facilities are constructed by or under direct and strict supervision of these drivers. The owners are the major players during ex post phase. They construct the remaining rooms and develop private spaces adjoining their houses. The urban local government of the town also drives the phase through plan and house design approval and providing regulations as well as planning and policy administration. Technical and professional guidance are provided to the owners during the construction of the remaining rooms of their house as well as the development of the adjoining green spaces.[5] Plan implementation activities, especially during ex post phase, determine whether the physical development and social character of the new housing scheme conform to the community envisaged in the plan. In fact, socio-cultural, economic and environmental sustainability of the community in terms of promoting the people's living standards in line with eco-city benchmarks is very much a function of activities during this phase (Orum and Chen 2003).

Site and service housing is one of the most visible urban housing sub-sector in Kenya's major towns. The sub-sector attained prominence as a means of housing delivery for the poor and low-income households in the 1970s and 1980s (UN-HABITAT 2006: 36–46). The Government of Kenya and urban local government councils in partnership with the World Bank and the United States Agency for International Development (USAID) implemented several site and service housing projects in Nairobi and other medium-size towns (Kenya 2004, UN-HABITAT 2009).

Different approaches were used to build in the site and service schemes in Nairobi. In Dandora, Kayole and Mathare North schemes, the Government of Kenya, Nairobi City Council and the World Bank were directly involved during the ex ante phase of housing construction (UN-HABITAT 2006). Outside government, non-governmental organizations were active in championing the rights of communities during the ex post phase as in the case of Mathare North site and service project. The households are actively involved in the construction of the remaining rooms after they move into their new homes.

9.3.1 Umoja 1 Residential Community

Umoja I project is a low-income site and service housing project initiated in 1974 by the Nairobi City Council and the USAID on successful completion of another similar project.[6] Umoja I, a comprehensive housing project, built 3,073 tenant purchase houses at a site located 7 km east of Nairobi Central Business District. The site which covers 126 hectares was planned for a density of 25 units per hectare; translating into 100 persons per hactare and 4 persons per household. The maximum population of this community is projected at 12,000 people. The soil cracks up to 50 cm deep when dry and expands by 50% when wet. Expert advice is therefore required when constructing the foundation for structures, and in landscape design so as to enhance the living environment, which is already preset by annual average temperature of 20°C, and which range between 15 and 25°C. This picture encapsulates sustainability benchmarks of the community and underlies eco-city attributes of residency in Umoja I. However, if this residential housing community has to accommodate a much larger low-income population, as indeed it has over the years, the resulting urban sprawl and additional new infrastructure and services costs are not desirable for two reasons. The first one is the inability of households to meet the costs for additional infrastructure and social service facilities. Secondly, the additional population would lead to higher densities and negatively impact the social and physical environment in the community.

The principles of sustainable urban development are secured in the concept of the plan and design of the houses (McManus and Haughton 2006). The development of the community is organized into a physical form of a “crescent”, which is built on a hexagonal grid and acts as a control in landscape design and a guide to development at both plot and community levels (Mwangi 1997). This has created a system of private and public green spaces, as well as a network for the mobility of people. Plots are 7 m by 18 m, with an effective area of 126 m^2. Houses are accessible through a hierarchy distributor, collector streets, pedestrian footpaths and a major road that link Umoja I with other urban districts in Nairobi. These define and secure front and rear private green spaces. The internal physical organization of the community is divided into 14 contagious sectors which are assigned formal reference letters A, C, D, E, F, G, H, J, K, L, M, N, P and Q.[7] Houses in each sector are given numbers 1, 2, 3,... *n,* where *n* is the number of the last house in a sector. There are no two sectors having the same number of houses. Public green spaces, services and infrastructure way leaves, adjoin the plots throughout the community. Sewerage for sanitation and run-off surface drains for storm water disposal and a network of streets are also fully developed (Kenworthy 2006).

Households are expected to manage domestic solid waste at the house level; with Nairobi City Council tasked to collect and transport the waste to a distant public garbage dumpsite. Two nursery schools and one primary school, a health centre, and a community market are also built. A site for the development of additional commercial centre is located close to the community market. Finally, land was allocated to religious and other social buildings in the community. Several churches and mosques were built.

The concept plan of the community and design principles outlined earlier makes three assumptions in order to meet the needs of first-time low-income homeowners with regular incomes. First, this category of urban households aspire to own and live in planned community of houses built around a formal design that meet the requirements of urban amenities and mobility standards. Participation of the owners in the construction of additional rooms using their own finances to complete the units according to the plan underscores this assumption. Second, structural constraints in accessing land and high investment costs of urban infrastructure development and service facilities hinder this category of households from owning homes in towns. It was assumed that some form of financial arrangement that facilitates them to pay for the initial cost of the houses during the ex ante phase would enhance the affordability of the houses, which they could pay for by monthly installments from formal employment. Furthermore, owning property through this route is considered a positive step in broadening and strengthening the urban economic base. Third, it is assumed that there exists a critical mass of potential households who aspire to own urban houses. These groups of people can afford and are committed and willing to service long-term tenant purchase credit. However, limited financial assistance to defray initial cost of land, investment in infrastructure and service facilities, as well as part of house superstructure that meet their initial shelter needs is required.

These three assumptions are the basis for ex ante and ex post phases of implementation of Umoja I residential development plan. The 3-room houses were built at three levels of completion during ex ante phase. In the first level, only the kitchen and wet-core and dining room were built. The owner was left with the task of constructing a further two rooms and a perimeter security wall. In the second level, one bedroom, wet-core and a kitchen were built. The owner was to construct one room and a perimeter wall only. In the third level, all the rooms were built and the owner was expected to construct the perimeter wall only. The tasks for ex post phase is to complete the remaining respective levels of the 3-room construction, and landscaping the front and rear private courts.

The goal of the plan is to encourage home ownership and discourage rental housing. The five specific objectives are: (i) to assist the Nairobi City Council implement a housing policy and to demonstrate that low cost houses for low-income social groups can be built cheaply; (ii) to influence the Nairobi City Council to adopt a housing strategy that focuses on the specific shelter needs of low-income families; (iii) to provide shelter to eligible low-income families; (iv) to increase housing stock of urban housing units; and (v) to improve the institutional capacity of the Nairobi City Council for implementation of low-income housing projects – planning, design and ex post implementation management[8] (Regional Housing and Urban Development Office 1983 and Nairobi City Council et al. 2007). Objective (v) in particular seeks to secure sustainable urban development and to foster eco-city practices in Umoja I community (Nairobi City Council et al. 2007, City of Alexandria 2008). However, weak planning and enforcement capacity and poor urban governance in Nairobi City Council has allowed owners to construct extension rooms on front and rear extension, as well as to encroach on reserved public green spaces since the project was first built in 1978.

9.3.2 Ex Post Plan Implementation by House Owners

A field study was conducted to appraise the 10-year ex post plan implementation of the Umoja I community in 1988 (Mwangi 1988). During this period, owners violated tenant purchase agreement and the plan of the community. The unauthorized construction of rooms at the front and rear, in addition to the permissible 3-room semi-detached house, reveal a grave weakness in Nairobi City Council's institutional capacity. Lack of enforcement of municipal planning and construction rules and regulations has emboldened the owners who have turned to construction of multi-storey rental apartments and commercial buildings from 1988 to 2008. Cases where original 3-room semi-detached houses are demolished and replaced by multi-storey structures covering the entire plot of land are rising by the day. The unfolding urban development chaos has transformed Umoja I from a community premised on social and environmental sustainability and implicit eco-city living; to a tragedy of urban planning in Kenya (Hardin 1968, Brightcloud and Hancock 2001, Mwangi 2009, Eco-city World Summit 2009).

9.3.2.1 Ten Years Ex-Post Plan Implementation Phase, 1978–1988

The small size of the 3-room house is the most cited reason by the owners for the violation of planning and development provisions in tenant purchase agreement. However, it has since been established that 95% of house owners who have built extension rooms rented the rooms to other tenants to get additional income. Family members used only 5% of houses with extensions for additional shelter. Sixty per cent of house owners without extension rooms in 1978 intended to construct the rooms, of which 60% of this proportion would rent out for additional income. This confirms that construction of unauthorized extensions is not an automatic response to any perceived inadequacy in plot size and the design of the house. The majority of households consider 126 m^2 areas of plot for one house to be adequate, although the violation of the plan and design is a potential source of serious threats to the environment now and in the future. Fifty-four per cent of all respondents in the study were aware of the consequences of unauthorized development on the environment in their community, while 46% said they were unaware. Similarly 68% of the respondents who live in houses without extensions are aware of the consequences of unauthorized development on the environment, while 32% of them expressed lack of awareness. It is therefore evident that the dangers of unplanned residential structures in the community are not appreciated.

The failure by the Nairobi City Council to enforce urban land use planning and development regulations at least confirms its objective for ex ante phase was not fully realized. The 29.8% level of staffing at the City Planning Department, against a 70.2% gap in pending staff recruitment at the time point to the Council's weak human resource capacity to enforce planning and development regulations. It also partly explains the unusually weak ex post plan implementation management regime. Following these weaknesses, the trend of construction of unauthorized accelerated during the latter years of 1978–1988; i.e. between 1982 and 1986. Distribution of the constructed unauthorized rooms during the second half of the 5 years is 2, 3, 5, 12 and 13 respectively, in the 50 housing units with rooms, were in the sample that was collected for appraisal. Indeed, of housing units with extensions, 70% had extensions built between 1982 and 1986, out of a total of 668 or 21.7% of houses with extension rooms in 1988.

The main cause for the unauthorized development is poor urban governance at the Nairobi City Council, which is reinforced by understaffing in the City Planning Department. There are six major outcomes from unauthorized development at the plot level. First, additional connections of the water supply system means that the 216-litre water storage tank installed in each of the 668 extensions held an additional 144,288 l of water at any one time. This was obviously not planned for. Second, additional sewer connections not only reinforced the problem associated with new water demand on existing water supply but also introduced stress on the designed capacity of the sewerage network. Third, at a rate of 1.3 kg per person per day in solid waste generation, households that lived in extensions produced additional 24,315.2 kg of solid waste during a 4-week period. Fourth, the construction of extension rooms on private open space at the front and rear of the main house led to an increase in the number of physical structures on each plot and a reduction of recreation and amenities space. Mobility on the plot was also severely curtailed. Finally, the trend in overcrowding the houses eroded the quality of social environment in the community. Besides, it altered micro-climate of community physical environment. These six outcomes initiated the degradation of the living environment of the 3-room house and the 668 extension rooms built by 1988. It was projected at the time that the remaining 78.1% of the 3-room houses without extensions would have had at least an extension room built, or the existing structure altered by the end of 2010 if the trend continued uninterrupted.

9.3.2.2 Twenty Years of Ex Post Plan Implementation by House Owners, 1988–2008

A field survey was carried out in early 2009 to confirm whether the construction of unauthorized extension rooms had stopped or continued during the last 20 years – 1988 to 2008. The survey that covered 30 housing units, about 1% of the 3,073 units, revealed that the trend has continued. In 2008, tenants consisted of 63.3% of sample residents and 36.7% were landlords, which reveal that tenancy housing is the dominant form of residency. Twenty per cent of the units covered in the survey have front and back extensions demolished and rental multi-storey buildings constructed in their place. The tallest multi-storey building has 5 floors built, and the highest number of rental rooms per floor is 8. Occupancy has also increased with 36.7% of household sizes in the units with 5 members each, while 26.7% of households have 4 members each. The original house is designed for a household size of 4 members. The size of 5 in the 36.7% of households is obviously higher than the recommended figure.

The survey covered 63.3% of tenant and 36.7% of landlord households. The lowest and the highest monthly rental by the tenants are Ksh 4,000 and Ksh 15,000, respectively, with 30% of the tenant paying Ksh 9,000.[9] Forty-three per cent of households have lived in the community for less than 10 years, while 56.7% of them have lived in the community for more than 10 years. The main reason for choosing Umoja I for rental housing is affordability (40%) and proximity to work place (10%). Construction of unauthorized extensions during 1988–2008 ex post phase of plan implementation has therefore continued. Of the newly constructed extension rooms, 96.7% and 93.3%[10] are connected to existing water supply and sewer line network respectively. This implies that direct provision of water to each tenement in the newly-built multi-storey apartments is considered to be more important than sanitation. Some tenant households living in these apartments do not have private sanitation rooms, – and several of them have to share the few rooms that are annexed to the apartments. This confirms that interference with service facilities, especially water supply, sanitation facilities, as well as open green spaces, has continued unchecked. As discussed earlier, this confirms the connection between continued construction of additional extension rooms with the problems of water supply and seepage of raw sewer into the portable waste systems that are frequently reported in the local media.

9.4 Conclusion

This chapter has discussed the ex post plan implementation of Umoja I low-income residential housing development. Implementation activities by house owners present daunting challenges to the sustainability of urban development, and has impacted negatively on the community. The house owners continue to construct unauthorized buildings that are used for rental and commercial activities. The inability of the Nairobi City Council to enforce land use and construction rules has led to these ex post implementation factors from 1978 to 2008, with deleterious implications on the built environment. Six factors have been noted here to have severely undermined the sustainability of Umoja I as an urban residential community poised to promote eco-city lifestyles (Alexandria City Council 2008).

The first is the altered and changed structural forms, character and functions of buildings. Second, planned capacities of roads, streets, water supply and sanitation networks and social facilities are overstretched, undermined and can no longer cope with increased loading without expanding the existing capacity. The third is the degradation of the social and physical environment, which is a gigantic daily challenge. These are hindering the realization of the benefits that come with planned urban development. Fourth, almost all houses are built beyond original planning and design level of completion. Densification of built structures with high-rise rental and commercial buildings dominating sites previously designated for a mere 3-room low-income house has spilled into public green spaces. A major physical and ecological change has therefore taken place in the community. Fifth, households from outside Umoja I have moved to live and work in new multi-storey apartments and commercial buildings; and this trend is on the rise. The community population has increased several times beyond the initial 12,000 people. Finally, service facilities such as those of water and sanitation stand out as areas of grave concern because they are overloaded and have turned out to be as a source of constant water supply shortage besides the health risks presented by the constant leaks from sewer pipes.

Overall, densification of unplanned for built structures and the additional population that has moved into Umoja I pose the greatest challenge for sustainable urban development in the community. The rental extension rooms built between 1978 and 1988; and the multi-storey buildings that were constructed in last 20 years, have severely weakened the principles of sustainability that were in-built in the concept plan and design of the houses along eco-city perspectives. The outcomes from this change include new factors in local microclimate, threats to social order, as well as increased potential risks from environmental neglect and poor services delivery. Lack of civic measures to stem, arrest and manage these outcomes undermine the promotion of eco-city standards of living in Umoja I community.

Traditional site and service schemes and low-density site and service housing may contribute to sustainable urban development on two interrelated grounds. The first involves elevating the importance of urban development to the level of regional and national economic and social development and environmental management. Clearly articulated goals in regard to employment creation and income generation in cities and towns should inform this concern. The second concern has to do with effective urban planning and implementation management of urban development projects during ex ante and ex post phases. Specifically, to overcome current challenges, effective and accountable urban governance, which is lacking in the Umoja I community, should focus on planning and implementation management process for sustainability and promotion of eco-city attributes if life in the community is to be reclaimed and retained.

Notes

1. This group is entirely excluded from access to urban shelter by a host of problems ranging from lack of employment, unreliable sources of income and collateral that would guarantee them access to housing finance from banks and non-banking financial institutions. This problem, combined with the inability to access urban land, hinders households in this group from owning formal urban housing.

2. Proof that at least one head of household has a formal job and earns a regular monthly income that qualifies her/him to join the scheme.
3. Ex ante implementation is when all works are done on site and service schemes and housing structures built at agreed levels of completion. It also involves the laying out of the requisite infrastructure and provision of essential services *before* target households move into their new houses. Ex post implementation is the building of additional rooms by households themselves according to the pre-approved house designs and site development plans in line with the provisions of the tenant purchase agreement between the owners and the development agency *after* target households have moved into their new housing structures. In this case study, the agency is the Nairobi City Council.
4. Apart from housing structures that are built to varying levels of completion, site and service housing communities benefit from an approved community development plan that caters for a transportation network, service facilities and drainage network and social services.
5. This development control involves policing and enforcing building and site development regulations so that they meet the expected land use development and habitat standards.
6. The project, Kimathi Housing Estate, comprises of 343 medium income housing units that were built in 1970, 4 km to the east of Nairobi Central Business District. A bank loan of US$10 million was secured in 1973 to finance Umoja I project.
7. There were no designed letters B, I and O.
8. The Housing Guaranty Programme (HGP) of USA guaranteed the bank loan of US$10 million that was secured by Regional Housing and Urban Development Office for the project. Nairobi City Council was given the loan and the Kenya Government guaranteed the loan.
9. Official exchange rates at the time was US$ = Ksh 76.
10. These figures are for the households surveyed.

References

Alexandria City Council (2008). *Eco-city charter 2008*. Alexandria, VA: Alexandria City Council. Environmental Policy Commission, Alexandria City Council; and The Urban Affairs and Planning Program, Virginia Polytechnic and State University – Alexandria Centre. http://www.alexandriava.gov/up/loadedfiles/tes/oeq/EcoCity/Charter2008.pdf. Accessed 26 September 2009.

Brightcloud, S. & Hancock, A. (2001). Tools for understanding and integrating sustainability: definition, concepts, framework, accountability, and planning. http://www.sustainablesonoma.org/resources/toolsforunderstanding.html. Accessed 17 September 2009.

Ecocity (2009). Welcome to ecocity world summit 2009, Instabul!. http://www.ecocity2009.com. Accessed on 18 September 2009.

Ecocity World Summit (2009). Announcing ecocity world summit 2009. http://www.ecocityworldsummit.org/index2.htm. Accessed on 18 September 2009.

Hardin, G. (1968). The tragedy of the commons. *Science*, 162(3859): 1243–1248.

Hope, K. E., Sr. (1997). The political economy of policy reform and change in Africa: the challenge of the transition from statism to liberalization. *Regional Development Dialogue*, 18(1): 126–140.

Kenworthy, J. R. (2006). The eco-city: ten key transport and planning dimensions for sustainable city development. *Environment and Urbanization*, 18(1): 9–22.

Kenya, Republic of (1978). *Human settlements in Kenya: a strategy for urban and rural development*. Nairobi: Physical Planning Department, Ministry of Lands.

Kenya, Republic of (2004). *Sessional Paper No. 3 on National Housing Policy for Kenya*. Nairobi: Ministry of Housing.

McManus, P. & Haughton, G. (2006). Planning with ecological footprints: a sympathetic critique of theory and practice. *Environment and Urbanization*, 18(1): 113–127.

Mwangi, I. K. (1988). *An appraisal of plan implementation and development control in Nairobi: a case study of Umoja I Estate.* Nairobi: Department of Urban and regional Planning, University of Nairobi. M.A. (Planning) Dissertation.
Mwangi, I. K. (1997). Nature of rental housing in Kenya. *Environment and Urbanization*, 9(2): 141–160. http://eau.sagepub.com/cgi/content/abstract/92/141. Accessed 4 September 2009.
Mwangi, F. (2009, September). Dagoretti area in a Shs 12 bn eco-city-plan. *The People Daily*, 28(2009): 3.
Nairobi Chronicle (2008a). Water rationing in Nairobi officially announced, Nairobi Chronicle, May 16, 2008. http://www.breakingnewskenya.com/2008/05/16/nairobi_water/. Accessed 3 August 2009.
Nairobi Chronicle (2008b). Water restored to Umoja amidst demolitions row. Nairobi Chronicle, June 21, 2008. http://www.breakingnewskenya.com/2008/06/21/water-restored-to-umoja-amidst-demolition row/. Accessed 23 August 2009.
Nairobi Chronicle (2009). Sewerage seeps into Nairobi water again. Nairobi Chronicle, January 15, 2009. http://www.breakingnewskenya.com/2009/01/15/sewarege-seeps-into-nairobi-water-again/. Accessed 3 August 2009.
Nairobi City Council, Ministry of Local Government and UNEP (2007). *City of Nairobi environment outlook: executive summary.* Nairobi: City Council of Nairobi.
National Housing Corporation (2008). *Presentation on NHC to Kenyans in the diaspora.* Nairobi: National Housing Corporation. http://www.nhckenya.co.ke/download/Presentation_to_Kenyans_in_the_Diaspora_Aug08.pdf, Accessed 23 October 2009.
OXFAM (2009). Urban poverty and vulnerability in Kenya: the urgent need for coordinated action to reduce urban poverty. http://www.oxfam.org.uk/resources/policy/conflict_disasters/bn-urban-poverty-vulnerability-kenya.html. Accessed 8 October 2009.
Orum, A. M. & Chen, X. (2003). *The world cities: places in comparative and historical perspective.* Malden, MA: Blackwell.
Regional Housing and Urban Development Office (1983). *Final evaluation report: Umoja I Estate.* Nairobi: Nairobi Regional Housing and Development Office.
Srinivas, H. (2009a). Sites and services. http://www.gdrc.org/uem/squatters/s-and-s.html. Accessed 20 September 2009.
Srinivas, H. (2009b). Urban development and poverty. http://www.gdrc.org/uem/squatters/urban-poverty.html. Accessed 20 September 2009.
Suzuki, H., Dastar, A., Moffatt, S. & Yabuki, N. (Eds.). (2009). *ECO^2 Cities: ecological cities as economic cities.* Washington, DC: The World Bank.
UNDP (2005). *Linking industrialization to human development.* Nairobi: UNDP Kenya. http://www.ke.undp.org/UNDP-4thKHDR.pdf. Accessed 26 October 2009.
UNDP (2006). *Kenya human development report.* Nairobi: UNDP.
UN-HABITAT (2005). *Urban-rural linkages approach to sustainable development.* Nairobi: UN-Habitat. HS/765/05E.
UN-HABITAT (2006). *Case study of sites and services schemes in Kenya: lessons from Dandora and Thika.* Nairobi: UN-Habitat. http://www.unhabitat.org/documents/docs/3659_64527_HS-111.pdf. Accessed 18 September 2009.
UN-HABITAT (2008). *The state of African cities, 2008: a framework for addressing urban challenges in Africa.* Nairobi: Economic Commission for Africa and UN-Habitat.
UN-HABITAT (2009). *Planning sustainable cities: Policy directions.* Global report on human settlement. London: Earthscan.
White, M. J., Mberu, B. U. & Collinson, M. A. (2008). African urbanization: recent trends and implications. In G. Martine, G. McGranahan, M. Montgomery, & R. Fenandez-Castilla (Eds.), *The new global frontier: urbanization, poverty and environment in the 21st century* (pp. 301–316). London and Sterling, VA: Earthscan.
World Commission on Environment and Development (1987). *Our common future.* Oxford: Oxford University Press. A World Commission on Environment and Development Report.

Chapter 10
Towards a Sustainable Regional Development in Malaysia: The Case of Iskandar Malaysia

Chin-Siong Ho and Wee-Kean Fong

Abstract Urban planners by duty play the role of developing functional and aesthetically pleasing cities with the most optimal use of land. At the same time, they ensure that the cities planned for are ecologically friendly, and that they develop low CO_2 emissions. The planning of the proposed high growth region, South Johor Economic Region (SJER), now commonly known as the Iskandar Malaysia, provides a good opportunity to achieve such a purpose. This chapter reexamines the concept of low carbon cities and explores the perspective and scenarios towards transforming the Iskandar Malaysia into an environmentally sustainable urban region.

10.1 Introduction

With more than 3.2 billion people living in the cities for the first time, the world's urban population now exceeds the number of people living in rural areas (Sandrasagra 2007). By the year 2030, it is expected that over 60% of the world's population will be living in towns and cities (UN 2002). These urban environments are responsible for over 70% of overall global carbon emissions. Hence, in order to tackle the issue of carbon emission, there is a need for global and national strategies for sustainability in urban environments – in both existing and new developments, and from inception to occupation.

Planning of a sustainable region involves creation of low carbon society (LCS) by promoting low carbon emission. In order to achieve a low carbon society, effort to reduce CO_2 emission is most important as CO_2 is the most significant anthropogenic greenhouse gas (GHG) emitted in urban areas. The increases of CO_2 concentration are due primarily to fossil fuel use and land use change. Urban planning through

C.-S. Ho (✉)
Universiti Teknologi Malaysia, Johor Bahru, Johor, Malaysia
e-mail: ho@utm.my

T.-C. Wong, B. Yuen (eds.), *Eco-city Planning*, DOI 10.1007/978-94-007-0383-4_10,

land use planning and planning control can play a vital role in implementing the idea of low carbon cities, particularly during the formulation of development plans. Spatial strategies in development plans should adopt sustainable development principles such as compact cities, eco-cities and Transit Oriented Development (TOD) and other concepts of energy efficient city. Some of these ideas are currently gaining popularity and have been incorporated in development plans of many newly planned cities.

Apart from spatial planning strategies, reduction in CO_2 emission can be done through non-spatial strategies such as fuel or vehicle legislations used in the Low Carbon Fuel Standard (LCFS), and through standards for CO_2 limits in vehicle engine emissions, for example, the use of hybrid vehicles. This chapter focuses on the spatial strategies and aims to examine the concept of low carbon cities and explore the scenarios towards the realization of sustainable cities in the newly planned Iskandar Economic Region in southern part of West Malaysia.

10.2 Sustainable Regional Development – Population and Economic Growth and CO_2 Emissions

The rapid increase of CO_2 emissions has caused many concerns among policy makers. In discussing ways to curb CO_2 emissions, most attention tends to focus on the role of affluence and population increase (Dietz and Rosa 1997). Studies have showed that population and economic growth are the major driving forces behind the rise in CO_2 emissions worldwide over the last two decades. It is particularly true in developing countries where the impact of population on emissions has been more pronounced. On average, it is found that a 1% increase in population is associated with a 1.28% increase in CO_2 emissions (Shi 2001). With such magnitude, global emissions are likely to grow substantially over the next decades. Thus, international negotiations and cooperation on curbing the rapid growth of CO_2 emissions should take into consideration the dynamics of future population growth.

The reduction of global emissions will become a more challenging task as most developing countries and newly industrializing countries will be experiencing rapid economic growth in the next decades. Rising income levels are associated with a large upward trend in emissions. Thus, another potential policy intervention on the reduction of emissions could also be in the area of increasing the energy efficiency of economic production both in developed and developing countries. Without these policy considerations on the role of energy efficiency, economic growth alone could be leading to a further worsening of global CO_2 emissions.

Planning of sustainable region needs to incorporate the ideas of low carbon society and low carbon economy in urban areas. Researchers and policy makers responsible for our climate change and energy modeling have used the term low carbon society in 2003 when developed nations announced a target for reducing CO_2 emission in order to stabilize the world's climate. Low carbon society project has been initiated by Japan/UK collaboration to draw out comprehensive vision and definition of low carbon society (NIES 2006). Several scientific research works have been carried out involving reviewing GHG emission scenarios studies, studying methodologies in achieving low carbon society and sharing best practice and information between countries.

10.3 Global and Malaysian – Carbon Dioxide Emissions

Every country contributes different amounts of heat-trapping gases to the atmosphere. Table 10.1 shows that in general, total amounts of CO_2 continue to increase in most world regions. Developed countries have been leading in total carbon emissions, accounting for more than 50% of the world's total. Countries in North America and Europe that consist mainly of developed countries emitted most of the anthropogenic GHGs. Over the last few decades, industrial development in Asia and the Middle East has resulted in rapid increase of CO_2 emissions, with an annual percentage change of more than 5%, as compared with global average of 1.6% during the period 1990–2003. Developing countries such as China, India, Russia and Brazil which are in the fast transitional stage of industrialization have contributed to this rapid increase.

Table 10.2 shows that the world average CO_2 emission was 4.1 metric tons per capita in 2003. Per capita emissions in developed nations such as in Europe and North America are higher than the world average, while developing countries

Table 10.1 Total CO_2 emissions by region, 1990–2003

Region	1990 million metric tons	2000 million metric tons	2003 million metric tons	Annual % change 1990–2003
World	21,283.38	23,832.70	25,575.99	1.6
Asia (excluding Middle East)	5,014.89	7,272.53	8,477.90	5.3
Central America & Caribbean	379.32	467.09	500.58	2.5
Europe	–	6,002.02	6,277.17	1.5
Middle East & North Africa	926.96	1,474.34	1,645.98	6.0
North America	5,274.41	6,232.06	6,257.98	1.4
South America	537.47	757.03	740.45	2.9
Developed countries	–	14,623.79	15,043.57	1.0
Developing countries	5,839.34	8,475.59	9,810.41	5.2
High income countries	10,452.47	12,123.43	12,420.82	1.4
Middle income countries	–	9,204.17	10,486.71	1.1
Low income countries	912.89	1,494.26	1,631.11	6.1

Source: WRI (2007a)

Table 10.2 Carbon dioxide emission per capita by region, 1990–2003

	1990 metric tons per capita	2000 metric tons per capita	2003 metric tons per capita	Annual % change 1990–2003
World	4.0	3.9	4.1	0.2
Asia (excluding Middle East)	1.7	2.1	2.4	3.2
Central America & Caribbean	2.7	2.8	2.9	0.6
Europe	10.1	8.1	8.5	−1.2
Middle East & North Africa	3.0	3.9	4.1	2.8
North America	18.6	19.8	19.3	0.3
South America	1.8	2.2	2.0	0.9
Developed countries	12.0	11.0	11.1	−0.6
Developing countries	1.5	1.9	2.1	3.1
High income countries	11.8	12.8	12.8	0.7
Low income countries	0.6	0.7	0.8	2.6
Middle income countries	3.3	3.2	3.5	0.5

Source: WRI (2007a)

are still less than the world average, in the range of 2–4 metric tons per capita. Obviously, this uneven distribution of CO_2 emissions is a big challenge to the world community in finding effective and equitable solutions for global warming and climate change issues.

Malaysia is a newly developing nation and one of the 172 countries that have signed the Kyoto Protocol in the United Nations Framework Convention on Climate Change (UNFCCC) on 12 March 1999. Clauses of this Convention were further ratified on 4 September 2002, aimed at combating global warming. However, ratification does not imply a country has agreed to cap their emissions, and Malaysia is not within the 35 countries that have agreed to cap their emissions.

Figure 10.1 shows the comparison of CO_2 emission per capita of Malaysia and other countries. Malaysia, with an average CO_2 emission of 6.2 metric tons per capita, is considered to be higher among the newly industrializing nations, and it is higher than the world average.

In spite of the absence of cap on emission, Malaysian government has been continuously promoting environmental stewardship in all its development plans. Since the Eighth Malaysia Plan (2001–2005), the incorporation of environmental consideration into planning and development was intensified (EPU 2001). With continuous efforts to promote sustainable development, Malaysia was ranked 38 out of 146 countries worldwide in 2005 (second in Asia after Japan), scoring a high 54 in the Environmental Sustainability Index (ESI) (Yale University 2005). Other

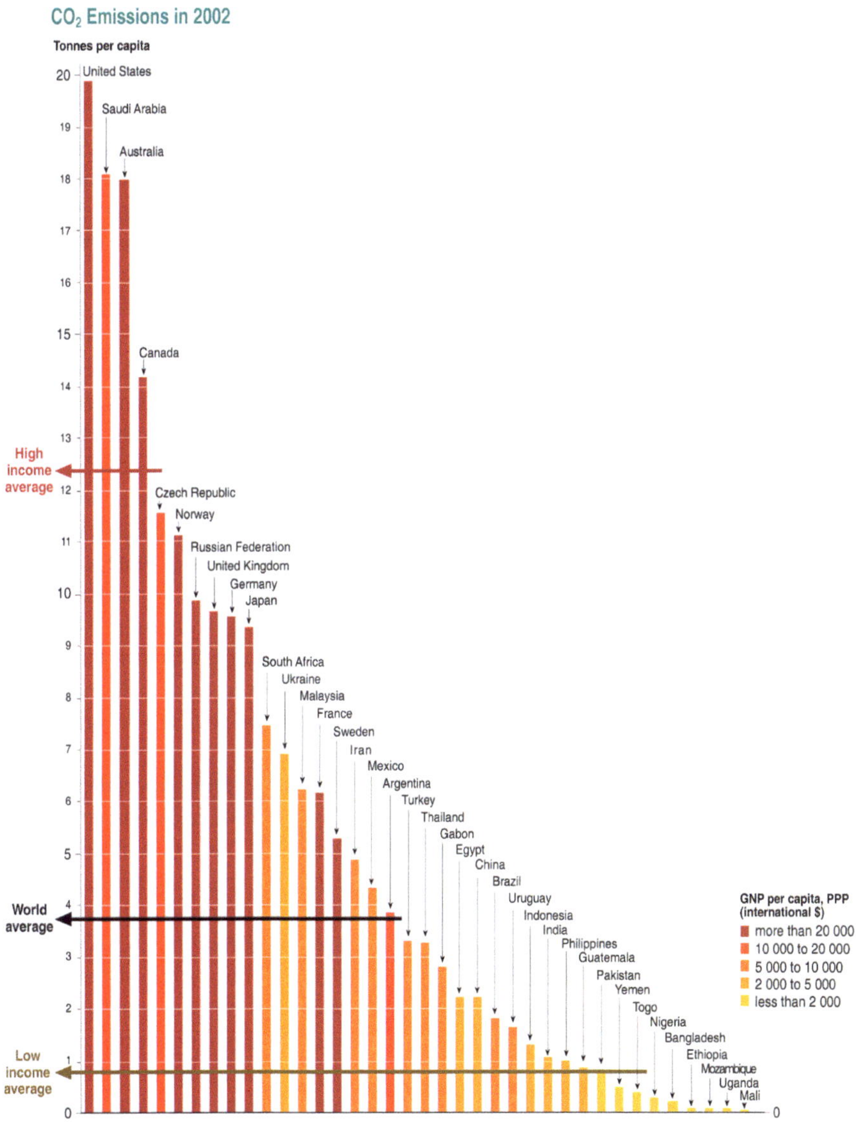

Fig. 10.1 Comparison of Malaysia and other countries in terms of CO_2 emission per capita in 2002. *Source*: United Nations Environment Programme/GRID-Arendal (2007)

multilateral environmental agreements and related amendments signed and ratified include the Stockholm Convention on Persistent Organic Pollutants, Montreal Protocol, on substances that deplete the ozone layer, Basel Convention on the transboundary movement of hazardous waste and their disposal, Rotterdam Convention on prior consent procedure for hazardous chemical and pesticides in international trade and the Cartagena Protocol on bio-safety.

Most developing countries and newly industrializing countries including Malaysia would consider that economic development must come first before handling environmental issues. Many of these countries are still building coal-fired power plants and still promoting predominately private transportation. Coal releases more CO_2 into the atmosphere than any other energy sources. Several automakers in these fast growing developing countries are competing to provide affordable cars to the country's increasing number of middle class population. The rapid increase of private car ownership and falling percentage of public transportation users will further increase the country's CO_2 emissions in the future.

In Malaysia, emphasis is placed on improving environmental quality through better management in major areas of concern particularly air, water quality and solid waste management as well as the utilization of cleaner technologies (EPU 2006). Concerted effort of the Malaysian government in formulating National Environmental Policy adopted in 2002 outlined strategies to propel the direction of national growth towards sustainable development. Accordingly, the new Ministry of Natural Resources and Environment to consolidate 10 environmental and natural resources agencies under one administration was set up to facilitate and manage environment and natural resources. However, environmental management is mainly carried out as environmental quality regulation such as measures to reduce occurrence of haze and reduction of pollutants (NO_x, CO, etc.). Comprehensive low carbon emission policy is not mentioned officially. Although such environmental quality regulations and protecting forest resources and other initiatives do indirectly reduce CO_2 emission, it is necessary to look into low carbon society scenario more holistically.

As a developing country, Malaysia responsibly attaches great importance to the issue of climate change, and has taken several initiatives to reduce carbon emissions and promote energy efficiency. Under the Ninth Malaysia Plan (2006–2010) (EPU 2006), policy strategies are outlined to increase energy efficiency and promotion on the use of renewable energy. In terms of sustainable energy development, the energy sectors aimed to enhance its role as an enabler towards strengthening economic growth. Source of fuel will be diversified through greater utilization of renewable energy. Emphasis was given to further reduce dependency on petroleum products by increasing the use of alternative fuels. A more integrated planning approach was undertaken to enhance sustainable development of the energy sector. During the Eighth Malaysia Plan (2001–2005) (EPU 2001), development of the energy sector was focused on ensuring a secure, reliable cost-effective supply of energy, aimed at enhancing competitiveness and resilience of the economy. Efficient utilization of energy resources as well as the use of alternative fuels particularly renewable energy, was further promoted. Energy related strategies were streamlined to moderate the impact of escalating oil prices on the economy.

Table 10.3 shows that transport sector was the largest consumer of energy in Malaysia, accounting for more than 40% of total during the period 2000–2005. This was followed by industrial and commercial/residential sectors at about 38% and 13% respectively. The overall energy demand at the national level is projected to

Table 10.3 Final commercial energy demand by sector in Malaysia, 2000–2010

	Peta Joules[c] (PJ)			Percentage of the total			
Sources	2000	2005	2010[d]	2000	2005	2010[d]	Annual growth rate (%)
Industrial[a]	477.6	630.7	859.9	38.4	38.6	38.8	6.4
Transport	505.5	661.3	911.7	40.6	40.5	41.1	6.6
Residential/ commercial	162.0	213	284.9	13.0	13.1	12.8	6.0
Non energy[b]	94.2	118.7	144.7	7.6	7.3	6.5	4.0
Agriculture/ forestry	4.4	8.0	16.7	0.4	0.5	0.8	15.9
Total	1,243.7	1,631.7	2,217.9	100.0	100.0	100.0	6.3

Note:
[a]Include manufacturing, mining and construction
[b]Include natural gas, bitumen, asphalt, industrial feedstock and grease
[c]1 P J = 1,000,000,000,000,000 J (10 to the power of 15)
[d]Projected
Source: EPU (2006)

increase by about 6.3% annually during Ninth Malaysia Plan period (2006–2010) to 2,217.9 PJ. Similarly, the per capita energy consumption is projected to increase from 52.9 GJ in 2000 to 76.5 GJ in 2010 (EPU 2006). The energy intensity (ratio of total primary energy consumption to gross domestic product) had also showed an increasing trend from 5.9 GJ in 2000 to 6.2 GJ in 2005 (EPU 2006). Although all the above demand parameters showed increasing energy demand to sustain economic growth, energy efficiency initiatives particularly in industrial, transportation and commercial sectors as well as government buildings are taken by government to achieve the aim of efficient utilization of energy resources. Similarly efforts were continued to promote the utilization of renewable energy (RE) resources such as the Small Renewal Energy Power Programme (SREP) and the Malaysia Building Integrated Photovoltaic Technology Application Project (MBIPV). All these projects will help to reduce CO_2 emission.

In spite of these efforts, Table 10.4 shows that CO_2 emission in Malaysia is still relatively high in terms of percentage change (120%) as compared to Asia (35.1%) and the world (12.7%), lagging far behind developed countries. With the per capita emission of 5.4 metric tons which is much higher than the global average of 3.9 metric tons per capita and Asian average of 2.2 metric tons per capita in the year 2000, data show that Malaysia continues to experience a rapid increase of CO_2 emissions (Fig. 10.2). It is expected that the emissions will continue at a high rate with relatively high rates of population and economic growth. The subsequent section will examine the ways in which Malaysia attempts to use a newly proposed development region, the Iskandar Malaysia to showcase a prototype of a green and sustainable urban region to achieve carbon reduction.

Table 10.4 CO_2 emissions in Malaysia as compared to Asia and the world, 2000

Comparison		Malaysia	Asia	World
CO_2 emissions	Total 2000 (mil. tons)	123.6	7,837.0	23,895.7
	% change since 1990	120.3	35.1	12.7
	Per capita (2000)	5.4[a]	2.2	3.9
Cumulative CO_2 emissions (million metric tons)	Fossil fuels and cement	1,714	175,087	781,501
	From land use change	20,654	163,621	315,122
CO_2 emission by sector (as % of total emission)	Transportation	26.2%	13.3%	24.1%
	Industry and construction	23.1%	24.7%	18.5%
	Electricity	25.5%	40.1%	38.3%

Note: [a]The value is different from that indicated in Fig. 10.1 due to different source of information and base year
Source: WRI (2007b)

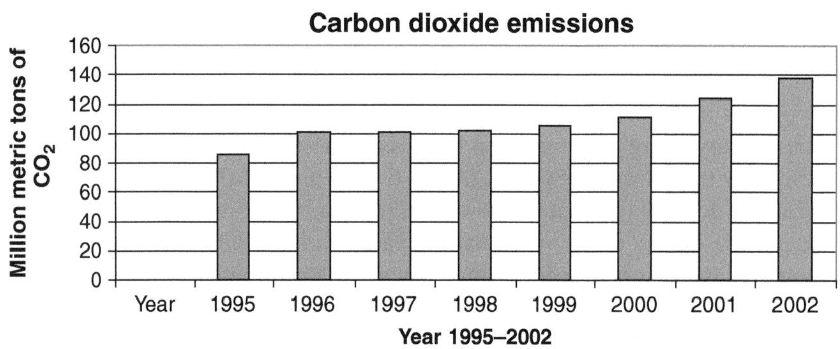

Fig. 10.2 CO_2 emission in Malaysia, 1995–2002. *Source*: Energy Information Administration (2007)

10.4 Iskandar Malaysia

10.4.1 Background

In order to plan for low carbon cities, it is more effective to look into the urban areas as they are engines of economic growth as well as main contributors to CO_2 emission. In the case of Malaysia, the natural resource management through spatial planning approach integrates environmentally sustainable development concepts. These strategies are incorporated into the National Physical Plan and then translated into structure plans where it also identifies and manages environmental sensitivity

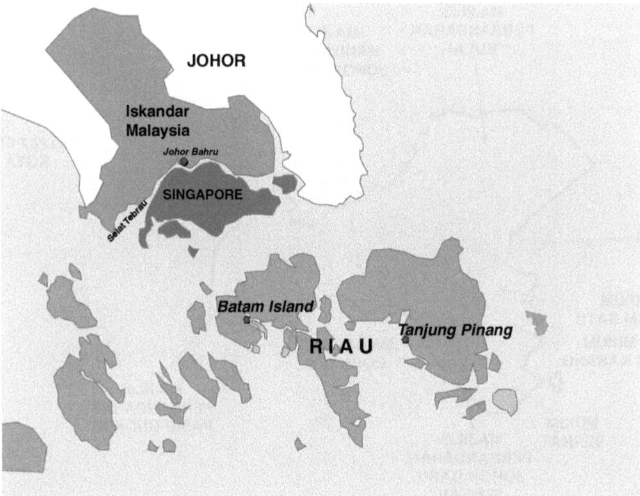

Fig. 10.3 Iskandar Malaysia and the surrounding region – Singapore and Riau of Indonesia

areas (ESAs) including forest and green lung reserves. Major urban conurbations are identified and three economic growth areas are demarcated in Peninsular Malaysia as regions or sub-regions where it will develop to be globally competitive. The South Johor Economic Region (SJER), which commonly known as the Iskandar Malaysia (IM), is one of these economic growth centres to be developed as an integrated global node of Singapore and Indonesia (cf. Fig. 10.3).

IM covers an area of about 2,216.3 km^2. The region encompasses an area about three times the size of Singapore. IM covers the entire District of Johor Bahru, and several sub-districts (*mukim*) of Pontian (cf. Fig. 10.4). The Planning Area falls under the jurisdiction of five local planning authorities, namely Johor Bahru City Council, Johor Bahru Tengah Municipal Council, Pasir Gudang Local Authority, Kulai Municipal Council and Pontian District Council. As shown in Table 10.5, there are a total of five flagship zones proposed as key focal points for developments.

Each of these flagships has a major urban centre. Among these urban centres are Johor Bahru City (financial district), proposed Nusajaya urban centre (new State administrative centre), Pasir Gudang/Tg. Langsat (port and industrial township) and Senai-Skudai/Kulai (transport and cargo hub). Four of the focal points will be located in the Nusajaya-Johor Bahru-Pasir Gudang corridor, also known as the Special Economic Corridor (SEC) (Fig. 10.5).

The planning of these urban centres in the 5 flagship zones provides opportunities for planners to explore the ideas of low carbon cities in these five proposed core areas. With the proposal of relatively high plot ratio of 3.0–7.0 and promotion of mixed land uses at Johor Bahru, Nusajaya centres and local centres, it allows development of a self-contained compact city. The corridor development along Johor Bahru – Nusajaya – Senai can also facilitate the TOD development in the region.

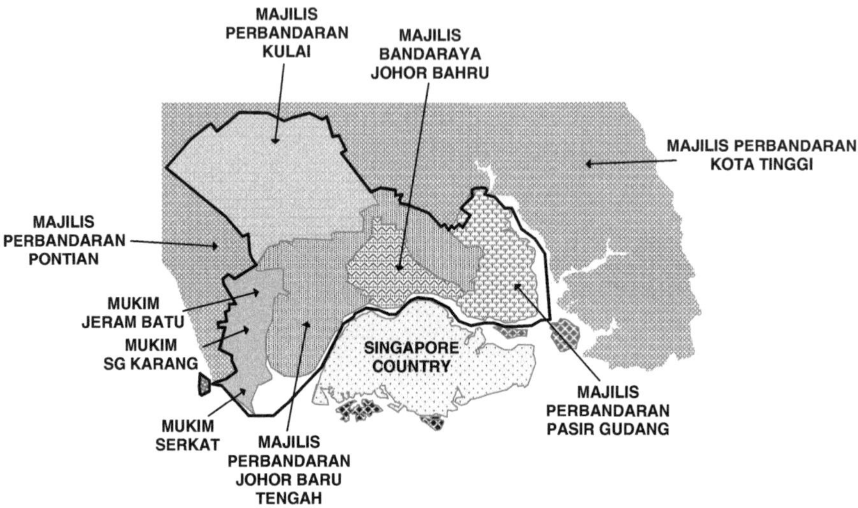

Fig. 10.4 Local planning authorities in IDR. *Source*: Khazanah Nasional (2006)

Table 10.5 The five flagship zones proposed in Iskandar Malaysia

Flagship Zone A:	Johor Bahru City Centre (new financial district, central business district, Danga Bay integrated waterfront city, Tebrau Plentong mixed development, causeway)
Flagship Zone B:	Nusajaya (Johor state administrative centre, medical hub, educity, international destination resort and southern industrial logistic cluster)
Flagship Zone C:	Western Gate Development (Port of Tanjung Pelepas (PTP), 2nd Link (Malaysia/Singapore), Free Trade Zone, RAMSAR World Heritage Park and Tanjung Piai)
Flagship Zone D:	Eastern Gate Development (Pasir Gudang Port and industrial zone, Tanjung Langsat Port and Technology Park and Kim-Kim regional distribution centre)
Flagship Zone E:	Senai-Skudai (Senai International Airport and Senai cargo hub)

Source: Khazanah Nasional (2006). Comprehensive Development Plan for South Johor Economic Region 2006–2025

Higher density and mixed land use are favourable for the implementation of compact city development as well as the use of district cooling (DC). All the above measures will help to improve energy efficiency as well as reduction in CO_2 emission in the planned region.

Broadly, the economy in the IM may be divided into three main sectors, namely primary (agriculture, fishing, forestry/wetland and mining), secondary sector (food processing, basic metal processing, non-metal processing, wood processing) and tertiary sector (retail and transport). Currently, the two main economic growth sectors in IM are manufacturing and services. The key manufacturing sectors that drive the economy are electrical and electronic (E&E), chemical and chemical products

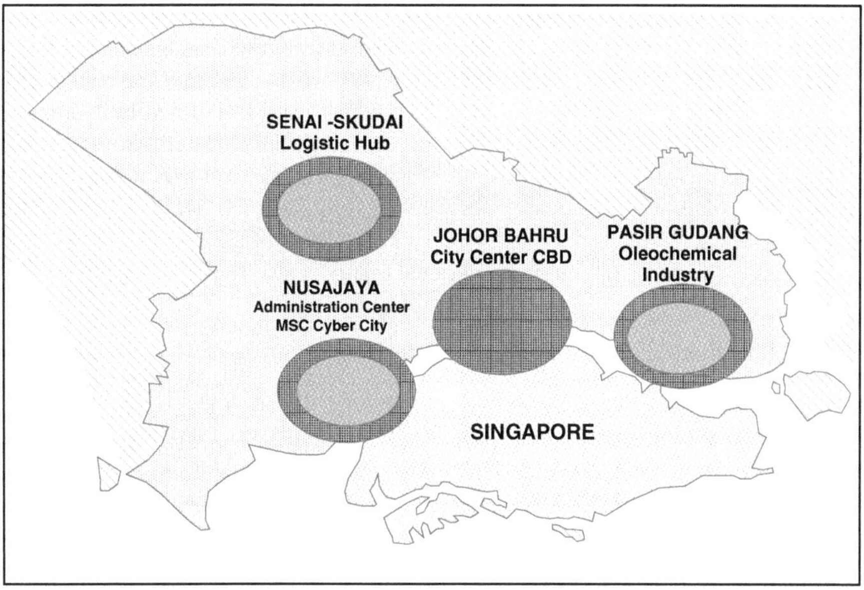

Fig. 10.5 Flagship zones and compact cities development in IDR. *Source*: Khazanah Nasional (2006)

(petrochemical, plastics, oleo chemicals) and food processing. They contribute 60% of the total value-added in manufacturing, and they lead to the emergence of supporting or induced sectors such as retail, wholesale, hotels, restaurants and finance. In manufacturing, the induced sectors include fabricated metal products, non-metallic products and transportation equipment.

10.4.2 Development Policies Related to Sustainable Region

There are three main policies stated in the master plan for IM, known as the Comprehensive Development Plan for South Johor Economic Region, 2006–2025 (hereinafter referred as "CDP"), which have direct impact on low carbon scenario of the IM development. These polices are energy efficient building, sustainable land use and transportation, and natural and green environment, which are elaborated below.

10.4.2.1 Energy Efficient Building and Sustainable Neighbourhood Design

In creating livable communities, energy efficient building and sustainable design guidelines are proposed. In order to encourage construction of energy efficient buildings, "green building rating" will be used for residential units, and to introduce energy efficient mechanisms on older or existing buildings in the development region. Green building is the practice of creating healthier and more resource-efficient models of construction, renovation, operation, maintenance and demolition.

Sustainable neighbourhood design will be used and implemented to encourage developers to plan neighbourhood with self-contained facilities to reduce the use of private vehicles and hence reduce transportation energy and CO_2 emission. One of the key thrusts of IM is to create livable communities that encompass quality housing, adequate facilities, quality services and a healthy, safe and lively environment. To this end, the CDP plans not only for the current needs of the population but also for the future, ensuring that inter-generational equity is also sensitively addressed.

10.4.2.2 Sustainable Land Use and Transportation

Land use planning helps to integrate environmentally sustainable development concepts by promoting mixed land use and public transport (and non-motorized vehicles) and compact city development. The use of zoning district system (base zoning district and special overlay zones) allows appropriate and compatible mixed use development by combining retail/service use with residential or office use in the same building or on the same site can help to reduce in between space movement. Hence it can also reduce transportation energy and CO_2 emission.

Transit Planning Zones is also introduced in city centre areas such as within Johor Bahru City Centre and Nusajaya City Centre to promote a combination of commercial and housing on the same site. It allows developments with increased intensity especially the residential component. This aims to support the strategy of encouraging city living and transit oriented development. Transit Planning Zone is area within the 400 m radius of rail stations where transit oriented development can be pursued. This form of development will help to promote the use of rail transport.

In addition, incentives are also given to encourage sustainable pattern of urban regeneration development through Brownfield development in the existing urban centres of Johor Bahru, Senai and Skudai. It provides a broad range of uses and is intensified in terms of commercial plot ratio and densities to reflect its role as the centre of administration, business, commerce, and employment of IM and the new growth centre within the Special Economic Corridor (SEC). This high density development will provide critical mass to support vibrant activity.

10.4.2.3 Sustainable Natural and Green Environment

The natural and green environment in IM covers a total of more than 150,000 ha of green spaces. This include RAMSAR site (9,483 ha), Pulai State Park (5,570 ha), regional park (3,178 ha), district park (1,514 ha), town park (941 ha) and local parks (204 ha) as well as the agriculture areas. All these green spaces play an important role as a carbon sink for this region.

RAMSAR site is wetland of international importance which are of rare and unique and for conserving biological diversity. The three RAMSAR sites in IM are Pulau Kukup, Sungai Pulai and Tanjung Piai, which are the Rank 1 Environmental Sensitive Areas (ESA). This would be able to reduce CO_2 in the atmosphere. Other public green space amenities to the general public, the private open space (POS) which refers to private green areas particularly golf courses are another green space located in the urban areas. There are also substantial areas in IM still under the category of agriculture, predominantly oil palm plantations. Some of the areas are classified as Environmental Protection Zone where it requires further environmental control by virtue of their identification as Environmental Sensitive Areas (ESA). In addition, water catchments zone (catchments of Sultan Iskandar Dam) is a Rank 1 ESA and needs to be protected. All activities within the water catchments zone must be controlled and no industrial activities should be allowed.

10.4.3 Scenarios of CO_2 Emissions from Energy Use

In realizing the vision of low carbon city, besides the real efforts to cut down the emissions, it is necessary to establish a database of CO_2 emissions. In this respect, it is necessary to develop a standard method for estimating CO_2 emission and benchmark of the present emission should be established, also, projection of the possible future emission trends should be carried out.

In this study, CO_2 emissions from energy use in IM have been estimated based on an integrated approach, using the System Dynamics Model (SD Model). A computer programming software known as STELLA was used to construct the SD model for the complicated urban energy consumption system, to estimate the CO_2 emissions from energy use in IM and to forecast the future emission trends.

The model consists of six sub-models, namely the residential, commercial, industrial, transportation, agriculture and carbon sequestration sub-models that representing the main sources of energy consumptions and CO_2 emissions as well as carbon sequestration by vegetations in the study area (Fong et al. 2009). These sub-models are interrelated with a number of variables such as population, economics, etc. and they are related to each other by various equations and assumptions.

In developing the SD Model, 2005 was used as the base year for this study on CO_2 emissions from energy use in IM, with the essential data mainly obtained from the CDP and complemented by various other sources. The underlying assumptions adopted in this SD Model are as follows:

- Only CO_2 from energy consumption were taken into consideration. Emissions from primary sectors were omitted;
- CO_2 emissions were calculated based on the consumptions of electricity, diesel, fuel oil, liquid petroleum gas (LPG), coal and coke, petrol and kerosene;
- Population growth rate, economic structure and economic growth rate were based on the values adopted in the CDP;
- One household per residential unit;
- Energy consumptions by commercial and industrial sectors were calculated based on energy consumption per unit of Gross Domestic Product (GDP); and
- Vehicles were classified into four categories, namely motorcycle, car, bus and lorry.

For the forecast of CO_2 emissions, an initial simulation using the SD Model was carried out based on the "Business as Usual" Scenario (BaU Scenario), according to various population and economic targets adopted in the CDP. This scenario was assumed to be a "high growth scenario" with high growth in all sectors (population growth at 4.1% and GDP growth at 8% per annum as envisaged by the Malaysian Government on the IM region over the next 20 years.

As mentioned above, it is challenging to maintain the high economic and population growths while reducing energy consumption and CO_2 emission. In order to examine the possible CO_2 emission reduction that can be achieved under the present target of high population and economic growths as reported in the CDP, besides BaU Scenario mentioned above, another two scenarios have been developed, namely Moderate Measure Scenario (MM Scenario) and Drastic Measure Scenario (DM Scenario). Table 10.6 presents the main assumptions made for each simulation scenario. Under MM Scenario, it was assumed that moderate measures such as improvement in energy efficiency, change of lifestyle, improvement of public

Table 10.6 Simulation scenarios

Scenario	Business as usual (BaU)	Moderate measures (MM)	Drastic measures (DM)	Measures
Population growth	4.1%	4.1%	4.1%	BaU
GDP growth	8%	8%	8%	BaU
Residential sector	BaU	Energy saving up to 20%	Energy saving up to 40%	Improved energy efficiency of home appliances Energy saving lifestyle Passive building technologies
Commercial sector	BaU	Energy saving up to 20%	Energy saving up to 40%	Improved energy efficiency of equipments Passive building technologies
Industrial sector	BaU	Energy saving up to 20%	Energy saving up to 40%	Improved energy efficiency of equipments/machinery Shifting to low energy consuming industries Usage of renewable energy
Transportation sector	BaU	Energy saving up to 20%	Energy saving up to 40%	Reduced vehicle population growth rate Reduced travel distance by motor vehicle through measures such as land use planning, improved public transportation system, car pooling and other transportation policies e.g. tolls, taxations, parking charges etc.

Source: Compiled by authors

transportation system, control of car ownership, and other environmental measures will be taken to reduce the energy consumption in all sectors. While in DM Scenario, it was assumed that similar measures as per MM Scenario will be implemented, but with more drastic moves to cut down energy consumption and CO_2 emissions.

Table 10.7 presents the simulation results. From the simulations, it was found that the total CO_2 emissions from energy use in IM in 2005 was about 5.6 million metric tons, with 53% from industrial sector, 34% from transportation sector, 11% from commercial sector, 2% from residential sector and less than 1% from agricultural sector. CO_2 emission per capita is estimated to be 4.1 metric tons, which is lower than Tokyo and Greater London (cf. Table 10.8).

For the coming 20 years, it is expected that the population and economy in IM will be growing at very high rates as envisaged by the Malaysian Government. The GDP per capita in 2025 is expected to be double of 2005 value.

With the high economic growth rate, it is likely to entail in high productivity and high consumptions in all aspects including energy. Under BaU Scenario, by 2025 it is expected that the total CO_2 emissions from energy use in IM would be about 21 million metric tons with per capita emission of 6.9 metric tons, which is much higher than the present emission levels of Tokyo and Greater London. As shown in Table 10.7, comparing to 2005, the total CO_2 emissions in 2025 is more than three times of 2005 level, and the per capita emission rate is also very high compared to 2005 level.

Table 10.9 shows the CO_2 emission trends over the 20-year projection period under BaU Scenario, MM Scenario and DM Scenario. For BaU Scenario, it can be seen that the proportion of industrial sector is getting smaller while commercial sector is getting larger compared to 2005. This is because the present manufacturing industry is focusing more on heavy industries such as oil and gas, cement, steel, etc. while according to the CDP, the future industry would gradually change to service industry, which has lower energy consumption rates.

Under MM Scenario, it was assumed that the population and economic growth rates would be maintained as in Scenario 1 (cf. Table 10.6). However, due to the implementation of various energy saving measures, CO_2 emission will be quite significantly reduced. The 2025 CO_2 emission under MM Scenario is expected to be about 86% of BaU Scenario and the per capita emission will be only 5.9 metric tons compared to 6.9 metric tons under BaU Scenario (cf. Table 10.7). However, comparing to 2005 level, the emission is still very high, over 300% growth over the 20-year period. Comparing the emissions from each sector, similar to BaU Scenario, the share of industrial sector is lowered due to shifting of industrial activities to service industry.

Under DM Scenario, it was assumed that more drastic measures than MM Scenario are to be implemented. Under this scenario, CO_2 emission in 2025 will be further cut down compared to BaU Scenario (cf. Table 10.7). The CO_2 emission is likely to be cut down 35% from 21 million metric tons in BaU Scenario to about 15 million metric tons. The per capita CO_2 emission is also possible to be cut from 6.9 metric tons under BaU Scenario to 5.0 metric tons. However, due to the high economic and population growth rate, there is still no reduction of CO_2 emission

Table 10.7 CO_2 emissions from energy use in Iskandar Malaysia (IM), 2005 and 2025

	2005			2025				
Scenario	GDP/capita	CO_2/cap, metric tons	Total CO_2, million metric tons (a)	GDP/capita	CO_2/cap, metric tons	Total CO_2, million metric tons (b)	CO_2 against BaU	CO_2 against 2005 (b/a)
BaU Scenario	50,251	4.1	5.6	105,740	6.9	20.8	100%	369%
MM Scenario	50,251	4.1	5.6	105,740	5.9	18.0	86%	318%
DM Scenario	50,251	4.1	5.6	105,740	5.0	15.1	73%	268%

Note: GDP in MYR

Source: Compiled by authors from different sources

Table 10.8 CO_2 emissions from energy use in Iskandar Malaysia (IM) in comparison with Tokyo and London, 2005

	CO_2 emissions from energy use (metric tons)		
City/region	Total	CO_2/capita	Remarks
IM	5,103,000	4.9	
Tokyo	71,300,000	5.8	Year 2003
Greater London	50,800,000	6.9	Year 2003

Source: TMG (2006)

Table 10.9 CO_2 emission from energy use in Iskandar Malaysia (IM) by sector, 2005 and 2025

		Share (%)				
Year	Scenario	Commercial	Industrial	Residential	Transportation	Agriculture
2005	Present	11.0	53.2	2.1	33.7	0.1
2025	BaU scenario	23.2	51.7	1.5	23.5	0.1
	MM scenario	22.1	49.2	1.5	27.1	0.1
	DM scenario	20.6	45.8	1.4	32.1	0.1

Source: Compiled by authors

compared to 2005, but an increase of about 270%. In terms of sectoral emissions, the share of each sector's emissions is similar to BaU Scenario and MM Scenario.

From the above simulations, it can be seen that if the present high growth scenario of IM prevails, CO_2 emissions from energy use in 2025 is likely to increase to more than three times of the 2005 level. However, if aggressive energy saving measures are taken by all parties, it is possible to slow down the growth in CO_2 emission to about 73% of BaU Scenario, although the 2025 emission would still be about 270% of the present level (2005) of emissions (cf. Table 10.7).

10.5 Concluding Remarks – Future Scenarios Towards Low Carbon Cities 2025

There are several strategies to achieve low carbon cities through sustainable development. In order to reduce CO_2 emission, the formulae in Kaya Identity may be used to achieve the target for a low carbon city. The Kaya Identity involved three main concepts, namely per capita activity, energy intensity and carbon intensity (cf. Fig. 10.6).

Hence in order to reduce CO_2 emission by reducing per capita activity is not feasible for developing region like IM. IM with current population growth rate of 4% per annum and economic growth rate of 6–8% per annum for the last 10 years may continue to grow until 2025 to achieve the objectives outlined in CDP to be

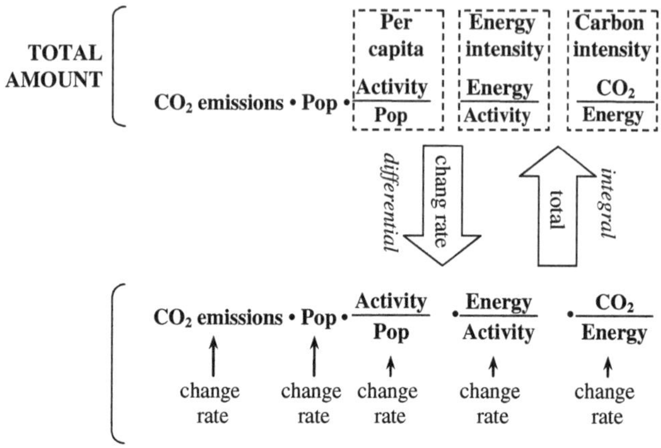

Fig. 10.6 Concept of Kaya identity. *Source*: NIES (2006)

strong sustainable conurbation of international standing (Khazanah Nasional 2006). Therefore to plan for a low carbon city, it is important to reduce CO_2 emission by reducing energy and carbon intensity.

Among the measures that may reduce energy intensity are low energy buildings, establishment of recycling system, transit oriented development, and Brownfield development, as explained in further detail below.

10.5.1 Low Energy Building

Low carbon buildings are defined as those that use 20–30 kWh/m^2 of energy. Low energy buildings typically use high level of insulation in cold countries, energy efficient windows, low level air infiltration and heat recovery ventilation. In the case of tropical countries, passive solar building design techniques are used. As space cooling and water heating are highest percentage of household energy consumption, reduction of air conditioning and water heating can reduce the total energy demand significantly. This measure is already adopted as policy strategies in livable communities for the IM region.

10.5.2 Establishment of Recycling System

Businesses generate a lot of wastes that require different types of refuse disposal facilities, such as landfills and incinerators. Japan has successfully reduced landfill wastes by about 75% per annum by recycling, reduction and reuse. In order to establish favourable recycling system in IM, it is important to reduce waste generation as well as to establish a system to recycle waste as resources. Public awareness and education on the importance of the recycling as well as setting up centres for collection of recyclable materials are important.

10.5.3 Transit Development

The linear spatial development pattern of IM along the Johor Bahru –Kulai – Pasir Gudang corridor provides great opportunities to for the region to develop rail system and develop the urban centres based on the TOD concept. The recent proposal for a Light Rail Transit along these corridors will help to promote the use of public transport to reduce dependence on private vehicle usage. This form of TOD development may reduce traffic on the roads and revitalize urban neighbourhood. The changing demographics and lifestyle of urban society of IM is similar to developed countries that indicates trend of growing number of smaller households and retiring baby boomers who opt to live in smaller homes in urban areas. There is a trend of demand to revitalize denser and more convenient living choices or decentralized concentrated urban centres. This yuppies and new urbanized society desires walkable communities with easy access to public transportation. This measure has also been adopted as one of the planning strategies of CDP (Chapter 16) to ensure viability of public transport.

10.5.4 Brownfield Development

The urban regeneration involve redeveloping brownfields either those abandoned or low density and uneconomic use of land for higher density development. Urban infill development is gaining popularity not only in developed countries but also in Malaysia. Apart from energy efficient and reduction in CO_2 emission, this form of development allows the existing infrastructure and amenities to be used and also prevent urban inner city decay.

Among the measures that may be able to reduce carbon intensity include the usage of alternative fuels such as bio-fuel, and prevent deforestation and promote carbon sink, particularly for the RAMSAR site and Sungai Pulai wetland.

10.5.5 Use of Alternative Fuel – Bio-fuel

In order to create low carbon cities in IM, it is important to look into the potential of reducing dependence on fossil fuel. This is where economy in which CO_2 emissions from the use of carbon based fuels (coal, oil and gas) can be significantly reduced. As Malaysia is a palm oil exporting nation, bio-fuel from palm oil provides a viable alternative for the government to consider. The optimal reduction is when the cities adopt zero carbon policies where any form of carbon emission is prohibited. According to the CDP, IM will experience high GDP growth of 8% per annum as compared with the Johor state average of 5.5% per annum (without SJER). In order to sustain such rapid economic growth, population will reach about 3 million by year 2025 and population growth rate is expected to be about 4.1% as compared with the annual state average of 2.1%. This will result in increases in energy demand due to industrialization and urbanization.

Low carbon intensity can also be achieved through the use energy efficiency measures. It also showed that energy efficiency could yield significant reduction in CO_2 emission at low cost and the substitution of renewable energy sources for fossils fuels and nuclear power including transport electrification. New technologies such as hydrogen solution power and carbon capture and storage should be explored in the near future.

Low carbon intensity may not always contribute to increase production cost for the enterprises. This is because a higher energy efficiency of low carbon energies, existing technological improvement, and increase cost of carbon fuels and the introduction of carbon taxes or carbon trading.

10.5.6 Prevent Deforestation and Promote Carbon Sink – RAMSAR Site and Sungai Pulai Wetland

Protection, preservation and enhancement of bio-diversity and natural green environment is an important policy. The total area of natural and green environment in IM is more than 150,000 ha including forests, mangrove areas, parks and open spaces as well as the agricultural areas. These green areas will play an important role as a carbon sink to absorb CO_2. Apart from the significance of Sungai Pulai Forest and wetland reserve of more than 9,000 ha as world renowned designated RAMSAR site, it is also one of the main catchment areas for water supply to Singapore and Johor Bahru. Sungai Pulai forest reserve consists of several areas including Kukup Island and Piai Cape (southernmost of the Asian Continent), National Park of Piai and Permanent Forest of Sungai Pulai.

All the above eco-friendly measures require a long-term thinking and they constitute an important technological, institutional, and social change to ensure successful implementation. In order to achieve the environmental goal for the IM it is important to set environmental targets for 2025. The key options to reduce CO_2 emission require possible combination of countermeasures on the energy demand and supply.

Apart from urban planning, the roadmap to achieve low carbon cities requires strong political will and decisive actions especially incentives of non-spatial policies such as to promote energy efficiency, renewal energy, recycling, and spatial policies such as TOD, regeneration/brownfield development, and energy saving building.

From the above System Dynamics simulations, it is expected that if the present high growth scenario prevails, and if aggressive energy saving measures were not taken, by 2025, CO_2 emissions from energy use is likely to increase to more than three times of the present level. In terms of emission per capita, the present level of emission rate is quite low, only about 4.1 metric tons per capita, which is well lower than the major cities in the developed countries such as Tokyo (5.8 metric tons per capita) and Greater London (6.9 metric tons per capita). However, should the high economic and population growths as envisaged by the Malaysian Government prevail, and the society is enjoying material affluence without giving much attention on energy saving, it is projected that the emission rate would significantly increase to 6.9 metric tons per capita in 2025. Nevertheless, if aggressive energy saving strategies were being drawn up and implemented in the government policies from national to local levels, and the general public are well aware of the importance of energy saving lifestyle, it is possible to bring down the CO_2 emission rate to 5.0 metric tons per capita in 2025 as shown in DM Scenario in Section 10.4.3 above. Therefore as a proactive measure, planning for low carbon city measures should be adopted in the planning and implementation of the CDP to ensure a more sustainable urban conurbation in south Johor where the Iskandar Economic Region is located.

References

Dietz, T. & Rosa, E. A. (1997). Effects of population and affluence on CO_2 emissions. *Proceedings of the National Academy of Sciences*, 94: 175–179.

Energy Information Administration (2007). http://ww.eia.doe.gov/oiaf/emission. Accessed 4 November 2007.

EPU (2001). *Eight Malaysia plan 2001–2005*. Putrajaya: Economic Planning Unit.

EPU (2006). *Ninth Malaysia plan 2006–2010*. Putrajaya: Economic Planning Unit.

Fong, W. K., Lun, Y. F. & Matsumoto, H. (2009). Prediction of carbon dioxide emissions for rapid growing city of developing country. *The HKIE Transactions*, 16(2): 1–8.

Khazanah Nasional (2006). *Comprehensive development plan for South Johor economic region 2006–2025*. Kuala Lumpur: Khazanah Nasional.

NIES (2006). *Aligning climate change and sustainability-scenarios, modeling and policy analysis*. Tsukuba: National Institute for Environmental Studies.

Sandrasagra Mitre, J. (2007). Climate change: cities getting serious about CO_2 emissions. http://www.psnews.net/news.asp?idnews=37765. Accessed 2 November 2007.

Shi, A. (2001). Population growth and global carbon dioxide emissions. Proceedings of the IUSSP conference. http://www.iussp.org/Brazil2001/soo/S09_Shi.pdf. Accessed 4 November 2007.

TMG (2006). *Tokyo environmental white paper 2006*. Tokyo: Tokyo Department of Environment. http://www2.kankyo.metro.tokyo.jp/kikaku/hakusho/2006/outline.html. Accessed 6 November 2007. (in Japanese).

UN (2002). *World urbanization prospects: the 2001 revision*. New York, NY: United Nations.

United Nations Environment Programme/GRID-Arendal (2007). National carbon dioxide (CO_2) emissions per capita. http://maps.grida.no/go/graphic/national_carbon_dioxide_co2_emissions_per_capita. Accessed 2 November 2007.

World Resource Institute (WRI) (2007a). http://earthtrends.wri.org/searchable_db/index.php?theme=3&variable_ID=470&action=select_countries. Accessed on 2 November 2007.

World Resource Institute (WRI) (2007b). http://earthtrends.wri.org/pdf-library/data_tables/cli_2005. Accessed 2 November 2007.

Yale University (2005). *Environmental sustainability index study –national benchmark environmental management*. New Haven, CT: Yale University.

Part III
Micro Local Planning: Design and Methods

Chapter 11
Presentation of Ecological Footprint Information: A Re-examination

Hoong-Chor Chin and Reuben Mingguang Li

Abstract In a short span of two decades, the ecological footprint concept as a framework for impact assessment and sustainability planning through the focus on earthly "capital" limits in the form of land resources has grown in popularity. A unique selling point of the concept is its focus on physical limits, thereby making it an "area-based analogue" of other popular impact assessment methodologies. In this chapter, a critical re-examination of the presentation of ecological footprint information is attempted with reference to past studies. In particular, the aspect of "spatiality" and "visualization" of the ecological footprint is explored by juxtaposing popular presentation techniques with the original goals of ecological footprint analysis. The result of the discussion is an identification of several shortcomings inherent in presentation techniques in ecological footprint literature and a subsequent suggestion of a standardized, spatial presentation technique that is in-line with present trajectories in the field of study. The ultimate aim of this chapter is to allow the various manifestations of ecological consumptions to be "mapped" in a comparable and meaningful manner (and traced dynamically), with a degree of flexibility among the different approaches to ecological footprint analysis.

11.1 Introduction

Nearly two decades have passed since William Rees and Mathis Wackernagel introduced the concept of *ecological footprint* (F_p) as a means to illustrate the impact of humans on the natural environment and resources (Rees 1992, Rees and Wackernagel 1994, Wackernagel and Rees 1996). The ecological footprint, not without its detractors, has shown much promise in terms of accounting for the overusage of earthly "capital" in the form of land resources. Literature on ecological footprint research and related case studies have taken on several trajectories as

H.-C. Chin (✉)
Department of Civil Engineering, National University of Singapore, Singapore
e-mail: ceechc@nus.edu.sg

T.-C. Wong, B. Yuen (eds.), *Eco-city Planning*, DOI 10.1007/978-94-007-0383-4_11,

the methodologies employed become increasingly complex. Among these are sub-national level footprint studies (e.g. Wackernagel et al. 1999, Hu et al. 2008, Scotti et al. 2009); sector-based studies (e.g. Bicknell et al. 1998, Herva et al. 2008); policy analysis and comparison (Liu and Kobayashi 2006, Browne et al. 2008); ecological footprint analysis of specific themes (e.g. production activity: Ferng 2001, commuting: Müniz and Galindo 2005, vehicle travel: Chi and Stone 2005; fuels: Holden and Hoyer 2005); remote sensing-aided studies (e.g. Cai et al. 2007); and research that employ time series (e.g. Lammers et al. 2008). These studies provide a form of measurement to gauge the sustainability of human activities within regions, countries and cities.

As the concept of ecological footprint is becoming increasingly popular and useful, in particular, towards environmental conservation and building of eco-cities, there have been attempts to improve the methodologies underlying the calculation of ecological footprint. It is not in the scope of this chapter to discuss the intricacies of computing accurate ecological footprints. Rather, the focus here is to add on to the process of refining ecological footprint research by re-examining the downstream portion of the research procedure – the *presentation* of ecological footprint information.

It is our opinion that ecological footprint research has progressed well, in terms of versatility and popularity, since the mid-1990s. However, several elements of the ecological footprint have to be scrutinized in detail to allow for the re-alignment of future research towards the original aims of the ecological footprint concept. In particular, we found that a certain lack of focus in the *spatiality* aspect of the ecological footprint, which in turn hampers the effectiveness of the presentation of results of various footprint studies conducted. In re-examining the available literature with relation to the original aims of the ecological footprint, some suggestions are made in the hope of further improving the efficacy of ecological footprint research.

11.2 The Evolution of the Ecological Footprint Concept

As previously mentioned, the ecological footprint concept was the brainchild of Rees and his then-PhD student Wackernagel. In a 1992 paper, Rees (1992: 121) first presented the idea of "appropriated carrying capacity" and "ecological footprint:" in response to the way mainstream urban economics dealt with sustainability issues. His view was that the conventional economic method of analyzing sustainable development was blind to issues such as environmental justice – where more developed nations consumed more of the Earth's natural capital than appropriate (e.g. consumption exceeds the availability of natural capital within the nation's political boundary). However, this remained a purely theoretical approach as little or no attention was placed on specific methodologies to measure ecological footprints.

Wackernagel's PhD thesis (1994) and subsequent papers by Rees and Wackernagel (1994, 1996) laid the foundations for the mathematical understanding of the ecological footprint as the basic equations were formulated. As the ecological footprint concept grew popular, many researchers (including Rees and Wackernagel themselves) began to refine the methodology behind footprint calculations. New techniques such as time-slicing and regional measurements improved on the capabilities of the ecological footprint concept. With such a positive burst of ideas with multiple trajectories, a re-examination of ecological footprint information presentation would enhance the usefulness of the ecological footprint.

11.3 The Spatiality of the Ecological Footprint

Some of the earlier pieces of work defining the ecological footprint, by pioneers Rees and Wackernagel, involved much spatial imagination. Phrases such as "180 times larger than its political area" (Rees and Wackernagel 1996: 233), "a modern city... enclosed in a glass or plastic hemisphere" (p. 227) and "the total land area required to sustain an urban region" (Rees 1992: 121), were used constantly in the description of ecological footprint. In essence, one of the main strengths (or as some may suggest, the main *purposes*) of the ecological footprint is its role as a "visually graphic tool" (Rees and Wackernagel 1996: 230), which accentuates the physical limits of our planetary home (e.g. Fig. 11.1) as opposed to rendering resources merely as another form of capital or as a carrying capacity.

Rees (1996, 2000) points out that the ecological footprint can be considered as an "area-based analogue" or indicator of Ehrlich and Holdren's I = PAT, where impact on the environment (I) is held as a product of population (P), affluence (A), and technology (T). However, many papers that employ the ecological footprint concept fail to capitalize on this particular strength, leaving spatial visualization out of the equation altogether. Among the studies that do take this into consideration, presentation styles are varied and non-standardized (cf. Wackernagel et al. 1999 and Chi and Stone 2005: Figs. 11.1 and 11.2).

11.3.1 The "Traditional" Ecological Footprint

Traditionally, ecological footprint calculations are conducted on the national-level scale that aggregate the energy and resource footprint of an entire economy (e.g. Austria: Haberl et al. 2001, Erb 2004; the Philippines and South Korea: Wackernagel et al. 2004; Ireland: Lammers et al. 2008). Such approaches are often geared towards the eventual output of per capita footprints (for the purpose of comparison between countries) and are not always comprehensive in accounting for every possible component of the footprints etched by the economies, as highlighted by Rees and Wackernagel (1996). The results are conclusive findings that allow for the attribution of "blame" in terms of identifying countries or regions "overshooting" their allotted biocapacity.

Even as the above style of research provides normative force for steering certain global and regional policies, the actual usefulness of the footprint figures within the economies is hampered by the inherent generalizations. Questions such as, "which

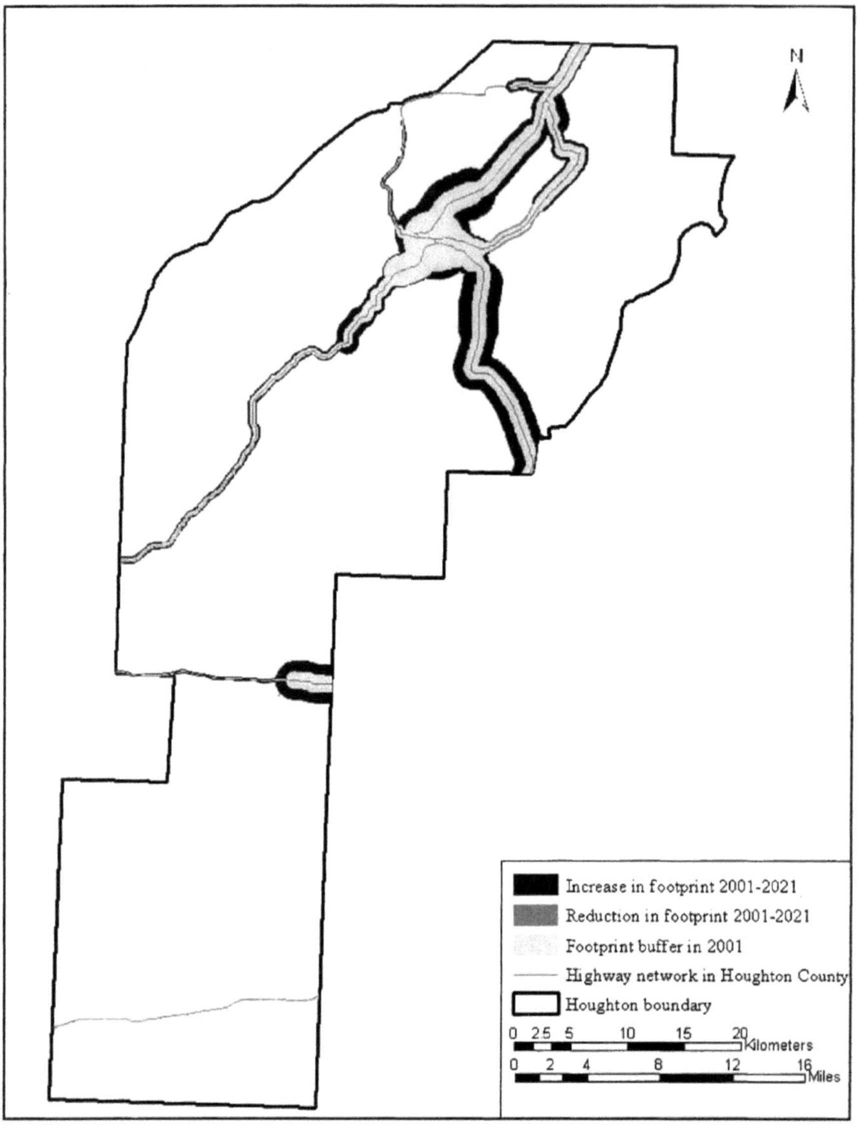

Fig. 11.1 Footprint change in Houghton County, Michigan, 2001–2021. *Note:* In this study, the authors used physical buffers from road networks to represent their footprint. The width of the buffer is a summation of physical footprint and energy footprint (both in m^2) over the actual length of the roadway segment (m). *Source*: Chi and Stone (2005)

sector of the economy is contributing most to the footprint?", "which policies will lead to a smaller footprint?" remain unanswered. While sub-national areal footprint studies such as on the municipal level (e.g. Scotti et al. 2009) or city level (e.g. Hu et al. 2008) have become increasingly popular, non-spatial sub-national scales such

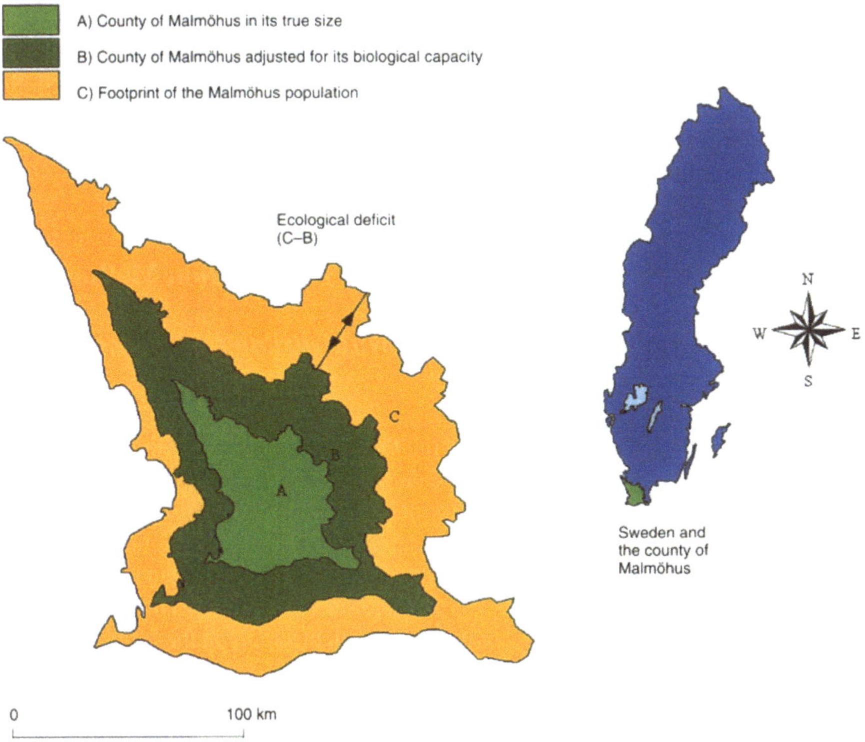

Fig. 11.2 Visual presentation of the footprint of Sweden and the county of Malmohus. *Source*: Wackernagel et al. (1999)

as sector-based approaches are limited in number due to the difficulty in portraying sectors as spatial entities, and the incompatibility of using per capita figures.

11.3.2 Non-spatial Scales and the "Footprint" Metaphor

One of our proposed suggestions is to employ a secondary form of sub-national ecological footprint calculation by a sectoral division of the economy within a land resource budget. A few benefits associated with a sectoral budgetary approach include: (a) the increased ownership of the "overshooting" problem, which translates to the ability to compare between sectors, as well as assesses sets of policies that affect certain sectors; (b) a better defined system for footprint comparison across countries rather than the inconsistent inclusion of variables from study to study; and (c) a more specialized look at each particular sector, thereby increasing the accuracy of the calculations. "Budget" is used here to further emphasize the limited nature of available land resources.

Recalling from the earlier works of Rees and Wackernagel, one of the selling points of the ecological footprint is its intuitiveness in the spatial measure shared by laymen and scholars alike. However, to achieve such a goal of effectively presenting footprints for a non-spatial scale of study, a method of spatializing these non-spatial scales is necessary. Sector-based studies by Bicknell et al. (1998) and Herva et al. (2008) are successful in developing ways to calculate the footprints of the sector, yet more could be attained if these methods could lead to the presentation of a "physical" imprint beyond merely presenting numbers.

11.4 Problems Associated with Current Presentation Methods

11.4.1 Single Aggregate F_p Value vs Multivariate Land Resource Types

Wackernagel et al. (1999) rightly pointed out the shortcomings of a single aggregate F_P value in their study on Sweden using data from 1994. Although the country had an aggregate per capita ecological footprint or demand (~7.2 ha cap^{-1}) that was smaller than its productive per capita biocapacity (~8.2 ha cap^{-1}), the figure did not take into consideration the "striking mismatch" (610) of demand for specific forms of biocapacity, or land types. The demand for fossil energy land for the case study was a staggering 2.6 ha cap^{-1}, with no equivalent supply in the form of carbon dioxide (CO_2) absorption land or renewable energy.

The resultant problem was a misrepresentation of the actual scenario. Using the analogy of a human needing nutrients, water and oxygen; the lack of any single one of components of land resource types will spell disaster – an implicit problem with the employment of aggregate values also pointed out by van den Bergh and Verbruggen (1999). A single footprint value is undoubtedly useful in quick comparisons, yet more has to be done to ensure that information is not lost altogether in the process of aggregation.

11.4.2 Ineffectiveness of Cartograms and Spatially-Dimensionless Line or Area Graphs

Cartograms – maps drawn to exaggerate land area based on certain themes (in this case F_p) – have been used in selected studies (e.g. Wackernagel et al. 1999: Fig. 11.1). Readers who encounter these exaggerated areal depictions of country outlines may respond in shock due to the visual impact created. Yet, the usefulness of cartograms is limited by several factor such as: (a) the difficulty of readers in estimating the actual scalar differences presented (gauging areal proportion is a skill difficult to master, especially for irregular shapes); (b) the inability of readers to extract any specific numerical figures; (c) the lack of any value in the created shape or outline; (d) the difficulty in comparing between countries which have significant differences in size, and (e) the need for cartographic skills to produce the cartograms, not to mention the understanding (by both readers and authors) of

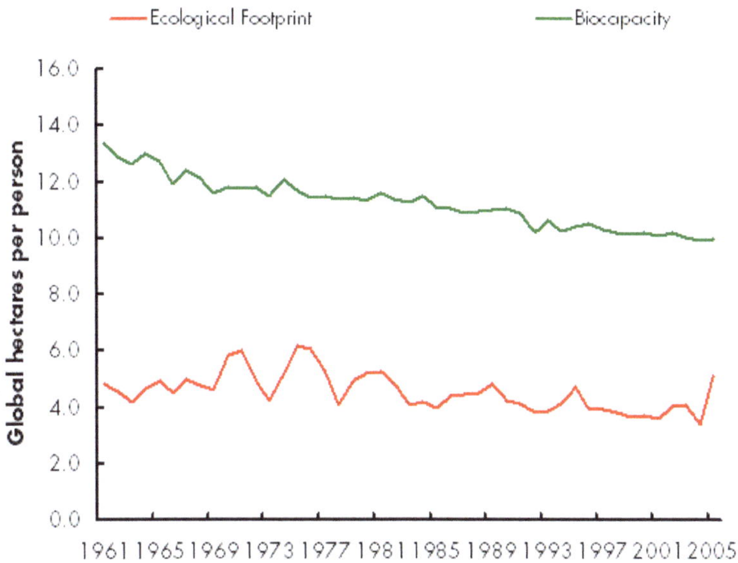

Fig. 11.3 Line graph showing the aggregate ecological footprint and biocapacity of Sweden over the year. *Source*: Global Footprint Network (2008)

geometric distortions created by the projections used (e.g. Mercator projection vs. Lambert azimuthal equal-area projection).

Of course, one could always turn back to the ever-reliable line or area graphs and bar charts to depict ecological consumption of a country. The only problem is that these charts tend to aggregate all the different land resources into a single line (re-creating the problem highlighted earlier) and lack a spatial dimension to them (presenting ecological "ceiling" more than a "footprint"). Figure 11.3 shows a line graph comparison between the ecological footprint and biocapacity of Sweden across several decades. The over-demand of fossil energy land by Sweden in 1994 (Wackernagel et al. 1999) is offset by the sheer surplus of forests, resulting in a healthy looking graph despite the obvious problem. Several studies, including Harberl et al. (2001), sidestepped this issue by presenting several graphs to account for each particular land resource type.

11.4.2.1 What Is Then Needed?

As such, an effective presentation scheme for ecological footprint has to meet several conditions that coincide with the strengths of present schemes yet negating their weaknesses. It has to:

- be simple and easy to interpret, i.e. intuitive;
- have comparability with other studies and be "standardizable";
- show a physical image of a footprint that can allow readers to relate with the spatiality of the problem;

- be able to reduce the problems that come with the aggregating of ecological demand for different types of land resources but still be concise enough to deduce a footprint at one look;
- preferably have numerical values that can be extracted; and
- be compatible with the sectoral approach mentioned earlier in this chapter.

11.4.3 Ecological Footprint Radar Charts

With the literature reviews and conditions in mind, our proposed solution for this problem is to provide a clear and standardized scheme that allows information to be quickly and easily extracted from the graphical output and yet remain comparable across sectors or even countries. This is achieved through the use of polygonal charts, named herein as Chin-Li or (C-L) Footprint Charts that comprise polygons fitted on axes representing biocapacities for individual land resource type. Figure 11.4 shows an *absolute C-L Footprint Chart* that displays the *absolute* ecological demand and supply for Sweden, in the 1994 case study by Wackernagel et al. (1999), broken down into the 6 main land resource types. This particular chart is useful for extracting exact figures of ecological demand and supply and to give an inkling of what the footprint is like.

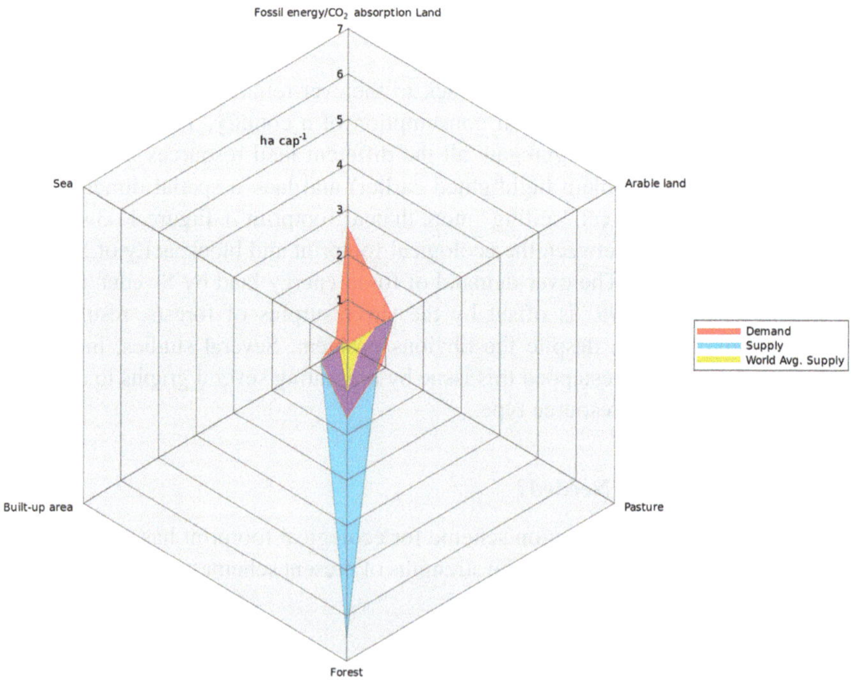

Fig. 11.4 Absolute C-L Footprint Chart showing the ecological demand and supply of Sweden in 1994. *Source*: Wackernagel et al. (1999)

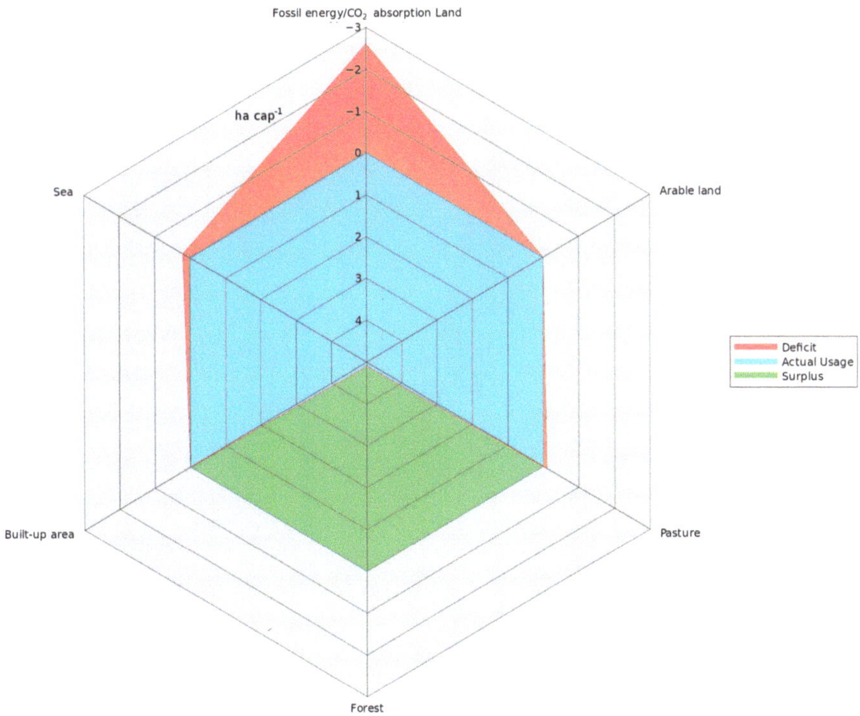

Fig. 11.5 Differential C-L Footprint Chart for Sweden, 1994. *Source*: Wackernagel et al. (1999)

Differential C-L Footprint Charts (e.g. Figs. 11.5, 11.6, and 11.7) are used for footprint analysis. Values plotted on this type of charts are the differences between the biocapacity/ecological supply of the country/city/region and the ecological demand. A perfect hexagon encased by the 0-line on the axes depicts scenario of a fully-utilised biocapacity, without deficit. The ease of interpretation and comparison can be seen in the following charts. Figure 11.4 shows the differential C-L Footprint Chart for same case study of Sweden as Figs. 11.2 and 11.3. Unlike the line graph in Fig. 11.2, we can immediately tell from the C-L Footprint Chart that there are imbalances between the biocapacity and ecological demand. Excess demand for fossil energy results in a deficit of ~2.6 ha cap^{-1} for CO_2 absorption land (as deficit), with minor deficits also visible in sea, arable land, and pasture. However, excess supply of forests means that a huge surplus of forest land is available (as surplus).

The hexagon representing the biocapacity of the studied area is clearly smaller in the case of Figs. 11.6 and 11.7 as compared to Fig. 11.5. This is due to the fact that Figs. 11.5 and 11.6 represent a city and a municipality, respectively, while Fig. 11.4 represents an entire country. Despite this, the ability to present ecological footprints is not diminished. Figure 11.6 clearly shows that Jiangyin city (Hu et al. 2008) is almost entirely utilising all of its biocapacity and creating deficits in several

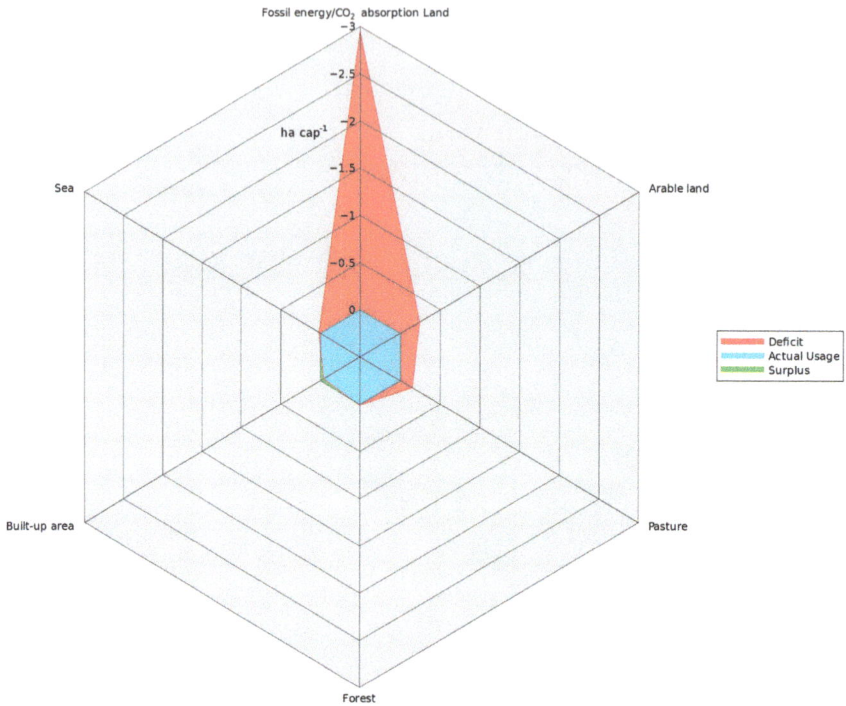

Fig. 11.6 Differential C-L Footprint Chart for Jiangyin city, China, 2005. *Source*: Hu et al. (2008)

of the land resource types. On the other hand, the Piacenza municipality in Italy (Scotti et al. 2009) experiences surpluses in arable land, built-up area and forests (Fig. 11.6).

11.4.4 Other Uses of the Polygonal Footprint Charts

There are other possible uses of the C-L Footprint Charts and these will be presented briefly in this chapter. Comparing time-slices and projecting footprints of a city as part of the biocapacity of an entire country are other creative ways in which the C-L Footprint Charts can be employed (e.g. Fig. 11.9). These will allow changes in footprints to be monitored visually. In the case of the suggested sectoral approach to footprint calculations, absolute footprints (as opposed to per capita) of individual sectors can be presented simultaneously as a footprint on the entire economy's biocapacity. These can be overlaid on the C-L Footprint Charts to show the contribution of different sectors (see Fig. 11.8).

Besides thematic variations, the components of the C-L Footprint Chart can also be changed. For example, the units used can be used with or without the equivalence factor. Percentages can be used as the axes of the charts instead of raw numbers.

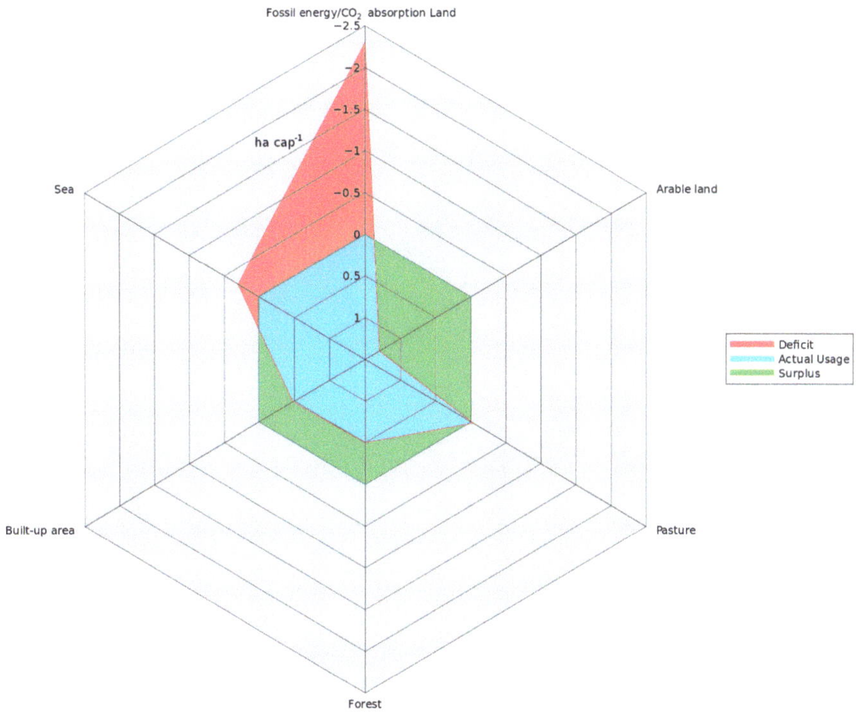

Fig. 11.7 Differential C-L Footprint Chart for Piacenza municipality, Italy, 2002. *Source*: Scotti et al. (2009)

The importance of the Footprint Charts lies not in its variable components but in its ability to show multivariate data while providing a quick summary of the overall pattern. As such, the ways in which the Footprint Charts can be used is extensive. The versatility of these charts can also be seen in Fig. 11.9, where global footprint over time can be analyzed.

11.4.5 Evaluating the C-L Footprint Charts

11.4.5.1 Technical Advantages of the C-L Footprint Charts

Some clear methodological advantages are delivered by the use of the C-L Footprint Charts. Among these are:

- The ability to provide an impactful visual imagery of the "footprint" created by a city, economy or country;
- The ability to include multivariate land resources thereby giving not just an aggregate footprint but one that showcases the distribution of different land types, yet can provide an indication of the overall footprint with one look;

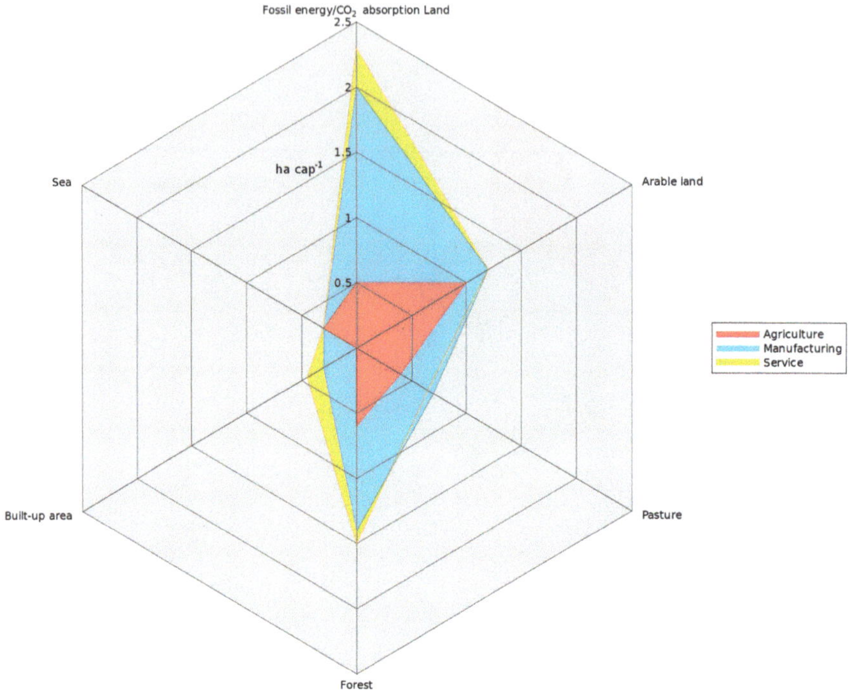

Fig. 11.8 Hypothetical C-L Footprint Chart displaying the contribution of different sectors to the overall footprint of an economy

- The ease of comparison between spatial scales (e.g. country to country) and temporal scales (e.g. across time slices);
- Its simplicity in creation and interpretation;
- Its commonalities with graphs such as the plotting of exact values which can be retrieved;
- Its compatibility with the sectoral approach to ecological footprint calculations;
- Potential enhancements and/or alternative uses of this approach are foreseeable.

11.4.5.2 Implementation of the C-L Footprint Charts

With an effective and standardized method of presenting ecological footprint information coupled by its ease of use and comparison, policies in urban planning and environmental management will no longer depend on arbitrary measures of ecological consumption. Within cities and the boundaries of a nation, domestic policies and their potential contribution or mitigation of natural capital consumption can be evaluated systematically. For example, in the transportation sector, decisions such as whether to increase funding for improving the public transport network or the multilevel roads can be better evaluated with respect to ecological cost. More "problematic" sectors may also be identified and efforts on reducing ecological impact may be focused there.

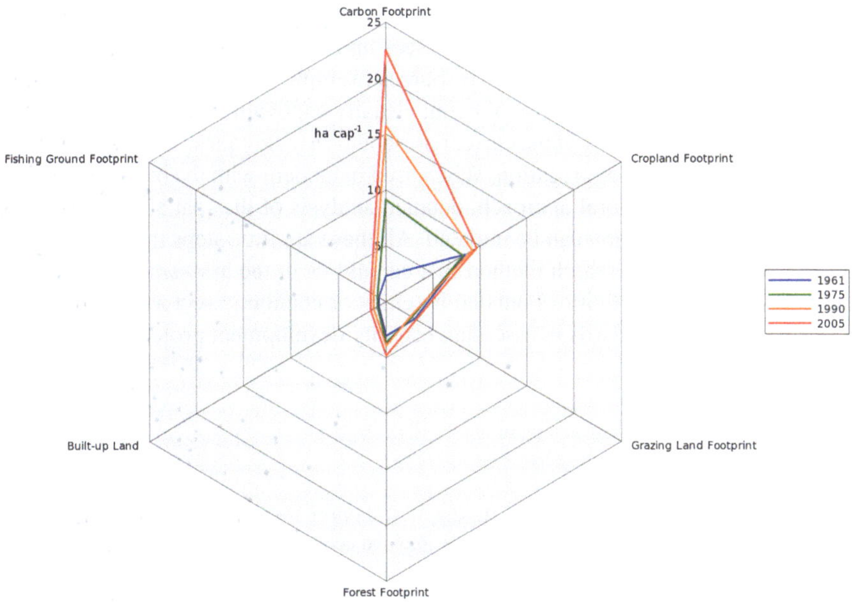

Fig. 11.9 Absolute C-L Footprint Chart of global ecological consumption from 1961 to 2005. *Note*: The categories used by the Global Footprint Network are different from the above examples. Units are also in billion global acres rather than hectares. *Source*: National Footprint Accounts 2008 edition

At the international level, various organizations such as the Global Footprint Network (http://www.footprintnetwork.org) currently exist. At present, the comparison of footprints of various nations are not commonly found and restricted to a comparison of numbers and simple line graphs. With the use of a graphically useful method in the C-L Footprint Chart, one can foresee a future where the footprint charts of different nations can be placed side by side and meaning can be made out of their sizes, shapes, and possibly, characteristic groups.

11.4.5.3 Limitations of the C-L Footprint Charts

While the shapes and sizes of the polygons on the C-L Footprint Chart are somewhat indicative of the magnitude of the footprint, they are not altogether proportionate. The rotation of axis may lead to a change in the shape under the graph. As such, this may prove to be a potential pitfall for readers who attempt to use the shape for visual interpretation.

11.5 Conclusion

All in all, it can be said that the purpose of this chapter is to refocus our current research practices in ecological footprint studies towards an embracing of the spatiality so crucial in the conception of the ecological footprint. The C-L Footprint

Charts introduced in this chapter offer such a possibility of re-establishing ecological footprint as "visually graphic tool" (Rees and Wackernagel 1996: 230) which continues to remind us of the limited resources exemplified by the limited land area of our world. Beyond the purpose of being visually captivating, the C-L Charts also solve some of the problems of the current presentation techniques, such as the loss of information due to data aggregation. When used in tandem with the other suggestion in this chapter of a sectoral approach, detailed analysis of the spatial manifestation of ecological consumption can be initiated. All these are little steps that we can take towards an enhanced research method that can aid us in the assessment of the sustainability of human activities bounded by cities or economic sectors of cities, and in setting required standards in assessing eco-city development projects.

References

Bicknell, K. B., Ball, R. J., Cullen, R. & Bigsby, H. R. (1998). New methodology for the ecological footprint with an application to the New Zealand economy. *Ecological Economics*, 27(2): 149–160.

Browne, D., O'Regan, B. & Moles, R. (2008). Use of ecological footprinting to explore alternative transport policy scenarios in an Irish city-region. *Transportation Research (Part D)*, 13(5): 315–322.

Cai, H. S., Zhang, X. L. & Zhu, D. H. (2007). Study of ecological capacity change and quantitative analysis of ecological compensation in a nature reserve based on RS and GIS: a case study on Po-yang Lake Nature Reserve, China. *New Zealand Journal of Agricultural Research*, 50(5): 757–766.

Chi, G. & Stone, B. (2005). Sustainable transport planning: estimating the ecological footprint of vehicle travel in future years. *Journal of Urban Planning and Development*, 131(3): 170–180.

Erb, K. H. (2004). Actual land demand of Austria 1926–2000: a variation on ecological footprint assessments. *Land Use Policy*, 21(3): 247–259.

Ferng, J. J. (2001). Using composition of land multiplier to estimate ecological footprints associated with production activity. *Ecological Economics*, 37(2): 159–172.

Global Footprint Network (2008). Sweden country trend. Online Article. 26 Oct 2008. http://www.footprintnetwork.org. Accessed 19 February 2009.

Haberl, H., Erb, K. H. & Krausmann, F. (2001). How to calculate and interpret ecological footprints for long periods of time: the case of Austria 1926–1995. *Ecological Economics*, 38(1): 25–45.

Herva, M., Franco, A., Ferreiro, S., Alvarez, A. & Roca, E. (2008). An approach for the application of the ecological footprint as environmental indicator in the textile sector. *Journal of Hazardous Materials*, 156(2–3): 478–487.

Holden, E. & Hoyer, K. G. (2005). The ecological footprints of fuels. *Transportation Research (Part D)*, 10(5): 395–403.

Hu, D., Huang, S., Feng, Q., Li, F., Zhao, J., Zhao, Y. & Wang, B. (2008). Relationships between rapid urban development and the appropriation of ecosystems in Jiangyin city, Eastern China. *Landscape and Urban Planning*, 87(3): 180–191.

Lammers, A., Moles, R., Walsh, C. & Huijbregts, M. (2008). Ireland's footprint: a time series for 1983–2001. *Land Use Policy*, 25(1): 53–58.

Liu, P. J. & Kobayashi, E. (2006). Health assessment and environmental impacts of modal shift in transportation. Oceans 2006 – Asia Pacific conference, May 16–19 2006, Volumes 1 and 2, pp. 127–132. Singapore.

Müniz, I. & Galindo, A. (2005). Urban form and the ecological footprint of commuting. The case of Barcelona. *Ecological Economics*, 55(4): 499–514.

Rees, W. (1992). Ecological footprints and appropriated carrying capacity: what urban economics leaves out. *Environment and Urbanization*, 4(2): 121–130.

Rees, W. (1996). Revisiting carrying capacity: area-based indicators of sustainability. *Population and Environment*, 17(3): 195–215.

Rees, W. (2000). Eco-footprint analysis: merits and brickbats. *Ecological Economics*, 32(3): 371–374.

Rees, W. & Wackernagel, M. (1994). Ecological footprints and appropriated carrying capacity: Measuring the natural capital requirements of the human economy. In A. M. Jansson, M. Hammer, C. Folke & R. Costanza (Eds.), *Investing in natural capital: the ecological economics approach to sustainability* (pp. 362–390). Washington, DC: Island Press.

Rees, W. E. & Wackernagel, M. (1996). Urban ecological footprints: why cities cannot be sustainable and why they are a key to sustainability. *Environmental Impact Assessment Review*, 16(4–6): 223–248.

Scotti, M., Bondavalli, C. & Bodini, A. (2009). Ecological footprint as a tool for local sustainability: the municipality of Piacenza (Italy) as a case study. *Environmental Impact Assessment Review*, 29(1): 39–50.

Van den Bergh, J. & Verbruggen, H. (1999). Spatial sustainability, trade and indicators: an evaluation of "ecological footprint". *Ecological Economics*, 29(1): 61–72.

Wackernagel, M. (1994). *Ecological footprint and appropriated carrying capacity: a tool for planning toward sustainability*. Ph.D. Thesis, School of Community and Regional Planning. The University of British Columbia. Vancouver, Canada.

Wackernagel, N., Lewan, L. & Hansson, C. B. (1999). Evaluating the use of natural capital with the ecological footprint: applications in Sweden and subregions. *Ambio*, 28(7): 604–612.

Wackernagel, M., Monfreda, C., Erb, K. H., Haberl, H. & Schulz, N. B. (2004). Ecological footprint time series of Austria, the Philippines, and South Korea for 1961–1999: comparing the conventional approach to an "actual land area" approach. *Land Use Policy*, 21(3): 261–269.

Wackernagel, M. & Rees, W. (1996). *Our ecological footprint: reducing human impact on the earth*. Philadelphia, PA: New Society Publishers.

Chapter 12
Towards Sustainable Architecture: The Transformation of the Built Environment in İstanbul, Turkey

Selin Mutdoğan and Tai-Chee Wong

But design, like sustainability, is a dynamic and living process... A design is sustainable, or it is not. If it is not sustainable, changes can be made to make it sustainable. If it is sustainable, by necessity it will be changing and evolving. Sustainability is not static – it is iteratively changing, based on evolving knowledge that connects science and design
(Williams 2007 *: 17).*

Abstract Towards the end of the twentieth century, new consumerist lifestyles coupled with technological innovations and rising environmental consciousness have changed the concept in environmental and building designs. Conscientious developers and city governments see green building and environmental designs as a useful approach to counter the adverse effects of global warming, fossil energy consumption and nature destruction that have threatened ecological and human health. Today, sustainable architecture and green design have become an important agenda of the built environment. The aim of this study is to examine the efforts made towards constructing sustainable office buildings in İstanbul, Turkey. In İstanbul, due to its unique location and historical background, innovative, modern and sustainable buildings have been built in its newly developed central city area, such as Büyükdere Avenue. Using a chosen set of green building rating systems as criteria, assessment was focused on site-environment, energy-water, materials-resources and indoor environmental quality. High-rise office buildings selected for this survey were Metrocity Office Building, İşbank Headquarters and the Akbank Tower along Büyükdere Avenue. Results showed that standards achieved were low but they were symbolic of a self-initiated will in line with the international agenda towards urban ecological conservation.

S. Mutdoğan (✉)
Hacettepe University, Ankara, Turkey
e-mail: selinse@hacettepe.edu.tr; smutdogan@gmail.com

T.-C. Wong, B. Yuen (eds.), *Eco-city Planning*, DOI 10.1007/978-94-007-0383-4_12,

12.1 Introduction

Cities are not merely centres of wealth accumulation but also of consumption, waste generation and pollutants production. In the twenty-first century, as technology, globalization and urban growth advance further, cities will become the centre of our environmental future where urbanites are anticipated to be more distanced from nature spaces which are being destroyed to make room for high-rise blocks and concrete infrastructures. Indeed, by the year 2020, 75% of the population of Europe and 55% of the population of Asia will be living in cities, consuming relatively little land surface of the planet yet the bulk of its resources (Yeang 1999, Sassen 2008).

If we treat cities as living organisms and a complex yet organized entity that should be integrated functionally within the ecosystems as a whole, the concept of sustainable architecture, innovative designs to minimize the problems of resource-consuming and environment-polluting impact of concrete structures will be highly useful for a more sustainable future in urban living. Thus, the fundamental approach towards building a sustainable architecture is complex yet simple – it is essential to ally its design, operations and management as closely as possible to the functioning logics of nature.

Given that high-rise urban buildings consume huge amounts of construction materials, and non-renewable energy sources which in turn produce massive volumes of waste discharge into the environment, the latter's adverse effects and damages are not environmentally sustainable. Indeed, buildings consume more energy and release more carbon dioxide than vehicles. In the United States, for instance, buildings use up 37% of all energy resources and two-thirds of electricity – six times more than automobiles in energy consumption and carbon dioxide emissions (Goffman 2006). As such, it is necessary to design buildings with minimum consumption of non-renewable sources of energy in terms of heating, hot water, cooling, lighting, power, ventilation and other internal functions. Indeed, with careful consideration to the climatic features of the project site and through bioclimatic design principles using climate-responsive building configuration, appropriate building orientation, buildings' energy needs to make internal mechanical and electrical environmental systems operational can be reduced (Winchip 2007, Yeang 1999, Cooper et al. 2009).

In Turkey, the concept of sustainability has become an important key element for architectural discourse during the last 20 years. Innovative, modern and sustainable high-rise office buildings have been tested out especially in İstanbul, due to its unique location and historical background. Buildings in this city have become very early experimental examples of sustainable architecture in Turkey. The objective of this research is to showcase a few high-rise office buildings in İstanbul and to examine their level of sustainability based on identified design criteria in line with architectural sustainability. The selected examples are the Metrocity Office building, Isbank Headquarters and the Akbank Tower. They are compared and analyzed using the sustainable design criteria focused mainly on site-environment, energy–water, materials–resources and indoor environmental quality on the basis of green rating systems. Nevertheless, before investigating the particularly identified building examples of İstanbul placed under scrutiny, the issue of sustainable architecture is discussed in the broad context of environmental sustainability.

12.2 Environmental Sustainability and Sustainable Architecture

The concept of sustainability has become a rigorously discussed phenomenon throughout the world during the last 40 years. The importance of sustainability and environmental sustainable development, whose foundations were laid in 1972,[1] has increased and it has been used to integrate humans, nature and building environments in a harmonious manner (Keleş and Harmancı 1993). The public acceptance for the term "sustainability" came when the World Commission on Environment and Development (WCED), which the United Nations General Assembly charged with formulating an "agenda for the future", based its proposals mostly on environmental sustainable development. As revealed in WCED's 1987 report, often called the Brundtland Report, the concept, in contemplating and reflecting on the environmental consequences of economic, political and environmental applications of the world, queried the extent to which national and international communities had considered the interests of the future generations to ensure that they could continue to live in a sustainable way (WCED 1987, Perkins et al. 1999). This Report also identified common concerns that threaten our future, the role of the international economy, equity, poverty and environmental degradation (WCED 1987: 24).

Similarly, the 1992 Rio Earth Summit organized by the United Nations alerted the alarming consumption of natural resources, the worrying rise in global warming and the rapid spectacular destruction of ecosystems. Participatory countries signed agreements that had been later translated into numerous regulatory measures affecting industry, transport, energy use and waste management. These measures were aimed at encouraging people in industrialized nations to conserve resources and to be aware of the potential consequences of their prevailing lifestyles. Indeed, global warming, land and sea degradation have been genuinely recognized as a consequence of human activities in their production methods, consumption patterns and choice of materials (Gauzin-Müller 2002, Ryder 2001, Wong et al. 2008). Soon after, came the debate and negotiations on similar issues that ended with the Kyoto Protocol of 1997 where participating nations designed more concrete measures including the reduction of greenhouse gas emissions. These measures have had wide-ranging implications in terms of land use, urban planning and especially architectural designs of city buildings.

Ten years after the Rio Summit in 2002, the United Nations sponsored another World Summit in Johannesburg. Global participants agreed to include the social issue of poverty eradication in the agenda as a basis of negotiation. Greater concern was also addressed over the broadened patterns of consumption and production, and as a result it was further appealed to protect and manage the natural resource-based economic, social, health-related activities so as to achieve a more sustainable lifestyle and living standards (Winchip 2007). Most recently in December 2009, the Conference of Parties (COP15) was held in Copenhagen, Denmark with an aim to establish an ambitious global climate agreement to begin from 2012 after the expiry of the Kyoto Protocol agreement.[2] Here, ministers and officials from 192 countries, together with a large number of civil society organizations, took part to exchange views and lay out principles for the new Agreement. Though the conference did not achieve a binding agreement or consensus for the post-Kyoto period, there was some "political accord" by approximately 25 parties including the United States and China. The accord was notable in that it required developed countries to commit new and additional resources, totalling US$30 billion for 2010–2012, including funds for forestry, to help developing countries. The 2020 goal is to cut deforestation by half (Light and Weiss 2009).

In principle, the Copenhagen Summit 2009 aimed to achieve the following targets (see United Nations Headquarters 2009):

(a) The most venerable and the poorest nations are assisted to adapt to the impacts of climate change;
(b) Substantial carbon emission reduction from industrialized countries;
(c) Developing countries to undertake nationally-appropriate mitigation actions with necessary international support;
(d) Financial and technological resources are significantly scaled up to meet the challenge of climate change; and
(e) To establish an equitable governance structure to ensure priorities of developing countries are recognized and financial auditing is transparent and effective.

Further negotiations have been planned for COP16 in Mexico, as there were unresolved issues from the Kyoto Protocol and a framework for long-term cooperative actions has yet to be established (UNFCCC 2009). All the above international negotiations have involved collective methods and measures to reduce emission of greenhouse gases, and consumption of fossil fuels, and use of green and renewable energy sources is encouraged. In this aspect, sustainable architecture has a great contributory role to play.

12.2.1 Sustainable Architecture

Sustainable architecture could be seen as a paradigm shift in conceptual thinking in architectural design history in recent decades. The shift has influenced values and professional practices of architects and landscape designers in their conceptualization, and reasoning about the relationship of the built environment and natural resources, energy use and the long-term future of living things (Williams 2007, Abley and Heartfield 2001). It is a key player in the environmental urban sustainable campaign due fundamentally to the rising energy consumption of buildings, particularly in urban places. In the 1990s, for example, buildings consumed globally 50%

of total energy and were responsible for 50% of the total ozone depletion; they had produced tremendously adverse environmental impacts. More significantly, modern buildings are heavily dependent on mechanical operations for heating, cooling, ventilation and lighting, using non-renewable forms of energy (Edwards 1999, Ryder 2001).

Arguably, structural designs of buildings, the ways buildings are serviced have all a direct influence over the volume of fossil fuels consumed leading to the quantity of carbon dioxide and other greenhouses gases released into the atmosphere. Such gases pollute the atmosphere, with potential damages to human and other biological health, and rise of global temperature. With an aim to counter buildings' adverse effects to the environment, sustainable architecture as a conceptual object may be defined as generally involving:

> environmentally conscious design techniques in the field of architecture. [It] is framed by the larger discussion of sustainability and the pressing economic and political issues of our world. In the broad context, sustainable architecture seeks to minimize the negative environmental impact of buildings by enhancing efficiency and moderation in the use of materials, energy, and development space. Most simply, the idea of sustainability, or ecological design, is to ensure that our actions and decisions today do not inhibit the opportunities of future generations (Wikipedia 2010).

Sustainable architecture is therefore a term used to describe an energy and ecologically conscious approach to the design of the built environment with the primary purpose of minimizing energy consumption and air pollution. Within sustainable architecture itself, building designs should be made compatible to fulfill the economic, social and environmental objectives of sustainability (Williams 2007: 13–16, Gissen 2003). Economic and social objectives of sustainability are equally crucial because buildings are meant to serve economic demand and social needs of human societies, without which buildings (commercial, residential, industrial, institutional, community etc) would not perform their intended functions and their demand is not summoned. Designers can be therefore in the position to use it as a powerful building process, and offer sustainable designs to function by adding environmental quality in the interest of the community. Green design is one of the sustainable design approaches.

12.2.2 Green Designs

Green designs are applicable to buildings to conserve energy, water and other material consumption and to protect the overall physical environment as well as building users. For the latter, creating a clean and green environment outdoor and a healthy building environment indoor implies the safeguarding of users' health and provision of a more productive working and living habitat. Health diseases inherent in problematic buildings include those related to the respiratory organs, asthma and allergy (Goffman 2006). In this sense, the green concept has a strong link to social and economic sustainability intertwined with environmental sustainability.

Green architecture or design indeed involves the approach that basically reduces energy, use of other resources and materials and regulates air flow into buildings targeted to deliver its intended outcome. Depending on local climatic conditions, temperate zones, for example, require forms of design that would capture well the passive solar light as against hot and humid tropic areas where solar exposure needs to be reduced to offer comfort. As widely acceptable, comfort contributes to health and productivity. The goals, inter alia, can be generally expressed in the following ways (Thomas 2003, Thomas and Ritchie 2009, Barnett and Browning 1999):

(a) Less demand for energy:
(b) Lower demand for space heating (for temperate zones), and cooling (for tropical zones) by using the orientation, form in relation to the solar and wind energy;
(c) Ensuring air ventilation is of a high quality, and materials within the building should not emit pollutants;
(d) Less hot and cold water consumption. It will be an advantage when waste hot water can be economically recovered;
(e) Optimal use of daylighting to cut down use of electrical energy; and
(f) Services that the building is able to put under its control should be energy-efficient, and materials used are environmentally friendly.

Today, national and city governments are increasingly conscientious of the importance of sustainable buildings for their cities, citizens and the environment as a whole. They have also passed legislations, regulations and environmental laws that require architects, engineers and designers to conceive and comply with sustainable architecture norms in both the building interior and exterior parts. Among the many international evaluation systems, four crucial green building rating systems have been used, namely (a) the BRE Environmental Assessment Method (BREEAM) in United Kingdom; (b) the Leaders in Energy and Environmental Design (LEED) in the United States; (c) the Comprehensive Assessment System for Building Environmental Efficiency (CASBEE) in Japan; and (d) the Green Star of the Green Building Council of Australia (GBCA) (WGBC 2010).

The above-cited systems have been used by many other countries as a reference, and with their own respective point grading system building approval certificates are issued. Despite certain differences, it is possible to identify four main categories out of these green building rating systems, with assessment criteria listed as follows:

(a) Site and Environment: assessing the relationship of the building to the existing topography and greenery application in the environment;
(b) Energy and Water: assessing energy consumption and saving capability, use of the most efficient methods to cut down energy and water consumption within buildings, as well as the efficient measures in the treatment of wastes and their storage;
(c) Indoor Environmental Quality: assessing the indoor air quality and the other comfort conditions like the heat efficiency, humidity ratio, natural ventilation etc; and

(d) Materials-Resources: assessing the sustainability value of building and service materials and how they would affect human beings, both physically and psychologically in the buildings.

The above four criteria of sustainable building design will be used in the assessment of the selected buildings in İstanbul below.

12.3 Sustainable Architecture in İstanbul, Turkey

In Turkey in general and in İstanbul in particular, urban-industrial growth first occurred during the 1950s, which saw massive numbers of rural migrants seeking urban-based jobs in emerging industries. This migration had a profound effect on the urban pattern and configuration characterized by unplanned land use and unhealthy living conditions. In poorly serviced spontaneous settlement areas filled up by migrants, it is not uncommon to witness serious degradation of the living environment.

With improved economic conditions and only after the 1980s that the city governments in Turkey have shown greater concern to the environmental problems of their urban habitat, and have become more aware of the sustainability concept. Concomitantly, the public has developed a greater consciousness and interest in environmental protection, and biodiversity. In city development programmes, urban sustainability issues and, for professional bodies and individuals, sustainable architecture, have started to gain momentum and capture greater attention. Turkey's association with sustainable architecture and green urbanism began in 1996 when İstanbul organized the Second UN Conference on human Settlements (Habitat II). This event helped not only put forward ways of applying sustainable principles in building designs but also adopted a specific approach to ensure urban groups were socially included in the decision-making process in that all urban residents have equal rights to the city. Such a concept of social inclusiveness of common people has helped lay the foundation of urban social sustainability and management of urban poverty in the country (Yılmaz 2008).

Indeed, the Second UN Conference on Human Settlements (Habitat II) of 1996 had reaffirmed the importance of the building sector by specifically identifying buildings to meet the needs of humans in the environmental, aesthetic, functional, security and ecological domains. According to Turkey's 8th Development Plan, the provision of environmental quality and building healthy communities in urban areas have been given high priority, and clean water, air and sustainable urban life are considered key elements of quality of life to be sought after (DPT 2001). Supplementary to this higher order development pursuit, a series of related regulations, laws and building codes were passed legislatively and came into force in 2002. The most important regulations of these are the Renewable Energy Sources Law, Energy Efficiency Law, Buildings Thermal Insulation Regulation, Building Materials Regulation and TSE 825-Thermal Insulation Requirements for Buildings.

These regulations and laws focused on the reduction of energy consumption and provided incentives to the use of renewable energy sources (DPT 2010).

Early examples of sustainable architecture are most commonly seen in İstanbul because it is a metropolitan area with a population of almost 13 million.[3] Also İstanbul has a significant role to play in this aspect, not only because of its historical and cultural background, but also due to its unique location that hosts the headquarters of many national and international companies. İstanbul is at the crossroads between Europe and Asia, and has a vast metropolitan area that accommodates the latest innovations of buildings and combines the old with the new.

12.3.1 Selected Buildings with Sustainable Architecture in İstanbul

As the financial and industrial centre of Turkey, it is symbolic and characteristically imperative that İstanbul's new built-up areas should respond and comply with the most-up-to-date building norms of sustainability demands, whether for corporate or residential communities. In site selection, the sustainability approach requires that new buildings be near the business district and accessible to the main transportation centres. Büyükdere Avenue is a perfect choice with its prime land characteristics. This avenue is located in a newly developed district and does not have any significant historical and cultural identity.

As early as the 1980s, land in this area was the least expensive which provided for unlimited renewal opportunities. Consequently, all of the prestigious high-rise office buildings for financial corporations were constructed there, replacing the small-scale residential settlements and factories with high-rise office and shopping blocks (Mimdap, 2007a).[4] Buildings constructed then in the 1980s had no legal obligations to comply with construction and design criteria on the basis of sustainable architecture. Generally landowners and architects constructed freely according to their own design principles and perceptions. Not surprisingly, the sharp increase in the number of buildings on this avenue has caused a lack of infrastructure, crowdedness, pollution and traffic jams. With the adoption of the green building criteria and with the İstanbul Building code revised many times (last revision was done in 2007), construction of environmentally friendly buildings has finally become a legal obligation[5] (IBB 2010).

İstanbul possesses buildings constructed by both national and multinational companies to reflect their corporate identities and they have become objects of prestige. However, even though the green design criteria were adopted in the 1990s, this new approach of development control for high-rise office buildings has not been effective, as buildings now standing amidst a disorganized pattern along Büyükdere Avenue clearly reflect. The sustainable design criteria for high-rise buildings as a means of development guidance have only started to see its effectiveness after the Building Code revision in 2007, thanks to the rising awareness of the architects involved in the building designs. For the purpose of this study, three building samples situated along the Büyükdere Avenue which have adopted sustainable design methods are selected for evaluation, namely the Metrocity Office Building, Isbank

Headquarters and the Akbank Tower (see Fig. 12.1). Each of these three buildings is discussed in turn.

12.3.1.1 The Metrocity Office Building

It is a complex of mixed use, consisting of a shopping centre, office and apartments, designed by architects Sami Sisa and Doğan Tekeli and built between 1997 and 2003. The first two storeys of the office building were designed as a shopping centre. The other two blocks are apartment units. The office block was designed facing the avenue after considering the orientation of the land plot which overlooks Büyükdere Avenue with a broad view (Boyut 2001). At the centre of the Metrocity Complex is a subway station, the most important facility provided for its users for easy access in a city such as İstanbul, known for its traffic congestion. The Metrocity Office Building has a total of 23 storeys and a height of 118 m. The floor area of each storey covers 750 m^2 and approximately 30–35 people work on each storey (Uluğ 2003).

12.3.1.2 The Isbank Headquarters

The banking headquarters was first designed by Doğan Tekeli and its construction was undertaken by Swanke Hayden Cornell Architects from 1996 to 2000. The decision to have the building facade facing the Büyükdere Avenue with its main entrance at the corner of the Bosphorus Bridge was to ensure that the building could be visible from all directions and that two tower blocks could be erected side by side parallel

Fig. 12.1 Overview of high-rise office buildings in İstanbul. *Source*: Mynet (2008)

to the bridge. This is solved with high blocks standing on a three-story horizontal mass. This horizontal mass houses the bank headquarters, provides a main entrance, and a central and exhibition hall. At the ground floor, there is a small shopping centre and a cafeteria which opens to a green garden facing the Büyükdere Avenue. The Isbank Headquarters building has 46 storeys, a height of 181 m, and a standard normal floor area of 1,407 m^2 (Boyut 2001).

12.3.1.3 The Sabancı Centre

The Sabancı Centre has two tower blocks, designed and constructed by Haluk Tümay and Ayhan Böke between 1988 and 1993. The Sabancı Centre comprises the 39 floor Akbank Tower, housing the central headquarters of Akbank, and another 34-floor Sabancı Holding Tower which houses the central and administrative headquarters of the Hacı Ömer Sabancı Holding and the Sabancı Group corporations (Sabancı, 1993) (see Fig. 12.2).

12.3.2 Comparison of the Buildings Using Sustainable Design Criteria

The three above-cited sets of buildings were designed by architects with different perceptions of design and were built at different times. With no legal obligations in Turkey 15–20 years ago, they applied certain green building standards to their projects. Some of the standards and technologies used were generally below the international norms at that time. It is however worthy of evaluating and comparing their applications using the four main criteria listed below.

Fig. 12.2 Metrocity, İşbank Headquarters, Sabancı Towers. *Source*: Mimdap (2007b) (Photo on *Left*). Photos (*centre* and *right*) by Selin Mutdoğan

12.3.2.1 Site-Environment

This criterion examines the building's relationship with its natural environment and green areas, in consideration of its site location, its position with surrounding buildings, as well as climatic and topographic characteristics. Besides, natural energy sources, such as solar and wind energy, should be exploited to decrease fossil-based energy consumption. On the other hand, the social relationship between humans and their environment in terms of human mobility and existing architectural heritage should also be considered (Karataş 2004).

The building sites of the three high-rise buildings on Büyükdere Avenue under study have no architectural heritage and valuable green areas, such as parks or orchards. This could have been an advantage because there were no physical restrictions in land use planning. Indeed, the developers had used this "freedom of expression" to design their own green areas in relation to the surrounding buildings constructed in recent years, within an architectural and design unity. Although an overall unity was created with the other buildings, the floor plans and form of the facades of the three buildings are different from each other. While the İşbank Headquarters and Sabancı Towers were designed as squares with a service core in the centre, the Metrocity office building is in the form of a quarter circle and the service core is oriented towards the north and west.

Climatic conditions affect the formation of the building envelope and materials used. Dark-coloured glass reflecting heat and light has been used for these three buildings; it is widely used in the İstanbul city region which has humid, windy and very hot summers. A panel system has been used for the wall cladding at the İşbank Headquarters (İşbankası Kuleleri 2006). The modular-sized panel system is significantly different from the other systems in that it is semi-finished at a local factory site, then transported to and completed at the building site. The building facade of the Sabancı Towers has been designed taking into account its intended function, aesthetics, safety and flexibility. The exterior surface of the building has been covered with a semi-panel system – the first example of this kind of utilization in Turkey. Reflective blue glass has been used on the outer facade and granite coating has been used on the concrete surfaces (Sabancı Center 1993) (see Table 12.1).

In the design of the facades and entrances of the three buildings, harmonization with the built-up environment, the building users and the established natural environment was given due consideration. To make up for the lack of green areas on the shell, an artificial natural setting was created at the main entrance and along the common spaces of different storeys. The entrance to the Metrocity office building and shopping centre looks onto Büyükdere Avenue. The subway station, a small pool and green areas were designed in this area (Fig. 12.3). Furthermore, the roof of the shopping centre is designed as a green roof.

The Isbank Headquarters, for security reasons, has a few entrances from different directions. The bank skyscrapers have a transport service system for both vertical movements and linking with the outside world, and a spacious car park. There is also greenery landscaping at the section overlooking the shopping centre entrance

Table 12.1 Comparison of three buildings using sustainable "site-environment" design criteria

Criteria	Metrocity office building	İşbank headquarters	Sabancı towers
Relation with outside environment	Connected with the main road axis	No connection with exterior	No connection with exterior
Entrance	Extroverted because of shopping mall	2 main entrances with security. Intraverted because of the security	2 main entrances with security. Intraverted because of the security
Building form			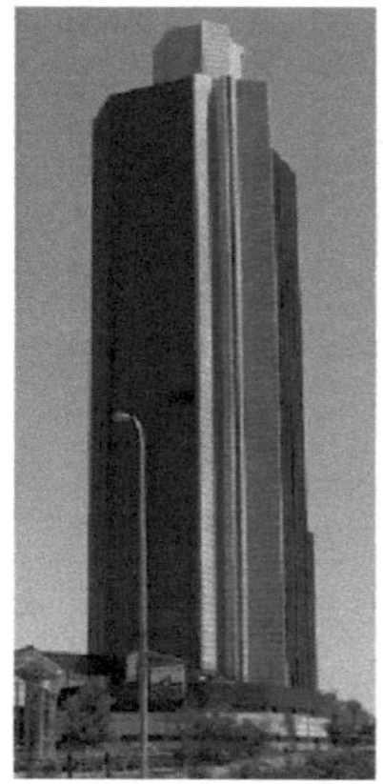

Table 12.1 (continued)

Criteria	Metrocity office building	İşbank headquarters	Sabancı towers
Typical floor plan	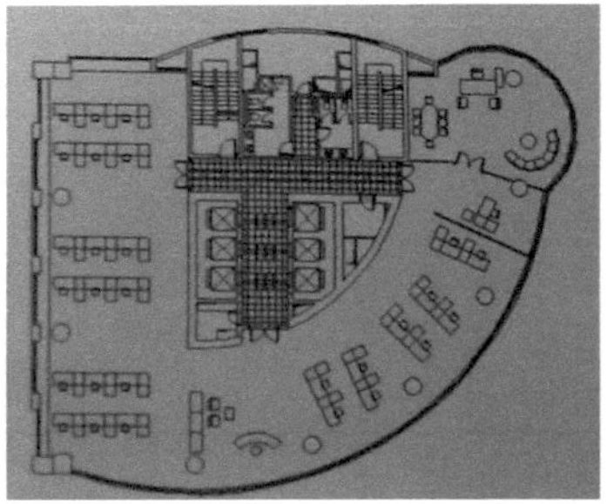	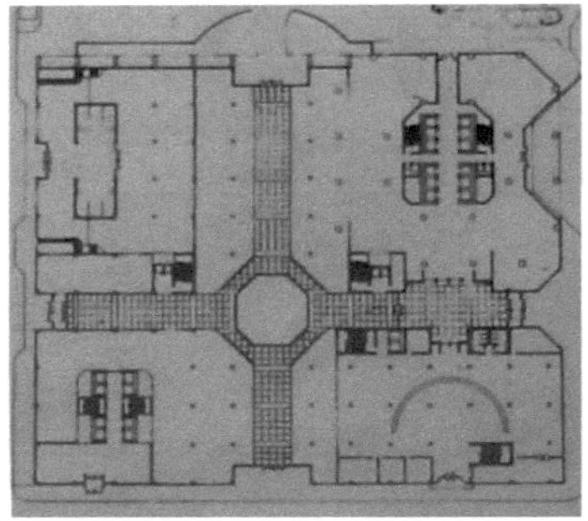	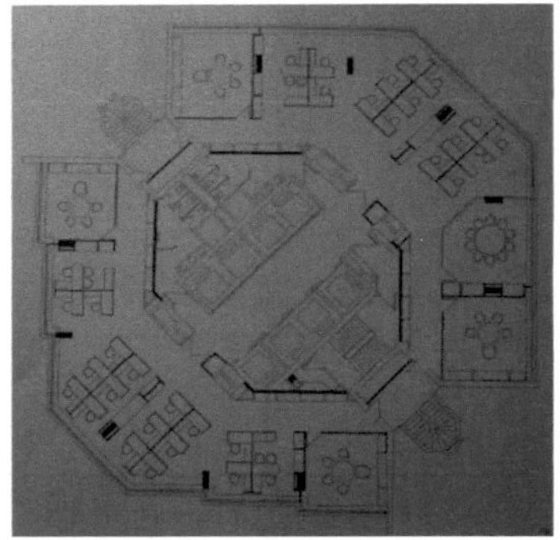
Orientation	South-East Service core at North and West	Uniform facade Service core in the middle	Uniform facade Service core in the middle
Density	30–35 people for each floor (23 floors)	65–75 people for each floor (46 floors)	38–48 people for each floor (33 floors)
Parking capacity	350	2900 (for all 3 towers)	600
Public transportation	Metro station within the tower	Very close to metro station	Very close to metro station
Open spaces and green areas	A pool and green area in front of shopping mall; a small roof garden designed for staff	A small recreation area for staff	Very small green spots designed in car parking area
Facades	Dark coloured glass	Dark coloured glass, panel system	Reflective blue glass, semi-panel system

Source: Morhayim 2003, Aydemir et al. (1992), Tekeli and Sisa (2003), and Tekeli (1999)
Note: Photos by Selin Mutdoğan

Fig. 12.3 Entrance of the Metrocity (Photos by Selin Mutdoğan)

of the complex and at the main entrance of the business towers. This green area, which covers the car parking entrance, has different species of plants (Fig. 12.4).

Similarly, the Sabancı Centre has green areas and gardens designed by landscape architects around the office blocks. All three buildings are located very close to the subway stations for easy access to public transport.

Fig. 12.4 Entrance and recreational areas of the İşbank headquarters (Photos by Selin Mutdoğan)

12.3.2.2 Energy and Water

For attaining sustainability, wise energy and water consumption has been given high priority in today's world, especially as buildings, as discussed earlier, consume 50% of total energy sources. High-rise buildings are targeted as they are a great consumer of energy as well as an emitter of pollutants and wastes. By green energy norms, buildings housing large numbers of workers and having an intense usage of equipment should be equipped with automated systems to ensure a well-coordinated operation and to provide maximum efficiency of all systems. Heating, air conditioning, ventilation, humidification, water purifying, heat recovery systems, fire detection, warning systems, stair pressure systems, lighting, security, garden irrigation, Uninterruptible Power Supply (UPS), generator and elevator systems available in the buildings can be centrally controlled with the automated systems to ensure the most efficient use of energy.

Unfortunately all three buildings under study have been equipped with traditional HVAC systems. Seen as high technology, the Freon gases (Freon 22 – Freon 134A – Freon 11) released by the buildings are harmful to the ozone layer, though the HVAC systems used in buildings are connected to the automated system for the most efficient operation. All three buildings have their own fan coil units. The Metrocity uses VRV (Variable Refrigerant volume), the Isbank uses 4 fan coils and VAV (Variable Air Volume) and the Sabancı uses parapet front heating systems and VAV (Aydemir et al. 1992, Karabey 2003, Morhayim 2003).

Theoretically, energy efficient equipment is preferred for lighting fixtures. All batteries should have photocell systems in order to save fresh water. The outdoor air is directly used for cooling the building and waste heat from the building is recovered (heat recovery) during the seasonal transition periods. The İşbank Headquarters and Metrocity Office building use stored storm waters for garden irrigation. In the treatment of waste waters, black water[6] is sent directly to the local drainage system and gray water in the whole building is filtered (Morhayim 2003, Pamir 2001).

The annual CO_2 emission amounts and energy consumption are measured at the İşbank Headquarters and Sabancı Centre, but not at the Metrocity Office Building. According to the survey data (Pamir 2001), the İşbank Headquarters consumes 120 kmw/m^2 of electricity per year. The Sabancı Towers has a total energy consumption of 161 kmw/m^2 per year. The energy consumption of buildings must be under 150–250 kwh/m^2 per year to pass an energy-efficient evaluation (Morhayim 2003). It appears that both buildings have met the target (Table 12.2).

The envelope, besides determining the outer appearance of the building and reflecting the institutional identity of the companies using the buildings as their office, is also effective in ensuring indoor comfort. Heat loss in buildings usually occurs at the shell. Hence, this is an important factor that tends to increase energy consumption. The surfaces at the connection points of the facade coating should be designed so as to prevent air leakage. Care should be taken in the fullness and emptiness rate on the facade to avoid heat loss and the northern side should be closed to the outer environment as much as possible (Williams 2007). These factors have been considered in the three buildings analyzed and the service cores

Table 12.2 Comparison of the three buildings using sustainable "energy-water" design criteria

Criteria	Metrocity office building	İşbank headquarters	Sabancı towers
Automation system for energy usage	Yes	Yes	Yes
Energy source for heating	Natural gas	Natural gas	Natural gas
Amount of CO_2 emission/year	Not measured	Measured	Measured
Amount of energy consumption/m^2	Not measured	120 kwh/m^2/year (just electricity)	161 kwh/m^2/year
Energy efficieny of lighting fixtures	Yes	Yes	Yes
Usage of gases harmful to ozone layer	Freon 22	Freon 134A	Freon 22 Freon 11
HVAC system	VRV	4 Fan coils, VAV	Paraphet front heater, VAV
Free-cooling in offices	No	Yes	Yes
Heat-recovery	No	Yes	Yes
Usage of storm water	Stored	Using garden irrigation	No
Usage of grey water	filtered	filtered	filtered
Usage of black water	No- Directly delivered to central sewer system	No- Directly delivered to central sewer system	No- Directly delivered to central sewer system

Source: Morhayim (2003) and Pamir (2001)
Note: HVAC = Heating, Ventilation and Air Conditioning; VRV= Variable Refrigerant Volume; VAV= Variable Air Volume

have been designed in the middle and northern parts of the building plan. The three buildings being investigated have used the best possible green technology in Turkey, and they can be served as pioneer examples for future reference on sustainability assessment.

12.3.2.3 Indoor Environmental Quality

Office workers spend a significant part of their working hours inside the buildings. As such, how the materials used in the buildings influence the ambient air; their impact on the human health, psychology, and comfort levels are very important. The impact is even more crucial if the buildings are high-rise and have no direct contact with the outside of the structure.

In an earlier study conducted by building owners and architecture firms of two buildings, questionnaires were prepared to survey the potential building users at the design stage. Their opinions were used as guidance for the design of a healthier indoor environment for the İşbank Headquarters and Sabancı Towers (Sabancı Center 1993, Tekeli 1999). This method could not be used at the Metrocity because of the rental contract restriction. The offices were sold after the completion of the building (Uluğ 2003). Special care was taken in the selection of materials in compliance with sustainable design criteria such as the indoor air quality (IAQ materials). For fire safety, materials not dispersing poisonous gases have been chosen.

To minimize the negative impact of low indoor air quality on workers, sensors measuring the indoor air quality in the three buildings have been used. However, according to the tests taken while the building was under construction, the indoor air quality at the İşbank and the Sabancı Towers passed marginally the standards (Morhayim 2003).

Another critical element of evaluating indoor environmental quality (IEQ) is to ensure that the building design allows adequate daylight to enter the office workplace. In the three buildings being examined, the focus of the service area is located in the central part of the building. Therefore, the design plan would need to make sure that the workers there can make use of natural light and natural air conditioning, where available. The orientation of the windows has helped perform such functions. A wide window opening on the façade of the İşbank building provides substantial amount of daylight into the building. Strip windows have been used at the Metrocity and Sabancı Towers which also enjoy a very broad angle of landscape view and high-quality natural light (Karabey 2003, Sabancı Center 1993).

The three office buildings have "flexible" office plan layouts. This means their office spaces have been designed to obtain the greatest possible window view and daylight penetration. Such design functions have been achieved at the Metrocity Building, especially by using local materials. Moreover, three different types of "Workstations" have been specially designed at the Sabancı Towers to achieve the most efficient operation of the energy distribution system (Tümay 1994). Indeed, the Isbank has been designed as an intelligent building, which consumes energy in the most rational manner and provides users with the maximum comfort. A computer-controlled system has been established to keep air quality, air conditioning (indoor heating and cooling conditions) at a high level of performance, supported by waste recycling and an open office system (Table 12.3).

12.3.2.4 Materials and Resources

By sustainable design, it is necessary that materials used do not have negative environmental consequences, even if values of aesthetics and costs have to be compromised. When selecting the materials, the following points have to be considered (Yeang 2006: 376):

- Materials are reusable and recyclable and dematerialization[7] principle is observed;
- Materials have a high potential for continuous reuse and recycling at the end of its useful life span;
- Materials have low energy impact;
- Locally produced materials are given priority of use; and
- Materials have no or negligibly low toxicity harmful to humans and the ecosystems.

Table 12.3 Comparison of the three buildings using sustainable "indoor environmental quality" design criteria

Criteria	Metrocity office building	İşbank headquarters	Sabancı towers
Precautions for healthy indoor environment	Yes	Yes (especially human-environment relation considered)	Yes
Based on questionnaire survey	No (rent)	Yes	Yes
– Natural ventilation	Yes	No	No
– Percentage of natural lighting	100%	85%	100%
– Facades with reflecting glass	Yes	Yes	Yes
– Low-e glass	Yes	Yes	Yes
– Heat sensor installed indoor	Yes	Yes	Yes
– Sensor for indoor air quality	Yes	Yes	Yes
– Measuring CO_2 indoor	Measurements with permission from office owners only	Yes	No
– Green areas indoor	No	No	No
– Indoor social facilities	Shopping centre Sport centre Restaurant	Concert hall Art gallery Shopping centre Cinema	Concert hall Art gallery Sport centre Restaurant

Source: Morhayim (2003), Sabancı Center (1993), and Tümay (1994)

In examining the three selected buildings under study, it is found that materials were largely chosen according to their quality for fire resistance since fire safety regulations were rigorously in force. Local materials were mostly used for furnishings and as finished products in Sabancı Towers. No recyclable materials were used, though preferred under the concept of sustainability. For the three buildings under study, architects had not considered about the reuse of construction waste in case of building demolition in the future. Using re-used and recycled construction waste and materials means extension of their life span (see Table 12.4). More recently, it has been reported that some local developers have used recyclable and sustainable materials, though to a limited extent.

12.4 Conclusion

Cities are centres of buildings, physical infrastructure and movements of economic activities of all kinds, the outcome of which produces tremendous amounts of waste harmful to the environment. The issue of common survival and long-term sustainability has called for concerted global efforts to save the Earth from being further degraded. Buildings, being a key culprit of carbon emission, have captured increasingly attention that sustainable architecture and green design are indispensable towards the ultimate end of controlling global warming and air pollution. On the top of financial and technological capabilities, moral responsibility is badly needed

Table 12.4 Comparison of 3 buildings using sustainable "material and resources" design criteria

Criteria	Metrocity office building	İşbank headquarters	Sabancı towers
PVC materials	Yes (fire resistancy)	No	Yes (fire resistancy)
Non-toxic materials	Galvanized materials used	Tested (especially for fire resistancy)	Yes (fire resistancy)
Source of materials used	Local	Local	Local
Usage of recycled materials	No	No	No
Usage of re-used materials	Not installed by tenants nor owners	Yes	Yes
Wastes (recycling and reuse)	Separate storage systems	Separate storage systems	Recycling (cooperating with another company)
Reuse of waste of construction materials in case of building demolition	No	No	No

Source: Morhayim (2003), Karabey (2003), and Sabancı Center (1993)

first, from developed countries, followed by developing countries which are fast industrializing and large consumers of raw materials as well as finished products. This is the key challenge towards a new international equitable management of the world's resources and environment on the basis of shared responsibility.

Turkey is a developing country but its pace of urbanization and the growth of its metropolitan centres are hardly any different from other high-growth developing countries. Over the last two decades, there have been efforts to set up examples of sustainable buildings in İstanbul. Unfortunately, none of them use renewable energy. However, pressure to adopt up-to-date norms to build sustainable buildings is being mounted. In the near future and as public awareness is enhanced and cost is reduced with the help of more advanced and innovative technology, green legislations will be put in place for implementation and enforcement. This study has demonstrated that the three modern buildings in İstanbul being surveyed have achieved little in sustainable design based on the criteria used to measure their annual carbon emission amounts and the energy consumption per square metre of floor area. Nevertheless, for Turkey, such low achievement should be evaluated as a good starting point.

Today, there are more advanced green technologies being adopted in other Turkish cities than the three buildings studied. Some of these buildings use geothermal energy, a cogeneration system or have double facades of higher standards. In Ankara and Erzurum, the Redevco Turkey opened two shopping centres in 2009 named respectively Gordion Shopping Centre in Ankara and Erzurumlu Forum in Erzurum. These two projects were awarded a BREEAM Certificate (rated "Very Good") – the first BREEAM-certified projects in the country. Up to now Siemens Office Building in İstanbul has been awarded a LEED certificate (rated "GOLD"). Also, Unilever and Philips Offices were awarded a LEED certificate for their renovation works in interior designs. Besides, some national organizations such as the

Association of Environmentally-Friendly Green Buildings (ÇEDBİK) are trying to create Turkey's own green building rating system. Results have been encouraging and one could claim such a self-initiated spirit will move Turkey a step forward in line with the international agenda towards urban ecological conservation.

Acknowledgements We wish to thank Associate Professor, Dr. Meltem Yılmaz of the Department of Interior Architecture and Environmental Design, Hacettepe University for the comments and feedback she has provided for the paper.

Notes

1. The first global environment summit was the UN Conference on the Human Environment in Stockholm, Sweden in 1972.
2. The Kyoto Protocol did not demand carbon reductions from developing countries. The Copenhagen Summit wants major polluting developing countries such as China, India, South Africa and Brazil to be binded to emission reductions. Both China and India had agreed to cut respectively 13 and 19% below business-as-usual carbon emissions from 2005 levels by 2020 (Light and Weiss 2009).
3. According to the Population Census of Turkey conducted in 2009, İstanbul's population was 12.915 million.
4. These opinions were discussed at a symposium entitled "Büyükdere Caddesi ve Mimarlik" (Büyükdere Avenue and Architecture) at the Taksim Campus of the Beykent University on 9 January 2008.
5. Without a comprehensive set of "Green Building Code" in Turkey, some companies use LEED or BREEAM to obtain a green building certificate.
6. Wastewater from toilets and urinals, which contains pathogens must be neutralized before water can be safely reused. After neutralization, black water is typically used for non-potable purposes, such as flushing or irrigation (Gissen 2003: 183).
7. Dematerialization means using fewer materials and less energy to produce goods and services. This idea fits in well with new technologies (Yeang 2006).

References

Abley, I. & Heartfield, J. (Eds.). (2001). *Sustainable architecture in the anti-machine age*. Chichester: Wiley-Academy.

Aydemir, C., Tümay, H., Dökücü, Ş., Resuloğlu, C. & Fındıkoğlu, S. (1992). Sabanci Center ve Akıllı Yüksek Yapılar. *Tasarim*, 20: 40–57.

Barnett, D. L. & Browning, W. D. (1999). Green building design. In F. Stitt (Ed.), *Ecological design handbook, sustainable strategies for architecture, landscape architecture, interior design and planning* (pp. 9–15). New York, NY: McGraw-Hill.

Boyut (2001). *Boyut Çağdaş Türkiye Mimarları Dizisi 2: Doğan Tekeli-Sami Sisa*. İstanbul: Boyut Press Group.

Cooper, R., Evans, G. & Boyko, C. (2009). *Designing sustainable cities*. Oxford: Wiley-Blackwell.

DPT (T.R. Prime Ministry State Planning Organization) (2001). The VIIIth development plan. http://plan8.dpt.gov.tr/. Accessed 02 July 2010.

DPT (T.R. Prime Ministry State Planning Organization). (2010). Legislation. http://mevzuat.dpt.gov.tr/. Accessed 02 July 2010.

Edwards, B. (1999). *Sustainable architecture: European directives and building design*. Oxford: Architectural Press.

Gauzin-Müller, D. (2002). *Sustainable architecture and urbanism: concepts, technologies, examples [translated from French by Kate Purver]*. Basel: Birkhaüser.
Gissen, D. (2003). *Big and green: toward sustainable architecture in the 21st century*. New York, NY: Princeton Architectural Press.
Goffman, E. (2006). Green buildings: conserving the human habitat. http://www.csa.com/discoveryguides/green/review/php. Accessed 10 June 2010.
IBB (2010). İstanbul İmar Yönetmeliği (İstanbul Building Code). http://www.ibb.gov.tr/tr-TR/kurumsal/Birimler/ImarMd/Documents/yonetmelik.pdf. Accessed 02 July 2010.
İşbankası Kuleleri (2006). *Tasarım*, 8: 92–94.
Karabey, H. (2003). Metrocity Konut ve Alışveriş Merkezi *Yapı*, 263: 72–80.
Karataş, B. (2004). *Sürdürülebilir Mimarlık Bağlamında Çok Katlı Ofis Binalarında Ekolojik Tasarım İlkelerinin İrdelenmesi*. Master thesis. İstanbul: Yıldız Technical University.
Keleş, R. & Hamamcı, C. (1993). *Çevrebilim*. Ankara: İmge Kitabevi Yayınları.
Light, A. & Weiss, D. J. (2009). Lesson learnt from Copenhagen. what you need to know following the Copenhagen climate summit. http://www.grist.org/article/2009-12-13.Accessed 21 June 2010.
Mimdap (2007a). Büyükdere Caddesi ve Mimarlık. http://www.mimdap.org/w/?p=3295. Accessed 18 April 2008.
Mimdap (2007b). Tekeli Sisa Mimarlık Ortaklığı. http://www.mimdap.org/w/?p=612. Accessed 4 June 2008.
Morhayim, L. (2003). *Ekolojik Mimari Tasarım Anlayışının İstanbul'daki Yüksek Ofis Yapıları Örneğinde Değerlendirilmesi*. Master thesis. İstanbul: Yıldız Technical University.
Mynet (2008) http://img2.mynet.com/emlakcontent/levent_detay.jpg. Accessed 23 July 2010.
Pamir, H. (2001). Kurumun ve Yapının Beraber Yapılması: *İş Kuleleri*. XXI, 8, 102–113.
Perkins, H., Thorns, D. C. & Field, M. (1999). *Urban sustainability: an annotated Bibliography*. House and Home Project, Canterbury and Lincoln Universities.
Ryder, P. H. (2001). If sustainable design isn't a moral imperative, what is?. In I. Abley & J. Heartfield (Eds.), *Sustainable architecture in the anti-machine age* (pp. 22–31). Chichester: Wiley-Academy.
Sabanci Center (1993). Istanbul: Sabanci Publisher.
Sassen, S. (2008). Cities: at the heart of our environmental future. In K. Feireiss & L. Feireiss (Eds.), *Architecture of change: sustainability and humanity in the built environment* (pp. 34–37). Berlin: Gestalden.
Tekeli, D. (1999). İş Bankası Kuleleri. *Arredamento Mimarlık*, 100(14): 37–40.
Tekeli, D. & Sisa, S. (2003). Metrocity Konut ve Alışveriş Merkezi, Levent-İstanbul (interview by Haydar Karabey). *Yapı*, 263: 72–80
Thomas, R. (2003). Building design. In R. Thomas & M. Fordham (Eds.), *Sustainable urban design: an environmental approach* (pp. 46–61). London: Spon Press.
Thomas, R. & Ritchie, A. (2009). Building design. In A. Ritchie & R. Thomas (Eds.), *Sustainable urban design: an environmental approach* (2nd ed, pp. 42–55). London: Taylor & Francis.
Tümay, H. (1994). Sabanci Center Planlama Hedefleri: Proje Konsepti. *Tasarım*, 49: 53–54
Uluğ, M. (2003). Bir Metropol Yapısı Olarak Metrocity. *Uluğ*, XXI(18): 54–59.
UNFCCC (Copenhagen Accord Advance Unedited Version) (2009). http://unfccc.int/files/meetings/cop_15/application/pdf/cop15_cph_auv.pdf. Accessed 26 March 2010.
United Nations Headquarters (2009, 22 September). Summit on climate change: power green growth, protect the planet. http://www.un.org/wcm/webdav/site/climatechange/shared/Documents/Chair_summary_Finall_E.pdf. Accessed 21 June 2010.
WCED (World Commission on Environment and Development) (1987). *Our common future*. Oxford: Oxford University Press.
WGBC (2010). World green building council. http://www.worldgbc.org/green-building-councils/green-building-rating-tools. Accessed 02 July 2010.
Wikipedia (2010). Sustainable architecture. http://en.wikipedia.org/wiki/Sustainable_architecture. Accessed 21 June 2010.

Williams, D. (2007). *Sustainable design: ecology, architecture and planning*. River St Hoboken, NJ: Wiley.
Winchip, S. M. (2007). *Sustainable design for interior environments*. New York, NY: Fairchild Publications.
Wong, T.-C., Yuen, B. & Goldblum, C. (Eds.). (2008). *Spatial Planning for a Sustainable Singapore*. Dordrecht: Springer.
Yeang, K. (1999). *The green skyscraper: the basis for designing sustainable intensive buildings*. Munich: Prestel.
Yeang, K. (2006). *Ecodesign: a manual for ecological design*. Great Britain: Wiley-Academy.
Yılmaz, M. (2008). *sustainable housing design consideration for Turkey: planning and design principles*. Ankara: Hacettepe University Publications.

Chapter 13
Urban Air Quality Management: Detecting and Improving Indoor Ambient Air Quality

T.L. Tan and Gissella B. Lebron

Abstract Current air pollution management and air quality control are primarily focused on outdoor and atmospheric issues. In major cities today with large numbers of shopping malls, offices and public administration centers which act as public spaces, contaminated indoor air could be public health hazards. In Singapore, diagnosing the causes of "sick building syndrome" is as important as treating outdoor pollution as its workforce is increasingly service-oriented and many of whom spend a substantial amount of time working in air-conditioned premises. It is known that indoor air quality (IAQ) can be easily and adversely affected by gas pollutants which are internally generated or infiltrated from external sources. One important and practical example is carbon monoxide (CO) which can be emitted at high concentration levels in an urban structure by burning of tobacco and incense, and by incomplete combustion from gas stoves and fuel engines used in renovation work. In this research, the decay rates of CO concentration (ppm) in air were measured accurately using the Fourier transform infrared spectroscopy in the 2,050–2,230 cm^{-1} wavenumber region. High levels of CO were obtained from sidestream environmental tobacco smoke (ETS). From the modeling of the decay curves of CO concentration with time, the air exchange rates in air change per hour (ACH) were derived for six different ventilation rates. They were found to be from 2.53 to 8.63 ACH. The ventilation rates for CO contained in a chamber were varied using different window areas. Half-lives of the CO decays at six different air exchange rates were also determined and found to decrease from 16.4 to 4.8 min as the air exchange rate increases. The implications of air exchange rate on the decay of indoor CO in ETS were discussed with reference to IAQ in air-conditioned buildings in Singapore, and to IAQ in general urban settings.

T.L. Tan (✉)
Natural Sciences & Science Education, National Institute of Education (NIE), Singapore
e-mail: augustine.tan@nie.edu.sg

T.-C. Wong, B. Yuen (eds.), *Eco-city Planning*, DOI 10.1007/978-94-007-0383-4_13,

13.1 Introduction

It has been established recently that the design of a building is becoming more important and crucial to the maintenance of good indoor air quality. In buildings where offices are used as air-conditioned workplaces, we are surrounded by artificial lighting, synthetic carpeting, furniture treated with chemicals, and high-tech equipment which can present various types of air quality problem. The World Health Organization has pinpointed that 30% of buildings in general are prone to sick building syndrome problems. Studies have shown that indoor air quality is influenced by two broad categories of building design: the "engineering" characteristics, and "quantitative" information on contamination levels in occupied buildings. The engineering characteristics refer to the information on air ventilation, infiltration standards, humidity, and levels of particles and pollutants in the building. The quantitative information on the building design includes the various monitoring devices and instruments used to measure all aspects of air quality within the building environment. The measurements are then compared to established and accepted standards. Therefore, it is important in the design of a modern building to include the engineering aspects of indoor air quality, and the quantitative information needed to maintain good indoor air.

Since the early 1970s, a considerable effort has been conducted by industrialized countries to measure the concentrations of air pollutants in the outdoor ambient air environment (Godish 2004, McDermott 2004, Yocom et al. 1971). In the United States, air monitoring became obligatory to determine the status of compliance with the National Ambient Air Quality Standards (NAAQs). These standards under the Clean Air Act (CAA) were set up by the US Environmental Protection Agency (U.S. EPA). The main objective of this Act is to improve ambient air quality by reducing emissions of pollutants which are detrimental to human health. The pollutants are listed in CAA as carbon monoxide (CO), nitrogen dioxide (NO_2), sulfur dioxide (SO_2), ozone (O_3), particulate matters, hydrocarbons, and lead (Pb) (Brooks and Davis 1992, Yocom and McCarthy 1991). Results of the monitoring of the concentrations of pollutants in outdoor or ambient air over the past 20 years has shown that emission controls have been effective in improving outdoor air quality (Brooks and Davis 1992, Spengler and Soczek 1984). However, a corresponding improvement in the general public health has not been observed in some epidemiological studies (Apte et al. 2000, Fang et al. 2004, Liu et al. 2001, Ohman and Eberly 1998, Seppanen et al. 1999, Smith 2002). One possible reason is that with urbanization, more people spend most of their time indoors where air quality differs from that of outdoors. Indoor pollutants may differ in type and concentration from some outdoor pollutants. For example, concentrations of pollutants such as CO from both outdoor and indoor sources may be higher indoors than outdoors when indoor sources are present (El-Hougeiri and El-Fadel 2004).

It is recognized now that indoor exposure to air pollutants comprises a significant component of total exposure to air pollution. Yocom et al. (1971) in their study of indoor-outdoor air quality in the United States found that the concentrations of CO were higher indoors than that outdoors if gas stoves were used. Indoor-outdoor air quality relationships were further studied by Yocom (1982). El-Hougeiri and El-Fadel (2004) investigated indoor and outdoor air quality at 28 public locations (such as restaurants, recreation areas, sport centres, school classrooms, kitchens, hotels, swimming pools, and movie theatres) in urban areas with proximity to main roads. In their work, the concentrations of CO, NO_2, and total suspended particulate matter (PM) were measured and studied. It is necessary that data be obtained on indoor exposures to pollutants if there is serious concern about the effect of air pollutants on the health of the occupants. The main reason is that most people spend over 90% of their time indoors in developed countries (Burroughs and Hansen 2008). Problems associated with poor indoor air quality caused by excessive pollutants in buildings are generally referred to as the sick building syndrome (SBS). To reduce the adverse health effects of SBS, the use of adequate ventilation has been found to be effective (Lam et al. 2006, Melikov et al. 2005, Seppanen et al. 1999, 2006).

In the last 15 years, the importance of research on indoor air quality (IAQ) issues in Singapore has been manifested in several studies (Chan 1999, Chan et al. 1995, Ooi et al. 1995, Sekhar et al. 1995, 1999, Tham 1994, Tham et al. 1996). Findings from these studies have guided the setting up of local IAQ guidelines for office premises by the Ministry of the Environment in Singapore (1996). Furthermore, Sekhar et al. (2003) developed an integrated indoor air quality energy audit methodology for Singapore with a purpose of providing an integrated multi-disciplinary model in obtaining measured data concerning different dimensions within the built environment such as ventilation, biological, physical, chemical, and occupant response characteristics. Their research work on five air-conditioned office buildings has led to the development of an indoor pollutant standard index (IPSI), ventilation index, energy index, and building symptom index (BSI). It was found that BSI values have some correlation among them as well as with IAQ and thermal comfort acceptability while there was no significant correlation between BSI and IPSI.

Recently, numerous studies on IAQ have been made on carbon monoxide (CO) because of its severe impact on public health and work performance in urban settings (Godish 2004, Harrison 1998). Statutory safe levels are that the average CO concentration for 8 h exposure must be less than 8.6 ppm (10 $mg.m^{-3}$) and the average CO concentration for 1 h exposure must be less than 26 ppm (30 $mg.m^{-3}$), according to WHO standards (World Health Organization 1999) given in Table 13.1. Concentrations of CO have been measured and studied in indoor parking facilities in France (Glorennec et al. 2008), in Greece (Chaloulakou et al. 2002), in Lebanon (El-Fadel et al. 2001), and in Hong Kong (Wong et al. 2002). Evaluation of IAQ in 10 school buildings in the urban area with major traffic roads in Athens, Greece was made in 2001 (Siskos et al. 2001). In their studies, CO concentration in classrooms was found to be lower than the WHO (1999) standard of 10 mg m^{-3} (8-h average). Determination of CO levels in 384 coffee shops in Ankara, Turkey was also made in 2009 (Tekbas et al. 2009). They found that 34% of the coffee shops had CO levels above the threshold level of 8.6 ppm (8-h average).

Further research on IAQ with measurements on CO concentrations was conducted in air-conditioned buses (Chan 2005) and offices and shops (Liao et al. 1991) in Hong Kong, in heavy traffic streets and tunnel in Santiago, Chile (Rubio

Table 13.1 WHO guideline values for CO (1999)

Pollutant	Annual ambient air concentration (ppm)	Guideline value (ppm)	Exposure time	Toxic effects according to ambient CO concentrations
CO	0.43–6.0 (0.5–7 mg/m^3)	87 (100 mg/m^3) 52 (60 mg/m^3) 26 (30 mg/m^3) 8.6 (10 mg/m^3)	15 min 30 min 1 h 8 h	35 ppm (8-h time-weighted average 200 ppm (maximum limit) 600 ppm (1-h exposure: headache, anxiety, irritability) 2,500 ppm (30-min exposure: loss of consciousness) 4,000 ppm (sudden death)

et al. 2010), in an air-conditioned building in New Delhi, India in 2002 (Prasad et al. 2002), in bathrooms using hot water boilers in Turkey (Tekbas et al. 2001), in indoor karting in Canada in 2005 (Levesques et al. 2005), in burning of joss sticks in London in 2005 (Croxford and Kynigou 2005), in cooking fuels in Pakistan in 2009 (Siddiqui et al. 2009), in commercial aircrafts (Lee et al. 1999), and in a bus terminal (El-Fadel and El-Hougeiri 2003). Emissions of CO from environmental tobacco smoke (ETS) in an indoor environment have serious effect on IAQ (Rickert et al. 1984, Sterling and Mueller 1988, Trout et al. 1998). In 2002, Dingle et al. studied the concentrations of pollutants including CO from ETS and ventilation in 20 social venues in Perth, Western Australia. Other studies on CO emissions from ETS include those of side-stream and second-hand smokes, conducted in 2010 (Daher et al. 2010) and of ETS in restaurants in Finland in 2000 (Hyvarinen et al. 2000).

To improve indoor air quality by reducing air pollutants such as CO in a building, the use of ventilation or air infiltration is the most suitable solution (Meckler 1996, Vitel 2001). Performance in office work was found to improve with higher ventilation rates in a 2006 study (Seppanen et al. 2006) and several sick building syndrome (SBS) symptoms were alleviated when the ventilation rate was increased in a 2004 study (Fang et al. 2004). Moreover, the purpose of ventilation for urban set-ups such as air-conditioned offices and lecture halls is to provide air of a level of quality that is perceived as acceptable (Fang et al. 1998). Since ventilation or air infiltration causes a stable gas at an elevated concentration in the building to be gradually removed from the building, the decay rate of the gas is usually used as a direct measurement technique for ventilation (Brooks and Davis 1992, Meckler 1996, Yocom and McCarthy 1991). The level of ventilation is usually measured in the form of air exchange rate with the unit of air change per hour (ACH). Tracer gases such as SF_6 and CO_2 are commonly used to study air exchange rate. Furthermore, measurements of the decay rate of a gas which is an indoor pollutant would provide useful information on the impact of ventilation on the IAQ of a building.

From a detailed literature review, studies on the impact of ventilation on indoor concentrations of a common pollutant, carbon monoxide (CO) appear to be quite limited. Possible sources of CO in a building can be environmental tobacco smoke

(ETS) from cigarettes, gas stoves and heaters, burning joss-sticks, or even leakage from nearby combustion engines of vehicles into the building (Yocom and McCarthy 1991). When a certain amount of CO is injected into a building whether from indoor or outdoor sources, it can be removed only by exchange with fresh CO-free air. Carbon monoxide is stable and often useful as an indoor tracer for air exchange determinations (Naeher et al. 2001, Yocom and McCarthy 1991). To understand the IAQ of a building, it would be advantageous to know how the concentration of CO changes with time for a certain ventilation or air exchange rate. In view of this, the decay rate of CO would provide information for the time needed for CO concentrations to reach safe levels for a given air exchange rate. Therefore, the objective of this study is to measure the decay of the concentrations of CO over time in an enclosed space with six different window sizes to allow six different values of air exchange rates for a given space. The air exchange rates in ACH would be derived by studying the exponential decay curves of CO concentrations. In this research, the CO concentrations at ppm level were measured using the Fourier transform infrared spectroscopy technique. The infrared technique for measuring the concentration of a gas in air is well established and is known to be sensitive, accurate and reliable (Hanst and Hanst 1994, Salau et al. 2009, Smith 1996).

13.2 Methodology

13.2.1 Experimental Method on CO Concentration Measurements

To contain the CO gas from side-stream environmental tobacco smoke, a chamber of 20 cm × 22 cm × 22 cm was constructed. It is made of transparent perspex, and airtight except for two small square open windows on opposite sides of the chamber to allow air infiltration. To allow the infrared beam to pass horizontally through the CO gas sample in the chamber, infrared windows (3-cm diameter) made of CaF_2 were placed at opposite sides of the chamber. A total of six open window sizes were used: $2 \times 2\ mm^2$, $3 \times 3\ mm^2$, $4 \times 4\ mm^2$, $5 \times 5\ mm^2$, $6 \times 6\ mm^2$, and $7 \times 7\ mm^2$. The air exchange rate is expected to increase as the window size increases. Side-stream tobacco smoke from a burning cigarette was injected into the chamber at the start of the experiment through the top cover of the chamber. The temperature of the smoke in the chamber was measured to be 25–27°C and its humidity was maintained at 62–65% during the experiments. These readings were measured using a humidity-temperature meter (model ETHG-912) from Oregon Scientific. The temperature of the ambient air in the room outside the chamber was 25–26°C while the humidity was 60–63%. The air in the chamber and outside it was under atmospheric pressure. A small fan placed at 2 m from one of the windows (inlet window) of the chamber provided a constant air draught of slightly higher pressure compared to that of the other window (outlet). The air speed at the inlet window was found to be less than about 1.5 m/s, using an air velocity meter (Thermo-Anemometer, AZ Instrument 8908). The use of fan was needed to activate a constant but low air infiltration rate into the chamber through the inlet window and out of the outlet window.

The chamber containing the smoke with CO is placed in the sample compartment of the infrared spectrometer. The infrared absorption spectra of the CO gas in the chamber were recorded using the Perkin-Elmer Fourier transform infrared (FTIR) spectrometer of model Spectrum 100 at a resolution of 0.5 cm^{-1} and in the 2,050–2,230 cm^{-1} wavenumber region. This infrared region contained the CO infrared signature (Smith 1996). A spectral run of 2 scans of total scanning time of about 60 s was sufficient in obtaining good signal-to-noise signal for each spectrum. The infrared absorption spectra of CO at different concentrations (ppm) are shown in Fig. 13.1. In the CO spectrum, there are 47 absorption peaks, and the absorbance *A* of the spectrum is determined by the total area under these peaks,

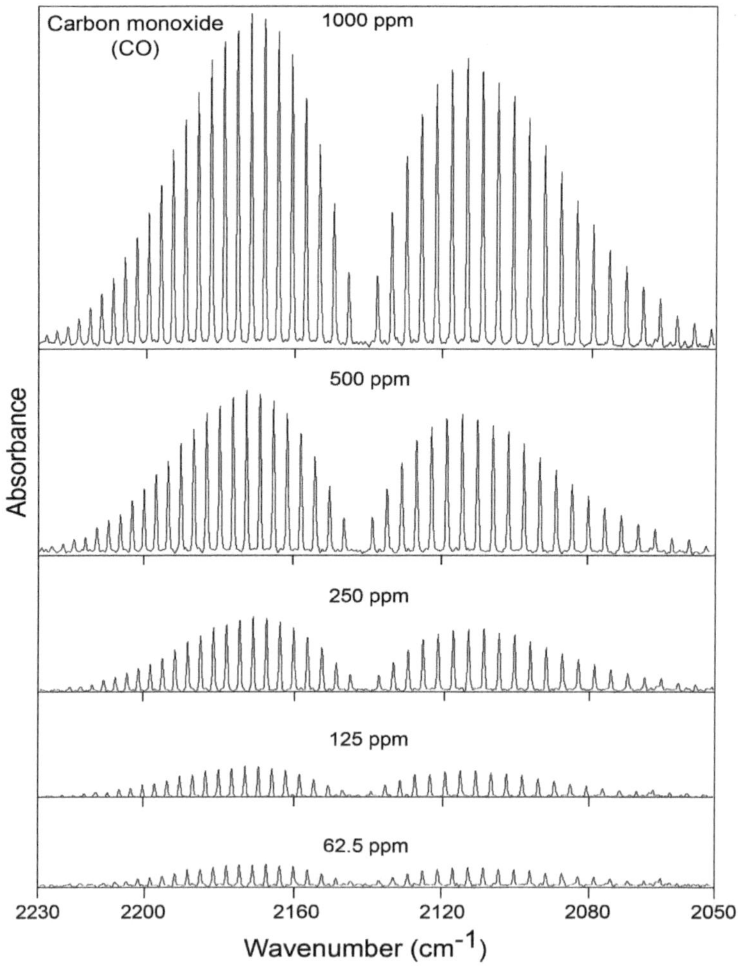

Fig. 13.1 The infrared spectra of CO in the 2,050–2,230 cm^{-1} region at different CO concentrations in ppm

using the SPECTRUM software in the infrared spectrometer. It is known that the absorbance value of the spectrum is directly proportional to the concentration of the absorbing gas (CO) in air (Smith 1996). As shown in Fig. 13.1, the intensities of the absorption peaks decrease (or the values of absorbance decrease) as the concentration of CO decreases from 1,000 to 62.5 ppm.

The CO infrared spectra were calibrated in terms of CO concentration in ppm using a Drager X-am 2000 digital detector meter which employs the electrochemical method. The detector has a range of 0–2,000 ppm with a measuring accuracy of less than 2%. Its reaction time is less than 6 s. The Drager meter was placed inside the gas chamber and CO reading (in ppm) was taken during the recording of the infrared absorption spectrum using the Perkin-Elmer spectrometer. The absorbance of the CO spectrum corresponding to the ppm concentration of CO was then measured. The calibration line of absorbance against CO concentration (ppm) from 0 to 1,700 ppm is plotted as shown in Fig. 13.2. The absorbance values ranged from 0 to about 2. The calibration line could be fitted well with the coefficient of determination $R^2 = 0.991$ or correlation coefficient $R = 0.995$. By using the calibration line, the CO concentrations in ppm could be determined from the absorbance values measured from the infrared spectra. The concentration of CO in ambient air in the room outside the gas chamber was 1–3 ppm.

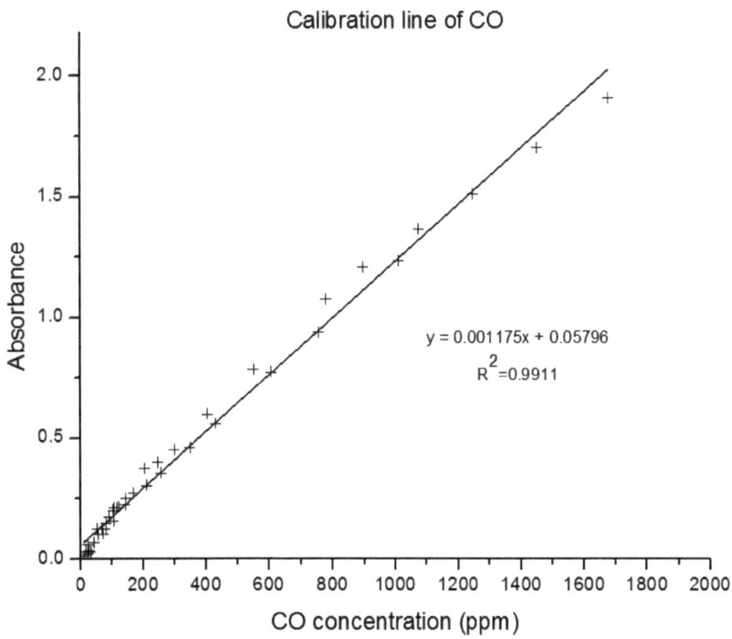

Fig. 13.2 The calibration line of concentration of CO measured in infrared absorbance and in ppm level

In the experiments, for each window size starting from 2×2 mm^2, a fixed amount of side-stream environmental tobacco smoke containing CO at 1,000–1,800 ppm was initially injected into the chamber. High levels of CO were used in the experiments because at 600 ppm with 1 h exposure, headache, anxiety, and irritability set in, followed by loss of consciousness at 2,500 ppm (30-min exposure), and sudden death at 4,000 pm (see Table 13.1 for WHO guidelines). The CO gas was allowed to mix well with the air in the chamber for about 5 min. At time $t = 0$, the first infrared absorbance spectrum of CO was taken, and subsequently, the spectra were taken at time intervals of every 80 s. The recording of the spectra and their corresponding time continued until the CO concentration reached a level as low as 30 ppm. These experiments were repeated for 5 other bigger window sizes. From these measurements, the decay rate of CO due to air exchanges for 6 ventilation rates in the chamber can be studied.

13.3 Modelling of Gas Decay Rate

The American Society of Testing Materials (ASTM) standard method for determining air leakage rate by tracer dilution in Annual book of ASTM Standards (1983) describes the protocol for measuring the air exchange rate using the tracer gas decay or dilution method. The air exchange rate I is calculated using the following exponential decay equation (ASTM 1983):

$$C = C_0 e^{-It} \tag{13.1}$$

where C = tracer gas concentration at time t; C_0 = tracer gas concentration at time = 0; I = air exchange rate, and t = time.

This relationship assumes that the loss rate of the initial concentration of tracer gas is proportional to its concentration. By applying natural log on the equation, Eq. (13.1) becomes linear as follows:

$$\ln C = \ln C_0 - It \tag{13.2}$$

If time t is measured in hours (h), the air exchange rate I is expressed in air change per hour (ACH). Values of I can be obtained by plotting the best-fit graph of measured values of C against t or of $\ln C$ against t.

The time taken for the concentration of the gas to decay by half is called the half life T of decay, and it is useful to apply it to estimate the total time needed for the concentration to reach a safe and accepted level. The half-life T of the exponential decay can be calculated using:

$$T = \frac{\ln 2}{I} \tag{13.3}$$

13.4 Results and Discussion

The decay curve of CO concentration (ppm) with time (min) for a window of 2 × 2 mm² is shown in Fig. 13.3. The starting concentration of CO was about 1,350 ppm which is at a health hazard level (Table 13.1). It decays to a safe level at 35 ppm after about 80 min. The curve could be fitted accurately with an exponential function with $R^2 = 0.995$ to give an air exchange rate $I = 0.04211\ \text{min}^{-1}$ in good agreement with the trace gas modeling given by the exponential decay Eq. (13.1). The value of I is multipled by 60 to give $2.53\ \text{h}^{-1}$ which is the value in air change per hour (ACH). The I value in ACH is given in Table 13.2. The plot of ln C against time (min) in Fig. 13.4 could be fitted accurately using a straight line which gives $I = 0.04542\ \text{min}^{-1}$ applying the natural log Eq. (13.2) with $R^2 = 0.974$. The values of the air exchange rate I determined from the best-fits of exponential decay curve (Fig. 13.3) and of linear ln C-time graph (Fig. 13.4) are in close agreement (within 8%).

From the other experiments, the decay curves of CO for window sizes of 3 × 3 mm², 4 × 4 mm², 5 × 5 mm², 6 × 6 mm², and 7 × 7 mm² are shown in Figs. 13.5, 13.6, 13.7, 13.8 and 13.9, respectively. The starting CO concentrations for various window areas ranged from 1,000 to 1,800 ppm. From the exponential decay fits, the corresponding air exchange rates (I) in ACH were obtained as given in Table 13.2. It is found that as the window area increases from 4 to 49 mm², the air exchange

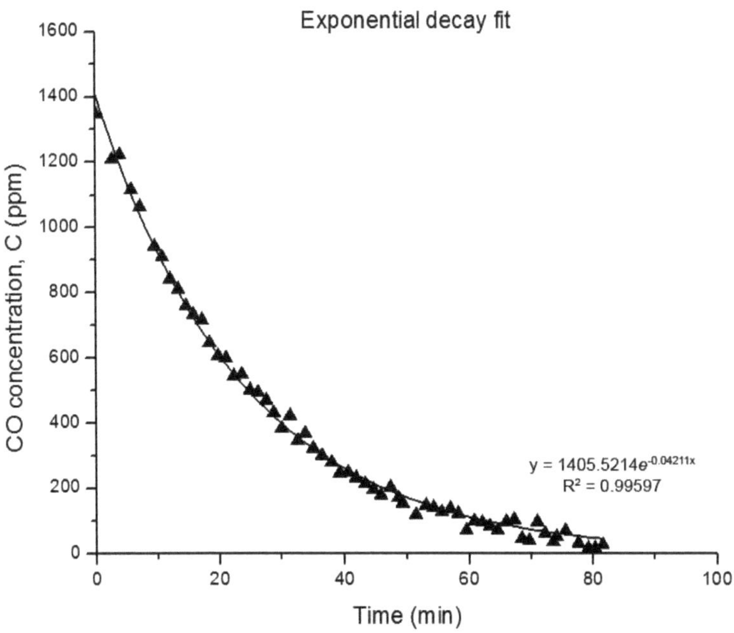

Fig. 13.3 CO decay curve for 2 × 2 mm² window

Table 13.2 Values of air exchange rate (ACH) and half-life T (min) obtained from CO decays for different window areas

Window area (mm^2)	Air exchange rate, I (ACH)	Half-life, T (min)
$2 \times 2 = 4$	2.53	16.4
$3 \times 3 = 9$	3.60	11.6
$4 \times 4 = 16$	4.16	10.0
$5 \times 5 = 25$	5.72	7.3
$6 \times 6 = 36$	6.64	6.3
$7 \times 7 = 49$	8.63	4.8

rate increases from 2.53 to 8.63 ACH. The values of half-life T calculated using Eq. (13.3) are also provided in Table 13.2. The value of T decreases from 16.4 to 4.8 min as the window area increases from 4 to 49 mm^2, showing that the decay rate increases as the ventilation rate increases. From the exponential decay curves, it can be observed that as the window area increases, the air exchange rate (I) increases, and therefore the time needed for CO to reach the safe exposure level at 52 ppm (30-min average exposure time) becomes shorter. Table 13.1 gives the CO guidelines for concentration values provided by WHO (1999) for health safety and toxic effect of CO.

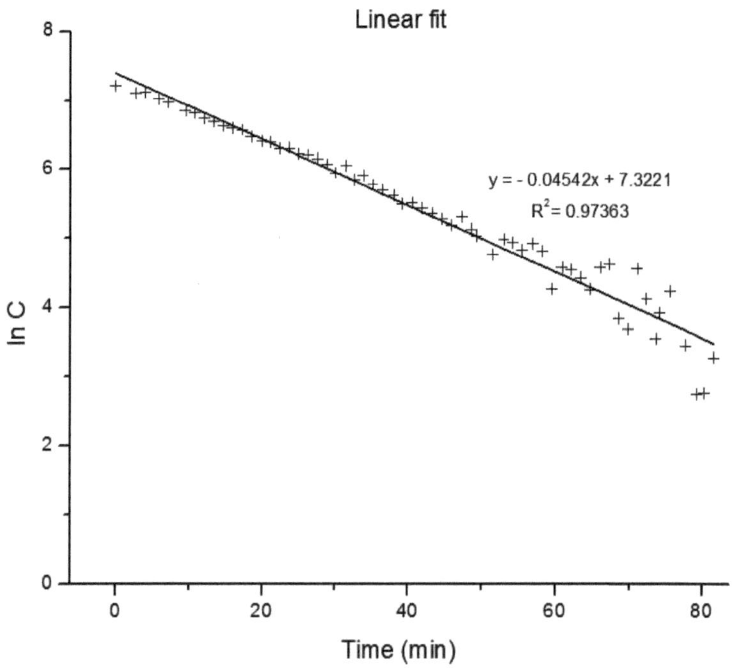

Fig. 13.4 Natural log graph of CO concentration, ln C against time for 2×2 mm^2 window

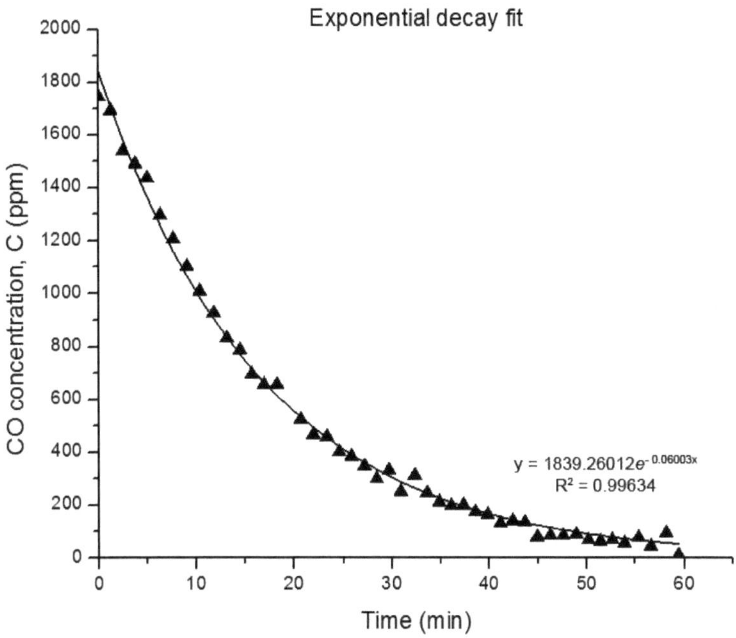

Fig. 13.5 CO decay curve for 3×3 mm^2 window

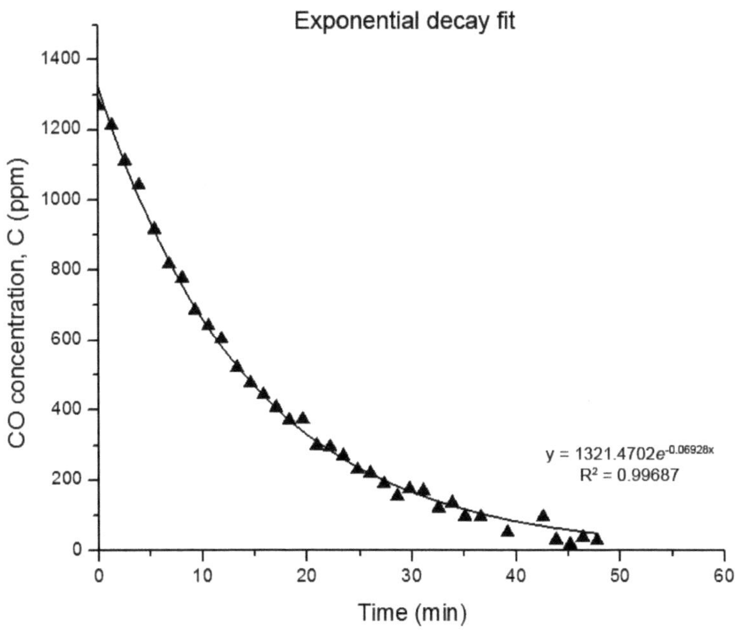

Fig. 13.6 CO decay curve for 4×4 mm^2 window

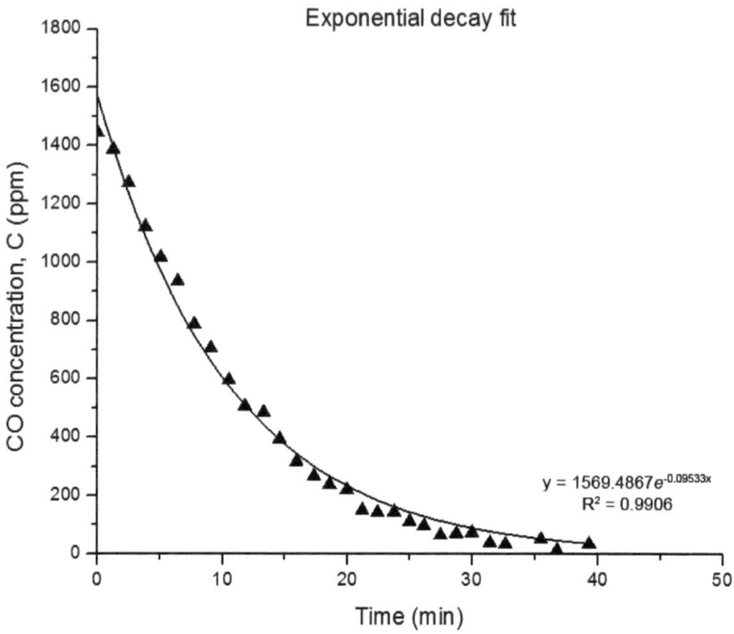

Fig. 13.7 CO decay curve for 5×5 mm^2 window

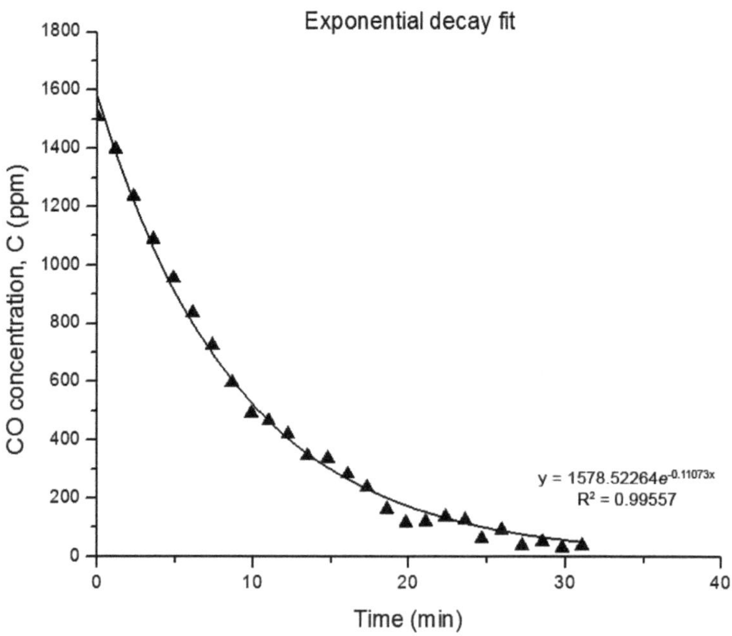

Fig. 13.8 CO decay curve for 6×6 mm^2 window

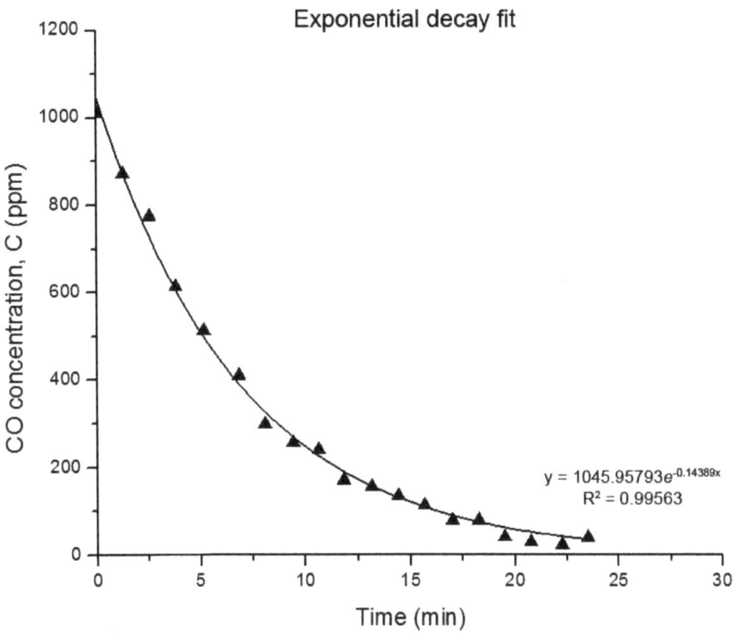

Fig. 13.9 CO decay curve for 7 × 7 mm^2 window

The application of the decay rate of gas concentration to measure the ventilation or air exchange rate has been well tested. For example, in an interesting study (Colls 2002), the particle concentration in a kitchen was artificially increased by burning some toast. Then the ventilation rate was set at a high constant level using an extractor fan, and the concentration of the particles was measured by an optical particle counter. The counter was able to measure particle size distributions as 1-min averages. It was found that the particle concentration increases to a sharp peak when the toast was burning. After the burning, the particle concentration was found to decay exponentially. The natural log fit of the concentration with time gave a straight line gave a gradient of 0.12 min^{-1} or 7 ACH. Using the values of floor area and volume, the deposition velocity for the particles was derived (Colls 2002). In another study (Naeher et al. 2001), CO was used as a tracer for assessing exposures to particulate matter in wood and gas cook stoves used in homes of highland Guatemala. The decay rates of high level of CO concentration produced by incomplete combustion were monitored so that exposure concentration and time to inhalable particles could be studied. In this work, we have used CO from side-stream environmental tobacco smoke (ETS) obtained from burning cigarettes of a popular brand. Therefore, the findings from the CO decay rates at various air exchange rates would be useful in assessing the exposures to the particulate matter and pollutants found in ETS. Recently, there has been an increased interest in understanding the serious health effects of ETS (Daher et al. 2010, Dingle et al. 2002, Hyvarinen et al. 2000, Rickert et al. 1984, Sterling and Mueller 1988, Trout et al. 1998). This is a useful application of this study, in addition to the assessment of IAQ with respect to CO concentrations in ambient air.

The graph of air exchange rate (I) in ACH against the window area (mm^2) using values from Table 13.2 is plotted as shown in Fig. 13.10. As shown, a proportional relationship is observed: as the window area increases, the air exchange rate increases almost linearly. Using values from Table 13.2, the graph of half-life T against the window area (mm^2) is plotted as shown in Fig. 13.11. As the window area increases, the air exchange rate increases resulting in a decreasing half-life or faster decay rate of CO. The trends observed in terms of the effect of window areas on the ventilation rate and decay rate of CO can be used for a better understanding of IAQ issues if CO is the main indoor pollutant.

In a study of five air-conditioned buildings in Singapore (Sekhar et al. 2003), the CO levels were found to be very low ranging from 0.2 to 2.64 ppm which are within the ambient air level allowed for CO by WHO (1999) and Ministry of the Environment, Singapore (1996), as given in Table 13.1. The air exchange rates for the five buildings ranged from 0.34 to 2.60 ACH. The findings that even at low air exchange rate (0.34–0.59 ACH) for one building, the CO concentration still remains low at 0.4–2.4 ppm showed that there were obviously no internal sources of CO such as from environmental tobacco smoke, and the infiltration of CO from external sources into the building was negligible. In their work (Sekhar et al. 2003),

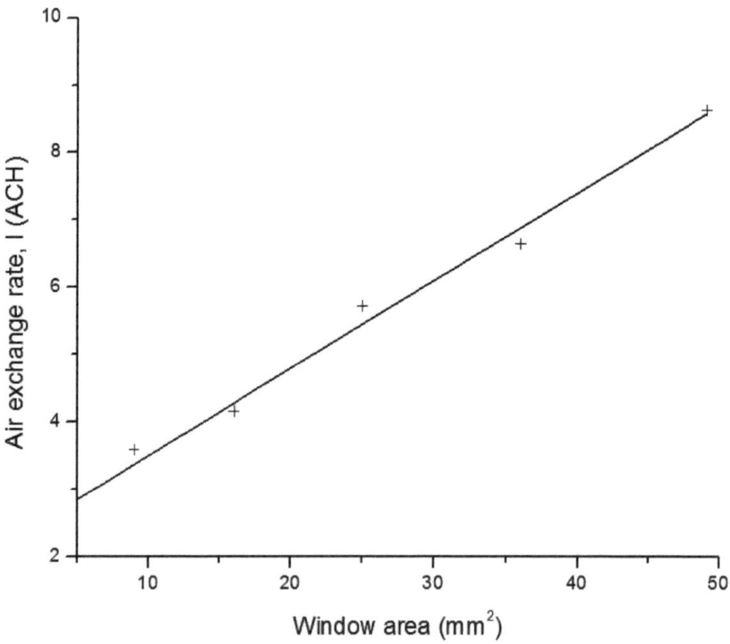

Fig. 13.10 The graph of air exchange rate (ACH) against window area (mm^2) for CO decays

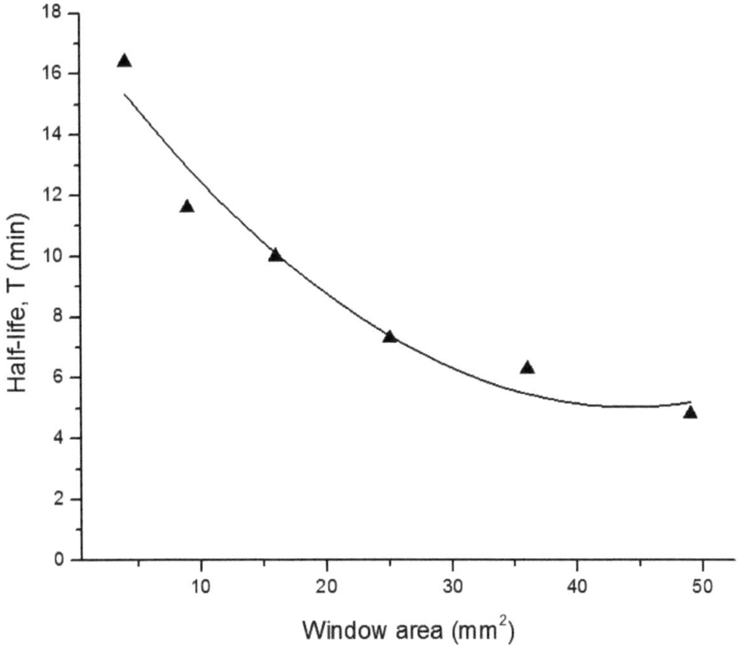

Fig. 13.11 The graph of half-life T (h) against window area (mm^2) for CO decays

the building with highest air exchange rate of 1.00–2.60 ACH was found to have the lowest maximum CO concentration of 1.23 ppm. This observation shows that higher air exchange rate reduces the concentration of CO even at very low levels in air-conditioned building. In agreement, the present results though at high levels of CO show that as air exchange rates increase, CO concentration decreases to a safe level in a shorter time. For example at air exchange rate of 2.53 ACH, the CO concentration level at a dangerous level of 1,400 ppm can be reduced to a safe level of 52 ppm (30-min exposure time) given in Table 13.2 within the time of 70 min, as shown in Fig. 13.3. The half-life of CO decay at 2.53 ACH was found to be 16.4 min, which is longer than 11.6 min at 3.60 ACH, as given in Table 13.2. From this work, it is found that as the air exchange rate increases, the half-life of CO decay decreases, showing that the CO concentration reaches a safe level in a shorter time (Table 13.2). The findings indicate that high air exchange rate not only keeps CO concentration at a low acceptable level, but also reduces elevated levels of CO in the building to safe levels, according to WHO guidelines (Table 13.1) in a shorter time.

From the results of this work, high ventilation rate (more than the usual 1 ACH) is needed to reduce the risk of CO toxic effect within the exposure of less than 1 h for high initial CO concentration (1,000–1,800 ppm). In a study (Tekbas et al. 2001), CO levels above 50 ppm (range of 54–300 ppm) were determined in 12 homes, out of 197 homes assessed, which used water boilers in the bathrooms running on liquid petroleum gas. Poor ventilation in these homes was observed. It was also mentioned that in a typical home air exchange rates would be expected to be near 1 ACH for the reduction of maximum CO concentration and also to increase decay rate for homes with internal sources of CO such as burning of joss or incense sticks (Croxford and Kynigou 2005). It was observed (El-Fadel et al. 2001) that the inadequacy of ventilation rates were often associated with high energy costs in an indoor environment, and the cost-benefit analysis was usually evaluated in terms of potential health impacts and decreased productivity. From practical observation (El-Fadel et al. 2001), typical ventilation rates of 5–10 ACH were commonly used to control IAQ if CO is selected as an indicator. These high rates are in good agreement with 2.53–8.63 ACH found in this work.

13.5 Conclusion

In managing urban air quality, the assessment of indoor air quality becomes a necessity as more people spend more time in an indoor environment. Since CO is an important indoor gas pollutant which can seriously affect IAQ, the decay rates of CO concentration for six different ventilation rates by using different window areas were measured and analyzed. Air exchange rates of 2.53–8.63 ACH and half-lives of 16.4–4.8 min were obtained from the decay of CO from ETS using high initial concentration of 1,000–1,800 ppm. The Fourier transform infrared technique was used to accurately measure the CO concentration in a gas chamber. The results would provide a better understanding of how the IAQ can be assessed if CO is of paramount concern.

Acknowledgements This study is supported financially by research grants from NIE, RS3/08TTL and RI9/09TTL. We are grateful for the support.

References

Apte, M. G., Fisk, W. J. & Daisey, J. M. (2000). Associations between indoor CO_2 concentrations and sick building syndrome symptoms in U.S. office buildings: an analysis of the 1994–1996 base study data. *Indoor Air*, 10: 246–257.

ASTM (1983). Standard test method for determining air leakage rate by tracer dilution. In Annual book of ASTM standards (Vol. 4.07, Designation E 741–783): American Society for Testing and Materials.

Brooks, B. O. & Davis, W. F. (1992). *Understanding indoor air quality*. Boca Raton, FL: CRC Press.

Burroughs, H. E. & Hansen, S. J. (2008). *Managing indoor air quality*. Lilburn, GA: Fairmont Press.

Chaloulakou, A., Duci, A. & Spyrellis, N. (2002). Exposure to carbon monoxide in enclosed multilevel parking garages in the central Athens urban area. *Indoor and Built Environment*, 11: 191–201.

Chan, P. (1999). Indoor air quality and the law in Singapore. *Indoor Air*, 9: 290–296.

Chan, M. Y. (2005). Commuters' exposure to carbon monoxide and carbon dioxide in air-conditioned buses in Hong Kong. *Indoor and Built Environment*, 14: 397–403.

Chan, Y. C., Tan, T. K., Char, L. K., Foo, S. C., Tan, T. C., Yap, H. M. et al. (1995). A study of bioaerosols within commercial office buildings in Singapore. Proceedings of the medical and public health session of the Asia-Pacific conference on the built environment, Singapore, 1995.
Colls, J. (2002). *Air pollution*. London: Spon Press.
Croxford, B. & Kynigou, D. (2005). Carbon monoxide emissions from joss or incense sticks. *Indoor Built Environment*, 14: 277–282.
Daher, N., Saleh, R., Jaroudi, E., Sheheiti, H., Badr, T., Sepetdijan, E. et al. (2010). Comparison of carcinogen, carbon monoxide, and ultrafine particle emissions from narghile waterpipe and cigarette smoking: sidestream smoke measurements and assessment of second-hand smoke emission factors. *Atmospheric Environment*, 44: 8–14.
Dingle, P., Tapsell, P., Tremains, I. & Tan, R. (2002). Environmental tobacco smoke and ventilation in 20 social venues in Perth, Western Australia. *Indoor and Built Environment*, 11: 146–152.
El-Fadel, M., Alameddine, I., Kazopoulo, M., Hamdan, M. & Nasrallah, R. (2001). Indoor air quality assessment in an underground parking facility. *Indoor and Built Environment*, 10: 179–184.
El-Fadel, M. & El-Hougeiri, N. (2003). Indoor air quality and occupational exposures at a bus terminal. *Applied Occupational and Environmental Hygiene*, 18: 513–522.
El-Hougeiri, N. & El-Fadel, M. (2004). Correlation of indoor-outdoor air quality in urban areas. *Indoor and Built Environment*, 13: 421–431.
Epidemiology, I. o. E. (1996). *Guidelines for good indoor air quality in office premises* (1st ed). Singapore: Institute of Environmental Epidemiology, Ministry of the Environment.
Fang, L., Clausen, G. & Ganger, P. O. (1998). Impact of temperature and humidity on perception of indoor air quality during immediate and longer whole-body exposures. *Indoor Air*, 8: 276–284.
Fang, L., Wyon, D. P., Clausen, G. & Fanger, P. O. (2004). Impact of indoor air quality temperature and humidity in an office on perceived air quality, SBS symptoms and performance. *Indoor Air*, 14: 74–81.
Glorennec, P., Bonvallot, N., Mandin, C., Goupil, G., Pernelet-Joly, V., Millet, M. et al. (2008). Is a quantitative risk assessment of air quality in underground parking garages possible? *Indoor Air*, 18: 283–292.
Godish, T. (2004). *Air quality*. New York, NY: Lewis.
Hanst, P. L. & Hanst, S. T. (1994). Gas measurements in the fundamental infrared region. In M. W. Sigrist (Ed.), *Air monitoring by spectroscopic techniques* (pp. 335–466). New York, NY: Wiley.
Harrison, P. T. C. (1998). Health effects of indoor air pollutants-CO. In R. E. Hester & R. M. Harrison (Eds.), *Issues in environmental science and technology – air pollution and health* (pp. 116–118). London: The Royal Society of Chemistry.
Hyvarinen, M. J., Rothberg, M., Kahkonen, E., Mielo, T. & Reijula, K. (2000). Nicotine and 3-ethenylpyridine concentrations as markers for environmental tobacco smoke in restaurants. *Indoor Air*, 10: 121–125.
Lam, K. S., Chan, F. S., Fung, W. Y., Lui, B. S. S. & Lau, L. W. L. (2006). Achieving “excellent” indoor air quality in commercial offices equipped with air-handling unit – respirable suspended particulate. *Indoor Air*, 16: 86–97.
Lee, S. C., Poon, C. S., Li, X. D. & Luk, F. (1999). Indoor air quality investigation on commercial aircraft. *Indoor Air*, 9: 180–187.
Levesques, B., Bellemare, D., Sanfacon, G., Duchesne, J. F., Gauvin, D., Homme, H. P. et al. (2005). Exposure to carbon monoxide during indoor karting. *International Journal of Environmental Health Research*, 15: 41–44.
Liao, S. S. T., Bacon-Shone, J. & Kim, Y. S. (1991). Factors influencing indoor air quality in Hong Kong: measurements in offices and shops. *Environmental Technology*, 12: 737–745.
Liu, K. S., Alevantis, L. E. & Offermann, F. J. (2001). A survey of environmental tobacco smoke controls in california office buildings. *Indoor Air*, 11: 26–34.
McDermott, H. J. (2004). *Air monitoring for toxic exposures*. New York, NY: Wiley.

Meckler, M. (1996). *Improving indoor air quality through design, operation and maintenance*. Lilburn, GA: The Fairmont Press.
Melikov, A., Pitchurov, G., Naydenov, K. & Langkilde, G. (2005). Field study on occupant comfort and office thermal environment in rooms with displacement ventilation. *Indoor Air*, 15: 205–214.
Naeher, L. P., Smith, K. R., Leaderer, B. P., Neufeld, L. & Mage, D. T. (2001). Carbon monoxide as a tracer for assessing exposures to particulate matter in wood and gas cook stove households of highland guatemala. *Environmental Science & Technology*, 35: 575–581.
Ohman, P. A. & Eberly, L. E. (1998). Relating sick building symptoms to environmental conditions and worker characteristics. *Indoor Air*, 8: 172–179.
Ooi, P. L., Quek, G. H., Chong, K. W. & Goh, K. T. (1995). Epidemiological investigations into sick building syndrome in singapore. Proceedings of the medical and public health session of the Asia-Pacific conference on the built environment, Singapore, 1995.
Prasad, R. K., Uma, R., Kansal, A., Gupta, S., Kumar, P. & Saksena, S. (2002). Indoor air quality in an air-conditioned building in New Delhi and its relationship to ambient air quality. *Indoor and Built Environment*, 11: 334–339.
Rickert, W. S., Robinson, J. C. & Collishaw, N. W. (1984). Yields of tar, nicotine, and carbon monoxide in the sidestream smoke from 15 brands of canadian cigarettes. *American Journal of Public Health*, 74: 228–231.
Rubio, M. A., Fuenzalida, I., Salinas, E., Lissi, E., Kurtenbach, R. & Wiesen, P. (2010). Carbon monoxide and carbon dioxide concentrations in santiago de chile associated with traffic emissions. *Environmental Monitoring and Assessment*, 162: 209–217.
Salau, O. R., Warneke, T., Notholt, J., Shim, C., Li, Q. & Xiao, Y. (2009). Tropospheric trace gases at Bremen measured with FTIR spectroscopy. *Journal of Environmental Monitoring*, 11: 1529–1534.
Sekhar, S. C., Tham, K. W. & Chan, P. (1995). Technical and legal issues of indoor air quality – the Singapore context. *International Journal of Housing Science*, 19: 301–314.
Sekhar, S. C., Tham, K. W. & Cheong, K. W. (2003). Indoor air quality and energy performance of air-conditioned office buildings in Singapore. *Indoor Air*, 13: 315–331.
Sekhar, S. C., Tham, K. W., Cheong, D., Kyaw, T. M. & Susithra, M. (1999) The development of an indoor pollutant standard index. Proceedings of Indoor Air 1999, the 8th international conference on Indoor Air quality and climate, Edinburgh, Scotland, 1999 (Vol. 2, pp. 272–277).
Seppanen, O., Fisk, W. J. & Lei, Q. H. (2006). Ventilation and performance in office work. *Indoor Air*, 16: 28–36.
Seppanen, O. A., Fisk, W. J. & Mendell, M. J. (1999). Association of ventilation rates and CO_2 concentrations with health and other responses in commercial and institutional buildings. *Indoor Air*, 9: 226–252.
Siddiqui, A. R., Lee, K., Bennett, D., Yang, X., Brown, K. H., Bhutta, Z. A. et al. (2009). Indoor carbon monoxide and $PM_{2.5}$ concentrations by cooking fuels in Pakistan. *Indoor Air*, 19: 75–82.
Siskos, P. A., Bouba, K. E. & Stroubou, A. P. (2001). Determination of selected pollutants and measurement of physical parameters for the evaluation of indoor air quality in school buildings in Athens, Greece. *Indoor and Built Environment*, 10: 185–192.
Smith, B. C. (1996). *Fundamentals of Fourier transform infrared spectroscopy*. New York, NY: CRC Press.
Smith, K. R. (2002). Indoor air pollution in developing countries: recommendations for research. *Indoor Air*, 12: 198–207.
Spengler, J. & Soczek, M. (1984). Evidence for improved ambient air quality and the need for personal exposure research. *Environmental Science & Technology*, 18: 268–280.
Sterling, T. D. & Mueller, B. (1988). Concentrations of nicotine, RSP, CO and CO_2 in non-smoking areas of offices ventilated by air re-circulated from smoking designated areas. *American Industrial Hygiene Association Journal*, 49: 423–426.

Tekbas, O. F., Gulec, M., Odabasi, E., Vaizoglu, S. A. & Guler, C. (2009). Determination of carbon monoxide levels in coffee shops in Ankara. *Indoor and Built Environment*, 18: 130–137.

Tekbas, O. F., Vaizoglu, S. A., Evci, E. D., Yuceer, B. & Guler, C. (2001). Carbon monoxide levels in bathrooms using hot water boilers. *Indoor and Built Environment*, 10: 167–171.

Tham, K. W. (1994). Integrated approach to indoor air quality investigation of office buildings. In L. Morawska, N. D. Bodfinger & M. Maroni (Eds.), *Indoor air – an integrated approach* (pp. 503–506). Oxford: Elsevier.

Tham, K. W., Sekhar, S. C. & Cheong, D. (1996) Integrated indoor air quality investigation of office buildings. Proceedings of Indoor Air 1996, the 7th International conference on Indoor Air Quality and climate, Nagoya, Japan, 1996 (Vol. 3, pp. 1027–1032).

Trout, D., Decker, J., Mueller, J., Bernert, J. & Pirkle, J. (1998). Exposure of casino employees to environmental tobacco smoke. *Journal of Occupational Medicine and Envrionmental Medicine*, 40: 270–276.

Vitel, C. (2001). The quality of the air in our buildings. *Indoor and Built Environment*, 10: 266–270.

WHO (1999). Environmental health criteria, no. 213: carbon monoxide. http://www.who.int/pcs/ehc/summaries/ehc_213.html. 03 May 2010.

Wong, Y. C., Sin, D. W. M. & Yeung, L. L. (2002). Assessment of the air quality in indoor car parks. *Indoor and Built Environment*, 11: 134–145.

Yocom, J. E. (1982). Indoor-outdoor air quality relationship. *Journal of Air Pollution Control Association*, 32: 500–512.

Yocom, J. E., Clink, W. L. & Cote, W. A. (1971). Indoor/outdoor air quality relationship. *Journal of Air Pollution Control Association*, 21: 251–260.

Yocom, J. E. & McCarthy, S. M. (1991). *Measuring indoor air quality: a practical guide*. New York, NY: Wiley.

Index

T.-C. Wong, B. Yuen (eds.), *Eco-city Planning*, DOI 10.1007/978-94-007-0383-4,

MIX
Papier aus verantwortungsvollen Quellen
Paper from responsible sources
FSC® C105338

FSC
www.fsc.org

If you have any concerns about our products,
you can contact us on
ProductSafety@springernature.com

In case Publisher is established outside the EU,
the EU authorized representative is:
Springer Nature Customer Service Center GmbH
Europaplatz 3, 69115 Heidelberg, Germany

Printed by Libri Plureos GmbH
in Hamburg, Germany